Edited by
Gösta Kjellsson
Vibeke Simonsen
Klaus Ammann

Methods for Risk Assessment

of Transgenic Plants

II. Pollination, Gene-Transfer and Population Impacts

supported by

Federal Office
of Environment,
Forests and
Landscape,
Switzerland

EUROPEAN COMMISSION
DG XI

MINISTRY OF ENVIRONMENT AND ENERGY; DENMARK

National Environmental
Research Institute

NATIONAL FOREST
AND NATURE AGENCY

ENVIRONMENTAL
PROTECTION AGENCY

Birkhäuser Verlag
Basel · Boston · Berlin

Editors:

Dr. Gösta Kjellsson
Dr. Vibeke Simonsen
Department of Terrestrial Ecology
National Environmental
Research Institute
Vejlsøvej 25
P.O. Box 314
DK-8600 Silkeborg
Denmark

Dr. Klaus Ammann
Botanical Garden
University of Bern
Altenbergrain 21
CH-3013 Bern
Switzerland

A CIP catalogue record for this book is available from the Library of Congress, Washington D.C., USA

Deutsche Bibliothek Cataloging-in-Publication Data

Methods for risk assessment of transgenic plants / ed. by Gösta
Kjellsson ... - Basel ; Boston : Berlin : Birkhäuser
   Bd. 1. verf. von Gösta Kjellsson und Vibeke Simonsen. -
   2. Pollination, gene-transfer and population impacts. - 1997
   ISBN 3-7643-5696-0 (Basel ...)
   ISBN 0-8176-5696-0 (Boston)

© 1997 Birkhäuser Verlag, P.O. Box 133, CH-4010 Basel, Switzerland
Printed on acid-free paper produced from chlorine-free pulp
Printed in Germany
ISBN 3-7643-5696-0
ISBN 0-8176-5696-0

9 8 7 6 5 4 3 2 1

Contents

List of contributors

Editors:

Gösta Kjellsson *
Address: Department of Terrestrial Ecology, National Environmental Research Institute,
          Vejlsøvej 25, P.O. Box 314, DK-8600 Silkeborg, Denmark
Phone: + 45 89 20 14 00
Fax: + 45 89 20 14 14
e-mail: gk@dmu.dk

Vibeke Simonsen
Address: Department of Terrestrial Ecology, National Environmental Research Institute,
          Vejlsøvej 25, P.O. Box 314, DK-8600 Silkeborg, Denmark
Phone: + 45 89 20 14 00
Fax: + 45 89 20 14 14
e-mail: vs@dmu.dk

Klaus Ammann
Address: Botanical Garden, University of Bern, Altenbergrain 21, CH-3013 Bern, Switzerland
Phone: + 41 31 631 49 37
Fax: + 41 31 631 49 93
e-mail: kammann@sgi.unibe.ch

Authors:

François Felber
Address: Institute of Botany, University of Neuchâtel, Chantemerle 18, CH-2007 Neuchâtel,
          Switzerland
Phone: + 41 32 718 23 39
Fax: + 41 32 718 30 01
e-mail: francois.felber@bota.unine.ch

Yolande Jacot
Address: Laboratory of Phanerogamy, Institute of Botany, University of Neuchâtel, Chantemerle
          22, CH-2007 Neuchâtel, Switzerland
Phone: + 41 32 718 21 11
Fax: + 41 32 718 21 01
e-mail: -

Rikke Bagger Jørgensen
Address: Resistance Biology and Plant Genetics, Department of Plant Biology and Biogeo-
          chemistry, PBK-301, Risø, DK-4000 Roskilde, Denmark
Phone: + 45 46 77 41 10
Fax: + 45 46 32 33 83
e-mail: rikke.bagger.joergensen@risoe.dk

* : Correspondence concerning the book should be addressed to the main editor.

Kathrine Hauge Madsen
Address: Weed Science, The Royal Veterinary and Agricultural University, Thorvaldsensvej 40, DK-1871 Frederiksberg C, Denmark
Phone: + 45 35 28 34 45
Fax: + 45 35 28 34 68
e-mail: kathrine.h.madsen@agsci.kvl.dk

Jens Mogens Olesen
Address: Department of Ecology and Genetics, Institute of Biological Sciences, University of Aarhus, Ny Munkegade, bld. 540, DK-8000 Aarhus C, Denmark
Phone: +45 89 42 31 88
Fax: +45 86 12 71 91
e-mail: jmo@pop.bio.aau.dk

Marianne Philipp
Address: Department of Plant Ecology, Institute of Botany, University of Copenhagen, Øster Farimagsgade 2D, DK-1353 København Ø, Denmark
Phone: +45 35 32 22 77
Fax: +45 35 32 23 21
e-mail: marianp@bot.ku.dk

Gitte Silberg Poulsen
Address: Climate and Biotechnology Division, Danish Environmental Protection Agency, Strandgade 29, DK-1401 København K, Denmark
Phone: + 45 32 66 01 00
Fax: + 45 32 66 04 79
e-mail: gsp@mst.dk

Pia Rufener Al Mazyad
Address: Botanical Garden, University of Berne, Altenbergrain 21, CH-3013 Bern, Switzerland
Phone: + 41 31 631 37 67
Fax: + 41 31 631 49 93
e-mail: rufener@sgi.unibe.ch

Mikkel Heide Schierup
Address: Department of Ecology and Genetics, Institute of Biological Sciences, University of Aarhus, Ny Munkegade, bld. 540, DK-8000 Aarhus C, Denmark
Phone: + 45 89 42 31 88
Fax: + 45 86 12 71 91
e-mail: mheide@pop.bio.aau.dk

Preface

The present work is a continuation of the work initiated in Autumn 1991, which resulted in the book, published by Birkhäuser Verlag in 1994, entitled: Methods for Risk Assessment of Transgenic Plants. I. Competition, Establishment and Ecosystem Effects. Already when the work on volume 1 started, it was obvious to the authors, that not only the physical establishment of a transgenic plant outside the cultivated area was important for risk assessment, but also the possible gene-transfer from transgenic plants to other plants had to be considered. It was then decided to write a second volume on test methods, as a complement to the first, covering the main topics: Pollination, gene-transfer and population impacts. The main user groups for this volume are scientists and students working with plant population genetics and risk assessment and administrators with responsibility for legislation of transgenic plants. In order to cover such a broad range of topics, specialist knowledge was required. Therefore, colleagues in Denmark and Switzerland, working in these fields in relation to the concerns of using transgenic plants, were asked to participate. The result was a Danish-Swiss cooperation. A list of contributors to the book and their addresses is shown on p. VII.

Financial support, which made the work possible, was given by: The National Environmental Research Institute, Denmark, the Federal Office of Environment, Forest and Landscape, Switzerland, the National Forest and Nature Agency, Denmark, the Danish Environmental Protection Agency and the European Commission, DG XI.

We would like to thank members of the steering group, Hans Erik Svart and Jan Højland from the National Forest and Nature Agency, Denmark, François Pythoud from the Federal Office of Environment, Forest and Landscape, Switzerland, Claus Frier from the Danish Environmental Protection Agency, Kaj Juhl Madsen, formerly DG XI, Bruxelles and Joanna Kioussi, DG XI, Bruxelles, for support and interest in the work.

Many colleagues have provided important information and helped with comments on early versions of the text: Arminio Boschetti, Freddy Bugge Christiansen, Christian Damgaard, Kirsten Engell, Hans Løkke and Helle Ravn.

Several people gave highly valuable technical assistance: Karin Friis Velbæk, Lilian Mex-Jørgensen, Charlotte Bech and Bodil Thestrup. Margrit Kummer is acknowledged for the drawing of the *Brassica* spp. in Chapter 6 and Käthe Møgelvang for the front cover. The proofreading and language control was made by the Danish editors and by Ian Cockburn-Curran from Eurotext International, Denmark.

Gösta Kjellsson, Vibeke Simonsen and Klaus Ammann
Silkeborg and Bern, Spring 1997

1. Introduction

Field releases,
commercialization and
environmental impacts

We live in a period where field releases of transgenic crops are becoming more and more numerous. Most of the approx. 2000 field releases have been done with transgenes with limited ecological relevance. Therefore, it is not surprising that up to now there are no cases reported where transgenic crops have caused serious ecological risk or any major impact on the environment. This might change in the near future, as plants with new traits are introduced and, consequently, we must develop methods for testing before release and monitoring the actual gene dispersal from field releases. We are also confronted with large-scale releases, in the course of general commercialization of major crop plants on a world scale, which will require both extensive monitoring and scientifically based assessment of environmental risks.

More closely watched, many of these field releases provide us with some confidence, but still a generalization seems to be premature. What we really need is a more differentiated attitude, which is based on a step-by-step philosophy, taking into account the crop plant and their various strains, the inserted traits, the weedy and wild relatives, the possible change in ecological characteristics and also the biogeographical situation. Thus, the procedures we need depend on the specific case studied. The risk assessment of transgenic cultivars involves some of the same problems as classical breeding, especially concerning resistance management. This needs to be discussed in the light of new insights due to the higher precision given by molecular genetics.

1.1. New developments

During the last three years, since 1994, when volume 1 was published (447), much has happened in the field of biotechnology and development of new transgenic crops and products. Many of the major crops have been genetically transformed and are now being commercialized as industrial products (especially from USA), e.g.: Rape, maize, soybean, cotton, tomato, tobacco, potato - and rice in a few years time. However, cultivation is still restricted in most countries and consumer reactions towards the modified products vary greatly between countries.

New traits and new
biotechnological methods

Until now, the biotechnologically inserted traits have mainly been herbicide resistance (especially Roundup and Basta resistance) and, to a lesser degree, insect resistance (e.g., Bt toxins) and virus resistance (see Chapter 9). In future years, new traits, such as tolerance to environmental stress, and biochemical products will become increasingly important. New biotechnological methods will hopefully lead to increased precision when new genes are inserted in target plants (see Chapter 8). The need to use certain marker types (e.g., kanamycin resistance), which are questionable from a human health point of view, will probably decrease in the future.

Test methods and risk assessment procedures

The need for relevant test methods is evident as new questions and environmental concerns arise. Furthermore, the importance of international agreement on risk assessment procedures is becoming more and more urgent (see Chapter 10). Hopefully, this book will contribute towards solving some of these important issues.

Biotechnology and the Internet

The immense popularity and growth of the Internet (the World Wide Web) have made many different sources of information on biotechnology easily available to both research scientists and environmental administrators. We have published a number of current Internet addresses (URLs) of potential interest to the reader in this book, hoping that they will not be obsolete too soon. Especially concerning field releases, regulations and commercialization they should provide useful information from much of the world.

1.2. Main purposes of the book

Risk assessment of transgenic plants

The main purposes of the present book are to:

-   **present a collection of methods which may be used for risk assessment of transgenic plants,**

-   **review useful methods for studying pollination, gene-transfer and establishment of alien genes in plant populations,**

-   **emphasize in which areas there is a need for development of new test procedures,**

-   **provide a review on genetic engineering techniques and on traits which have been inserted in plants,**

-   **identify concepts and procedures for risk assessment of GMPs.**

The book focuses on the topics pollination, gene-transfer including hybridization and population impacts. Pollination is an obvious prerequisite for gene-transfer, as genes have to be moved by an agent from one plant to another for effective sexual reproduction. In the next step, the alien gene is incorporated in the genome of the receiver plant through hybridization and introgression. The expression of the alien gene in the new host plant and the effects on plant survival, competitive ability and establishment are important issues covered in vol. 1 (447). The two method books should also be useful as **general references of methods for pollination ecology and for population ecology and genetics.**

Method selection

The methods included in the book have been selected to broadly cover the major topics. Some overlap to methods in vol. 1 could not be avoided, in so far as applications and new techniques made this necessary. The selection and description of methods is based on a literature search in several library databases covering the period 1985 to 1995 and supplemented with references from the authors.

1.3. How to use the method book

Several ways are used to identify relevant methods for specific research areas: 1. Subject key words (category and subcategory) in Chapter 3 to 5 refer to methods in lists. 2. The subject index, Chapter 12, may be used. 3. A synopsis, Chapter 6, covers the method applications. 4. Suggestions of relevant use in risk assessment are given in Chapter 12. When the final choice between methods is made, information on method applications and restrictions is available in the method descriptions in Chapter 7. A résumé of the content of the chapters is presented below:

Contents - Chapter 2-5

Chapter 2 provides a glossary covering the scientific terms used in the text. The definitions are based on discussions of current usage. **Chapter 3** presents the list of relevant **key words** covering the main subject. The list is divided into 12 **main categories** (e.g., breeding system, gene-flow, hybridization) each with a number of **subcategories** (e.g., cross-pollination, flowering phenology, heterostyly). The list of key words was initially created for the literature search. The placement of subcategories in respective categories is mainly made for structural reasons and may occasionally be tentative. **Chapter 4** and **5** are entries for **subcategories with corresponding methods** and **methods with corresponding subcategories** for selection of the relevant methods for a specific topic or finding the topics covered by a specific method.

Contents - Chapter 6-7

A **synopsis on subcategories and recommended methods** is presented in **Chapter 6**. For each subject, concise background information is given, a number of methods are recommended and comments on research need and need for new methods are provided. The **methods** are fully described in **Chapter 7** with category and subcategory key words given for each method. Information on assumptions and restrictions, application and advantages of the method is included.

Contents - Chapter 8-10

Chapter 8 deal with the **biotechnological techniques** that are used for production of transgenic plants. The different types of mechanical, chemical and biotic methods for transformation are described in terms of precision, efficiency and applicability. In **Chapter 9** the different **traits** that have been **inserted in transgenic plants** are reviewed. The biochemical and physiological mechanisms for their functioning are described with reference to the main types of traits, i.e.: Herbicide tolerance, pest and pathogen tolerance, changed

metabolic content and stress tolerance. **Chapter 10** presents suggestions for how to proceed - when ecological risk assessment of transgenic plants is performed - and describes the concepts currently
used. Different types of test systems are discussed in relation to the
methods covered in vol. 1 and 2 of the method book. The role of the
environment and the use of modelling are further discussed.

Contents - Chapter 11-12   **Chapter 11** contains the **references** for the book. Each reference is
identified by a number in parentheses in the text. The list was made
from a database in the bibliographic system Pro-Cite. The **subject
index, Chapter 12**, provides quick entrance to major subjects,
specific methods and plant species of interest. Both the key entrance
and full references are given.

2. Glossary of terms and abbreviations

References to specific methods are indicated by M-numbers, while reference to transfer methods are indicated by T-numbers.

Addition	Chromosome complement of the original species is supplemented by one or more chromosomes from another genotype or species.
AFLP	Amplified Fragment Length Polymorphism, see M1.
Agamospermy	Production of seeds without fertilization.
Agrobacterium	Soil-living bacteria genus which can transfer plasmids into plants and other organisms, see Plasmid.
Allele	One of two or more forms that a gene may take, differing in DNA sequence and normally affecting the function of the product (e.g., wild-type and mutant form).
Allogamy	Mating system where pollen is transported between individuals. See also geitonogamy and xenogamy.
Allopolyploid	Plant with more than two sets of chromosomes, some of which are non-homologous, originating from two or more parents. See also Autopolyploid.
Amphidiploid	Plant which is diploid for two genomes, each from a different species. See also Diploid and Allopolyploid.
Androdioecy	Mating system where some individuals have male flowers and others have hermaphroditic flowers.
Androecium	Male parts of a flower, i.e., stamens each consisting of anther, filament and connective.
Andromonoecy	Mating system where every plant possesses both male and hermaphroditic flowers.
Aneuploid	Individual where the chromosome number is not a multiple of the haploid chromosome number of the species.
Angiosperm	Plant bearing seeds which are enclosed in an ovary or carpel. The majority of seed plants are Angiosperms. See also Gymnosperm.
Anther	Terminal part of the male organs (stamen) of flowers, usually born on a stalk (filament) and containing pollen. Pollen grains are commonly released through longitudinal slits or through terminal pores (poricidal anthers).

Anthesis	Period from opening to withering of a flower, including the period when pollen is released and pollination is possible.
Antisense RNA	RNAs which can latch to mRNAs and prevent the production of a protein.
Apomixis	Reproduction without sex either through agamospermy or through vegetative propagation (cloning). See Pseudogamy.
Apospory	A kind of agamospermy where the embryo sac originates from a cell in the center of the ovule.
Artificial hybrid	Hybrid made by man which may or may not occur naturally. See Natural hybrid.
Autogamy	Mating system where a flower is pollinated and fertilized with pollen produced in the same flower. See also Self-fertilization.
Autopolyploid	Plant with more than two sets of chromosomes which originate from a single parent, or from parents which have homologous chromosomes.
Backcrossing	Cross between offspring and one of its parents or any individual identical to one of its parents.
Biparental inbreeding	Mating between related individuals. Includes all forms of inbreeding except self-fertilization, and is the only kind of inbreeding possible in dioecious species.
Breeding system	Processes (such as pollen release, pollination, pollen competition and fertilization) and structures (such as flower size, form and colour, nectaries and stigma surface) which influence transfer of gametes and fertilization of ovules.
Buzz-pollination	Type of pollination where bees vibrate flowers for collecting pollen.
Callose	Complex polysaccharide found in, e.g., pollen mother cell walls, in germinating pollen grains, within pollen tubes, and as a wounding reaction.
Callus	Mass of undifferentiated cells that initially arises from plant cells and often used as tissue in artificial culture.
Calyx	All the sepals of the flower.
C-banding	Specific type of chromosome-banding produced by a chemical treatment, see M5 and M33.
cDNA	cDNA is a DNA copy of mRNA, which contains the coding region (exons) of the genomic gene (i.e., introns are excluded), see also DNA and mRNA.

Chemical attractant	Any chemical elicited by a plant that attracts a pollinator to move towards it. Syn.: Chemoattractant.
Chromosomal aberration	Abnormal chromosomal complement resulting from loss, gain or rearrangement of genetic material. An aberration often changes or disrupts the function of a gene.
Cleistogamy	Breeding system where small and inconspicuous flowers, which never open, are self-pollinated and self-fertilized.
Cline	Graded series of a morphological or genetic trait, distributed along a spatial dimension.
Clogging	Phenomenon where a stigma is overcrowded with incompatible pollen grains, thereby reducing seed set.
Corolla	Petals which usually constitute the part of the flower that visually attracts pollinators.
Cotyledon	The first leave(s) of a seed plant, found in the embryo, which usually form the first emergent photosynthetic leaves. Sometimes they remain below the ground. Syn.: Seed-leaf.
cpDNA	Chloroplast DNA, which is present in chloroplasts. It is haploid, non-recombining and usually maternally inherited.
Cross-fertilization	Mating system where ovules are fertilized by gametes produced on another individual.
Cross-pollination	Mating system where stigmas are pollinated with pollen grains from another individual.
Dehiscence of anthers	Spontaneous opening of anthers to release pollen grains.
Deletion	Chromosomal aberration resulting in the loss of one or more base pairs (segment) from a chromosome.
Dichogamy	Mating system where female and male sexual organs in a plant mature at different times (e.g., protogyny, protandry).
Dicliny	Mating system where female and male sexual organs are found together in some flowers (hermaphroditic) and separated in other flowers (male or female) on the same individual (gynomonoecy, andromonoecy, monoecy) or different individuals (gynodioecy, androdioecy, dioecy).
Differential display PCR	PCR on cDNA by random primers, see M59.
Dioecy	Mating system where some individuals possess female flowers and others male flowers. See also Monoecy.

Diphasy	Mating system where individual genets can vary their sexual mode from year to year influenced by environmental conditions.
Diploid	Individual resulting from a single zygote with two homologous sets of chromosomes.
Diplospory	Kind of agamospermy where the diploid embryo sac is formed from a megaspore mother-cell and the embryo develops without fertilization.
Distyly	Heterostylous mating system in which there are two flower morphs (see heterostyly).
DNA	DeoxyriboNucleic Acid, a molecule containing the genetic information of an organism.
Duplication	Chromosomal aberration where a segment within a specific chromosome is present in two copies, usually caused by unequal crossing-over.
Electroporation	Method of genetical engineering where cells are exposed to an electric impulse which enables alien DNA to penetrate the cell wall and the membrane and become incorporated in the DNA of the host cell, see T2.
Emasculation	Removal of anthers in buds or open flowers to prevent self-pollination, see M12.
Embryo sac	Female gametophyte of angiosperms, most often composed of eight haploid cells, one of which is the egg cell.
Endosperm	Food-storing tissue inside a seed.
Epistasis	Interaction of alleles in different loci, e.g., expression of one locus masks the expression of another locus.
Exine	Outer layer of the two wall layers of a mature pollen grain.
Exon	Part of DNA which encodes a part of a functional protein, i.e., the coding sequence.
F-statistics	Short for fixation-statistics, coined by Wright as a way of describing the distribution of genetic variation within and between populations in a species, see M19.
Female choice	Phenomenon where the female functioning part of the plant rejects certain pollen donors either as pollen, pollen tubes or as part of a developing ovule.

Fertilization	Fusion of gametes. In angiosperms double fertilization occurs whereby the egg cell and the primary endosperm cell (polar cells) are fused with male gametes and the diploid embryo and the triploid endosperm are produced.
Fitness	Fitness of a genotype is commonly measured by the number of successful offspring compared to other genotypes (i.e., relative fitness), see M14 and M35. Fitness should always be defined in relation to environmental variables, e.g., habitat and climate.
Flower	Reproductive structure in angiosperms composed of calyx, corolla, androecium and gynoecium.
Fruit	Mature ovary containing seeds. A number of different types of fruits exists some of which are adapted to different kinds of seed dispersal, e.g., capsules (often wind-dispersal) and drupes (often bird- or animal-dispersed).
Functional gender	Quantitative sex expression of a flower or an individual weighted by the relative amount of male and female reproductive units in the population, see M70. See also Phenotypic gender.
Gametophyte	Pollen grain and the embryo sac (in angiosperms). The haploid phase in the reproductive cycle of a plant.
Gametophytic incompatibility	Mating system, where the alleles included in the pollen grains determine the compatibility type of the pollen grain. See also Sporophytic incompatibility.
Geitonogamy	Mating system where a flower is pollinated and fertilized with pollen produced in another flower on the same individual. Genetically identical to autogamy.
Gender	Quantitative measure of the sex expression of a flower or of an individual.
Gene	Functional unit of inheritance (a sequence of DNA) which is located on a certain position (locus) on a chromosome. Genes include structural genes, which encode for proteins, and regulatory genes, which control the function of structural genes.
Gene amplification	Process by which a DNA sequence is replicated more extensively than reflected by its representation in the parent DNA molecule.
Gene dispersal	Spread of genes within populations through pollen and seeds. The term is often used indiscriminantly for Gene-flow.
Gene expression	Physical or chemical characters produced by the translation of genes. Also the extent to which a gene is transcribed into m-RNA and translated into protein.

Gene-flow	Spread of genes among populations through pollen or seeds. Syn.: migration.
Gene pool	All the genes, and their different alleles, available to a plant by sexual reproduction.
Gene silencing	Absence of expression of a gene which is present in the genome.
Genet	Genetically identical unit that arose from a single zygote. A genet may produce genetically identical new units, ramets, by vegetative growth.
Genetic diversity	Amount of genetic polymorphism which is present within the genomes of all the individuals of a population or a species. It can be measured by the extent of polymorphism at the protein level or the DNA level in a species throughout its distribution.
Genetic drift	Changes in gene-frequencies in a finite population due to stochastic effects such as random sampling of gametes. Opposed to deterministic changes in gene frequencies caused by selection.
Genetic load	Average decrease in fitness per individual in a population caused by deleterious alleles, relative to a theoretical population without deleterious alleles.
Genetic marker	Marker which is based on molecular or biochemical properties, see also Marker.
Genetic polymorphism	Occurrence in a population or species of different forms of the same genetic locus.
Genetically Modified Plant (GMP)	See Transgenic plant.
Genome	Genetic constitution which includes a single representative of each chromosome pair.
Genotype	Genetic constitution of an organism. Often used for a single gene. See also Phenotype.
Gymnosperm	Plant bearing seeds which are not enclosed in an ovary or carpel. Mostly coniferous species. See also Angiosperm.
Gynodioecy	Mating system where female and hermaphroditic flowers are found on separate individuals.
Gynoecium	All female reproductive parts of a flower including ovary, style and stigma.
Gynomonoecy	Mating system where individuals possess hermaphroditic as well as female flowers.

Haploid	Cell or organism with one set of chromosomes.
Haplotype	Genetic constitution of a non-recombining, haploid piece of DNA (e.g., mitochondrion or a piece of one of the homologous chromosomes).
Hardy-Weinberg law	Basic law of population genetics to determine frequencies of genotypes in a panmictic population for a polymorphic locus under various dominance processes, see M19.
Hardy-Weinberg equilibrium	Stable frequency distribution of genotypes AA, Aa and aa, in the proportions p^2, 2pq, and q^2, respectively, following random mating in the absence of mutation, migration, selection and genetic drift. It applies to a locus with two alleles, but can be extended to multiple alleles.
Herbicide	Chemical substance designed to kill plants, which can be either selective or non-selective. Selective herbicides kill some plant species or group of plants (e.g., grasses) without stunting other plants (e.g., dicotyledonous plants). Non-selective herbicides are toxic to all plants.
Herkogamy	Mating system where female and male sexual organs are spatially separated within the flower. The most common condition in plants.
Hermaphroditic	Plant where every flower possesses both female (pistil) and male (stamens) sexual organs.
Heteromorphic incompatibility	Mating system where the mating types are morphologically different.
Heterosis	Increased vigour of growth, fertility, etc., in a cross between two genetically different lines, as compared with growth, etc. in either of the parental lines, generally with the best parent. It is the result of new genetic combinations associated with increased heterozygosity. Syn.: Hybrid vigour.
Heterostyly	Mating system where a genetic polymorphism occurs in which the length of styles and stamens varies between individuals of a species. There can be two or three morphs differing reciprocally in the heights at which the stigmas and anthers are positioned (see Distyly and Tristyly). Hence, self-fertilization is prevented. The system is most often combined with a sporophytic incompatibility system with two or three mating types in accordance with the morphological morphs. See also Homostyly.
Homogamy	Mating system where anther dehiscence and stigma receptivity occur simultaneously.
Homomorphic incompatibility	Mating system where the mating types are morphologically similar.

Homostyly	Flower morph in which anthers and stigmas are positioned at the same level. Hence, self-fertilization is not prevented. See heterostyly.
Horizontal gene-transfer	Non-sexual transfer of genes between organisms by way of vectors able to insert DNA (e.g., bacteria). Genes may be transferred between unrelated species by this process.
Hybrid	Plant resulting from a cross between parents that are genetically dissimilar; often restricted to the offspring of two different species or of well-marked varieties within a species. Hybrids may be fertile (capable of producing offspring) or sterile. The more distant the genetical relationship between parents the greater is the probability that hybrids will be sterile.
Hybrid depression	Decreased vigour in growth, fertility, seed production, etc., in a cross between two genetically different lines, in comparison with either of the parental lines.
Hybrid swarm	Interbreeding between two species and their hybrids, showing a full range of variation from one parental genotype to the other plus recombinations.
Hybrid vigour	See Heterosis.
Inbreeding	Production of progeny by the mating of related parents, e.g., in small populations, or by selfing. Less variation occurs in the traits of the offspring, which often possesses lower fitness than outbreed progeny. The breeding system is non-panmictic, hence frequencies of heterozygotes fall below that expected by the Hardy-Weinberg equilibrium, see M32.
Inbreeding coefficient	Probability that the two copies of a gene in an individual are identical by descent, i.e., come from a common ancestor.
Inbreeding depression	Decrease of fitness of the offspring after inbreeding. Inbreeding increases the number of genes which are homozygous for deleterious alleles, see M32.
Incompatibility	Mating system where the inability for cross-fertilization between two individuals (genotypes) within a species is exclusively genetically based.
Inflorescence	Ramifications of the flowering part of a plant. Descriptive types constitute a classical tool in plant morphology which functionally may partly be adaptations for pollination.
Insert	Specific base pairs (i.e., alien DNA-segment) inserted into the genome. See also Insertion.
Inserted trait	New trait a plant attains as a result of genetic engineering.
Insertion	Process where one or more base pairs are inserted into the genome.

Intine	The inner of the two wall layers of a mature pollen grain.
Introgression	Transfer of segments of a genome and genes between species, subspecies or populations. Hybridization and backcrossing to one or both of the parental types result in incorporation of alleles from one taxon into the gene pool of the other.
Intron	Intron is a part of the DNA which is not encoding a functional protein.
Inversion	Chromosomal aberration where the DNA sequence of a chromosome segment has been inverted by turning it 180° with respect to the rest of the chromosome.
Isozyme	Multiple forms of an enzyme that catalyses the same reaction. These proteins have the same catalytic ability but different electric charge.
Lambda vector	Lambda phage vector can clone alien DNA up to a size of 20 kbp.
Linkage	The phenomenon that two genes are physically close together at the same chromosome. This excludes free recombination between the genes, resulting in non-Mendelian segregation ratios. The extent of linkage is measured in recombination units, termed Morgan, see M34.
Locus	Physical position of a gene on a chromosome. See also Gene.
Male sterility	Flowers with no, or only little, viable pollen (may functionally lead to gynodioecy).
Marker	A marker indicates a specific trait of a genotype. Different types of markers include: Visual traits (e.g., plant and flower morphology), tolerance (e.g., antibiotic-, pest-, disease- and herbicide-), protein profiles, allozymes, antibodies and DNA (nuclear, extra-nuclear and construct). Markers are mainly used for detecting and monitoring genotypes and for measuring gene-flow.
Marker gene	Gene of known effect, which marks individuals having a specific trait. The existence of this trait can then be visualised by means of physiological, morphological, biochemical or molecular marker tests. Marker genes are often inserted in transgenic plants during the transformation process.
Mating system	Describes in which pattern gametes are actually combined in produced offspring.
Mating type	Group of plants sharing incompatibility alleles and which, consequently, are not able to breed with each other.

Metapopulation	Set of single populations interconnected by migration of genetic material. The metapopulation is relatively stable, whereas the single populations have a more limited life-span. An equilibrium is present between extinction of already existing populations and establishment of new populations in unoccupied patches.
Microinjection	Method of genetical engineering where alien DNA is transferred into the cell nucleus by injection, which may result in incorporation of the DNA in the host genome, see T5.
Microprojectile bombardment	Method of genetical engineering where alien DNA is transferred by tungsten or gold particles shot by a particle gun, see T4.
Micropyle	The opening in the ovule through which the pollen tube enters.
Microsatellites	Simple sequences of DNA repeat which are scattered all over the genome.
Migration	In plants, the extent to which plant genes are dispersed either through pollen dispersal (haploid) or seed dispersal (diploid).
Mixed mating	Mating system where some seeds produced by a plant are the result of outcrossing while others are the result of selfing.
Monoecy	Mating system where every individual possesses female and male flowers. See also Dioecy.
mRNA	Messenger RNA, see RNA.
mtDNA	Mitochondrial DNA, which is present in mitochondria, is haploid and usually maternally inherited.
Mutation	Random change of a DNA sequence located on a chromosome.
Natural hybrid	Hybrid existing in nature which has been produced without any human intervention.
Nectar	Secretion produced by flowers or by glands inside flowers; consisting of sugars, amino acids and other compounds of value for visiting animals.
Nectar production	Amount of nectar (including concentrations of solutes) produced per flower during a specified time period (e.g., a day), see M41.
Neighbourhood-area	Area within a population from which there is a 95% probability that both parents (pollen and ovule) of the focal plant come from. Under panmixia, the genetic neighbourhood-area is equal to the whole population.
Neighbourhood-size	Number of genets within the neighbourhood-area.
Northern hybridization	Technique used for detecting gene expression, see M71.

Outbreeding	Breeding system which is panmictic, giving rise to heterozygote frequencies as predicted by the Hardy-Weinberg equilibrium, see M32.
Outcrossing	Mating system where pollination and fertilization occur between different genets.
Outcrossing rate	The probability that a seed has two genetically different parents (i.e., a ratio). Is equal to one minus the rate (i.e., ratio) of self-fertilization.
Ovary	Part of the gynoecium containing ovules. It is developing into a fruit after fertilization of the ovules.
Ovule	Reproductive structure containing the female gametophyte (embryo sac) and developing into a seed after fertilization.
P/O-ratio	The pollen-ovule ratio. The number of pollen grains divided by the number of ovules produced in a flower.
Panmictic unit	Number of individuals between which mating is approximately random.
Panmixis	Theoretical concept of a breeding system which has an infinitely large number of genetical individuals which are equally likely to mate with each other.
Pathogen tolerance	Tolerance against diseases caused by bacteria, fungi or vira.
Pedicel	Stalk of a flower.
PCR	Polymerase Chain Reaction. A method which amplifies a certain DNA sequence, see M59.
PEG	PolyEthylene Glycol, a chemical which enhances the ability of a cell to incorporate alien DNA, see also T3.
Pest tolerance	Tolerance against attacks of invertebrate animals, e.g., insects or nematodes.
Petal	One of the leafy structures that make up a flower; present as one or several whorls called a corolla inside the calyx, often adapted to attract pollinators.
Phenology	Study of periodically occurring biological events in nature, e.g., the timing of flowering in relation to pollinator activity.
Phenotype	Appearance of a given genotype in a given environment.
Phenotypic gender	Quantitative morphological sex expression of a flower or an individual. See also Functional gender.

Pin	Term for the long styled morph in a heterostylous system. See also Thrum.
Pistil	Ovary, style and stigma.
Plasmid	Small, self-replicating circular DNA which is often used as vectors in genetic engineering.
Plastid	Cellular organelle in higher plants, e.g., chloroplasts and mitochondria.
Pleiotropy	Different unrelated phenotypic traits which are controlled by the same gene.
Pollen dispersal	Spread of pollen within populations in distance and space which determines the chances of cross-pollination.
Pollen flow	Spread of pollen among populations.
Pollen grain	Male gametophyte in the reproductive cycle of angiosperms and gymnosperms.
Pollen sterility	The level of non-fertile pollen produced.
Pollen tube	Outgrowth of the intine of a pollen grain growing down the style during the fertilization process.
Pollen vector	Pollen transporting agent used for transfer of genes from one species to another, e.g., wind, a bee or a bird.
Pollination syndrome	One or more adaptive traits affecting the success of pollination. Syndrome characters may include: Flower morphology (e.g., colour, shape and texture), type of inflorescence, breeding system (e.g., heterostyly and phenology), chemical attractants (nectar and scent) and pollen production. Hence a plant species or group of plant species may be adapted to different kinds of pollinators or physical agents (e.g., water and wind). See Chapter 6, p. 55.
Pollinator	Biotic pollen dispersal vector (agent), e.g., a bat, a bee or a bird.
Pollinator activity	Amount of movement of pollinators within and between individual plants in a population.
Pollinator foraging behaviour	Handling of flowers and their resources, and movement pattern between flowers by a pollinator.
Polymorphism	A trait which may occur in different forms either in a population or in an individual at different life stages. See also Marker.
Polyploid	Individual with more than two sets of chromosomes.

Population	1. Group of individuals separated in space or time from other groups of the same species. 2. Group of reproductive individuals sharing a common gene pool (genetic population). See also meta-population.
Promoter	Region of DNA involved in transcription of a gene with sites for gene regulatory proteins.
Protandry	Mating system where the male stage (anthers dehiscence) occurs before the female stage (stigma receptivity).
Protogyny	Mating system where the female stage (stigma receptivity) occurs before the male stage (anthers dehiscence).
Protoplast	Plant cell in which the cell wall has been removed.
Pseudogamy	Mating system where pollination of stigmas is necessary to trigger development of ovules, but no fertilization takes place.
ptDNA	Plastid DNA, see Plastid.
QTL	Quantitative Trait Loci,
Ramet	Physically independent plant. Often several neighbouring ramets have a common origin and thus are genetically identical. Together they constitute a genet or a clone.
RAPD	Random Amplified Polymorphic DNA, see M61.
Resistance	Resistant plants grow and reproduce normally when exposed to high doses (levels) of a specific agent (e.g., herbicide, insect attack or pathogen infection), which would harm non-resistant plants. Resistance and Tolerance are often used indiscriminately, but in this book the terms have separate definitions. See also Tolerance.
Restriction enzyme	Endonuclease that recognises a specific sequence of bases in the double-stranded DNA. The restriction enzymes used in recombinant DNA technology bind and cut within their recognition site. They can cut the two DNA strands exactly opposite one another and generate blunt ends or they may make staggered cuts to generate sticky ends.
Reverse transcriptase	Enzyme (from retrovira) which makes a DNA copy of an RNA molecule when primed with a suitable oligodeoxyribonucleotide. The enzyme is commonly used for making DNA copies of mRNA molecules in cDNA cloning.
RFLP	Restriction Fragment Length Polymorphism, see M64.
RNA	RiboNucleic Acid, a molecule which, in various forms, participates in the translation of the genetic information from DNA into proteins.

Seed	Product of a fertilized ovule, consisting of an embryo enclosed by protective seed coats. Most seeds contain endosperm with nutrition surrounding the embryo.
Segregating alleles	Pair of alleles at a locus which will segregate into different gametes and hence into different offspring at meiosis. See M9.
Self-compatibility	Mating system where individuals are capable of self-fertilization.
Self-fertilization	Fertilization of ovules with gametes from the same individual.
Self-incompatibility	Mating system where individuals are incapable of self-fertilization or fertilization by pollen expressing the same incompatibility alleles.
Self-pollination	Pollination of a stigma with pollen from the same or another flower (geitonogamy) on the same individual.
Selfing	Mating system where flowers are pollinated with pollen from the same genet.
Sepal	Individual segment of the calyx, present in a whorl in the flower outside the corolla, often green.
Somaclonal variation	Genetic aberrations in individuals from *in vitro* culture in comparison to the genotype of the donor tissue.
Southern hybridization	Technique used for detecting fragments of DNA, see M71.
Sporophyte	Diploid phase of the reproductive cycle of a plant.
Sporophytic incompatibility	Mating system where the alleles in the pollen donor plant determine the compatibility type of the pollen grain. Hence, all pollen express the same phenotype of incompatibility.
SSR	Simple Sequence Repeats. See Microsatellites and M38.
Stamen	Male structure of sexual reproduction, consisting of a filament that supports the anther, which produces the pollen.
Stigma	Distal part of the pistil on which pollen grains are deposited and are germinating.
Stigma receptivity	Period or stage in the life-span of a flower where its stigma is receptive to pollen grains. The deposited pollen which are compatible will then produce tubes penetrating the style.
Stress tolerance	Tolerance to changed environmental condition for growth, e.g., salt, drought, temperature or day length.
Structural chromosome changes	Changes in structure of chromosome (e. g. deletion, duplication, insertion, inversion, translocation) or genome (e.g., addition, amphidiploid, aneuploid, polyploid, substitution).

Style	Narrow part of the gynoecium between the ovary and the stigma; containing the transmission tissue important to pollen tube growth.
Subdivided population	Population subdivided into smaller units between which migration is possible but restricted. This allows for genetic differentiation to build up through genetic drift.
Substitution	One or more chromosomes of the original species are substituted by chromosomes from another species or genotype.
Tapetum	The secreting inner part of the anther wall involved in pollen development.
Thrum	Term for the short styled morph in a heterostylous system.
Tolerance	Tolerant plants are less susceptible to damage from a specific agent (e.g., herbicide, insect attack or pathogen infection) than a non-tolerant plant, but high doses (levels) of the agent will diminish growth and reproduction. Resistance and Tolerance are often used indiscriminately, but in this book the terms have separate definitions (see resistance).
Transformation	Incorporation of alien DNA into the genome of plants and other eucaryote organisms by various techniques (e.g., *Agrobacterium*-mediated gene transfer), see T1. Transformation may also occur naturally in plant-microorganism interactions.
Transgenic plant	Plant with genes inserted from another organism through non-conventional gene transfer (transformation).
Translocation	Chromosomal aberration resulting in a changed position of a chromosome segment within the genome.
Transposition	Transfer of transposons (i.e., jumping genes). The transposons may be enhanced in a genome, can change positions and cause a mutation.
Tristyly	Heterostylous mating system where there are three morphs differing in anther and stigma positions and compatibility.
Vector	DNA or RNA molecule used as a vehicle for transfer of genes from one species to another. Examples of vectors used for genetic transfer: Lambda vector, Plasmids, YAC vector (see separate entries). See also Pollen vector.
Vertical gene-transfer	Introduction of genes from one plant species into another through introgressive hybridization.
Visual attractant	Any visual clue in form of inflorescence, flower shape, colour (including UV-emission) or size which attracts a pollinator to move towards the flower.

Xenobiotic   Substance or chemical compound which does not originate from
                                    biological processes.

Xenogamy   Mating system where pollination and fertilization occur with pollen
                                    produced by another genet.

YAC vector   Yeast Artificial Chromosome vector. YAC vectors can clone alien
                                    DNA up to a size of 400 kbp.

Zygote   First cell of a new individual in sexual reproduction from the fusion
                                    of two gametes.

3. Categories and corresponding subcategories

For the key entrance words (see Chapter 1) the following main topics were chosen, which are relevant to the reader looking for specific information: **Reproduction** including pollen development and production, seed development and production, pollination and breeding system, **gene-flow** including hybridization, **genome structure** at chromosome and gene level, and **population**. The three first topics are connected to the dispersal of an inserted gene at the individual level, whereas the last focuses on the dispersal among plant populations. Furthermore, the major topics **genetic engineering techniques**, **inserted traits** and **test procedures for risk assessment** are included. The subcategory-words are listed in alphabetical order for each category. Explanations of the subcategory-words are available in the glossary, Chapter 2.

Category	Subcategory
Pollen development and production	Pollen competition
	Pollen germination
	Pollen production
	Pollen viability
Seed development and production	Agamospermy
	Fertilization
	Fruit abortion
	Ovule development
	Seed abortion
	Seed germination (see also vol. 1)
	Seed production (see also vol. 1)
Pollination	Chemical attractant
	Nectar production
	Pollen dispersal
	Pollinator activity
	Pollinator foraging behaviour
	Visual attractant
Pollination syndrome	Ant-pollination
	Bat-pollination
	Bee-pollination
	Beetle-pollination
	Bird-pollination
	Butterfly-pollination
	Fly-pollination
	Insect-pollination
	Mammal-pollination
	Water-pollination
	Wind-pollination

Breeding system	Cross-pollination
	Flowering phenology
	Heterostyly
	Incompatibility
	Phenology
	Reproductive allocation (see also vol. 1)
	Self-pollination
	Sex distribution
Gene-flow	Gene-transfer
	Horizontal gene-transfer
	Introgression
	Marker
	Outcrossing
	Vertical gene-transfer
Hybridization	Artificial hybrid
	Heterosis
	Hybrid
	Hybrid depression
	Hybrid vigour
	Natural hybrid
Genome structure	Epistasis
	Gene expression
	Gene stability
	Genotype
	Insert
	Insertion
	Pleiotropy
	Recombination
	Structural chromosome changes
Population	Cline
	Gene pool
	Genetic diversity
	Genetic drift
	Genetic load
	Genetic neighbourhood-area
	Metapopulation
	Polymorphism
Genetic engineering technique	Biological vector
	Chemical poration
	Electroporation
	Microinjection
	Microprojectile bombardment
	Transformation

Inserted trait

Altered flower colour
Altered metabolic content
Bacteria tolerance
Drought tolerance
Frost tolerance
Fungal tolerance
Herbicide tolerance
Insect tolerance
Male sterility
Nematode tolerance
Oxidative stress tolerance
Pathogen tolerance
Pest tolerance
Pollen sterility
Salt tolerance
Stress tolerance
Viral tolerance

Test procedure

Contained use (contained trial)
Data analysis
Deliberate release
Experimental design (narrow sense)
Field experiment
Greenhouse experiment
Growth chamber experiment
Laboratory experiment
Large scale release
Long-term monitoring
Model
Release
Risk assessment
Short-term monitoring
Small scale release

4. List of subcategories with corresponding methods

The different methods available for the study of specific subcategories are given below. Method numbers used in Section 7.2 are shown in parentheses. Subcategory words from the categories "genetic engineering technique", "inserted trait" and "test procedure" are not included. See Chapters 8, 9 and 10 for further information.

Subcategory	Method
Agamospermy	Bagging of flowers (M2) Emasculation of flowers (M12) Experimental pollination (M13) Marking plants and flowers: Flowering phenology analysis (M36) Ovule counts (M44) Pollen viability tests (M54) Seed development analysis (M66)
Ant-pollination	See Insect-pollination
Artificial hybrid	Experimental pollination (M13) Forced fertilization (M18) See also Hybrid
Bat-pollination	See Mammal-pollination
Bee-pollination	See Insect-pollination
Beetle-pollination	See Insect-pollination
Bird-pollination	Pollen carry-over between flowers (M49) Pollen deposition on stigmas: Pollen numbers and fertility (M51) Pollinator attraction: Visual cues (M56) Pollinator foraging behaviour (M57) Pollinator preference experiments (M58)
Butterfly-pollination	See Insect-pollination
Chemical attractant	Gas chromatography (GC) (M20) Nectar production: Chemical composition (M40) Pollinator attraction: Chemical cues (M55) Reproduction in alien plants: Community and invader analysis (M62) Reproductive allocation measures (M63) Thin layer chromatography (TLC) (M74)
Cline	Local adaptation analysis (M35) Transplantation experiment (M76)

Cross-pollination   Bagging of flowers (M2)
      Emasculation of flowers (M12)
      Experimental pollination (M13)
      Fluorescent dyes and marking of pollen (M17)
      Marking plants and flowers: Flowering phenology analysis (M36)
      Ovule counts (M44)
      Paternity analysis (M45)
      Pollen carry-over between flowers (M49)
      Pollen counts (M50)
      Pollen deposition on stigmas: Pollen numbers and fertility (M51)
      Pollen germination tests (M52)
      Pollen traps (M53)
      Pollinator foraging behaviour (M57)
      Reproduction in alien plants: Community and invader analysis
      (M62)
      Seed development analysis (M66)
      Seed germination tests (M67)
      Seed viability tests (M68)
      Spatial autocorrelation analysis and Moran's I (M72)
      Transplantation experiment (M76)

Epistasis   Diallel cross (M9)

Fertilization   Classification of fruits, seeds and ovules (M6)
      Experimental pollination (M13)
      Forced fertilization (M18)
      Pollen germination tests (M52)
      Seed development analysis (M66)
      Seed germination tests (M67)
      Stigma receptivity test (M73)

Flowering phenology   See Phenology

Fly-pollination   See Insect-pollination

Fruit abortion   Classification of fruits, seeds and ovules (M6)
      Marking plants and flowers: Flowering phenology analysis (M36)
      Ovule counts (M44)
      Reproductive allocation measures (M63)
      Seed development analysis (M66)
      Seed germination tests (M67)
      Seed viability tests (M68)
      Thinning of flowers (M75)

Gene expression   Immunological methods (M30)
      Protein electrophoresis: Isozyme analysis (M60)
      Southern and Northern hybridization (blotting) (M71)

Gene pool   Fitness measurement (M14)
      Genet identification (M25)
      Plant material preservation: Cryopreservation and freeze-drying
      (M47)
      Plant material preservation: Rapid drying (M48)

Gene stability	Immunological methods (M30) Protein electrophoresis: Isozyme analysis (M60) Southern and Northern hybridization (blotting) (M71)
Genetic diversity	Computer programs for analysis of genetic data (M8) DNA sequencing (M10) Fitness measurement (M14) F-statistics (M19) Gene-flow estimation with private alleles (M21) Genetic distance (M22) Genet identification (M25) Haplotype statistics (M26) Herbarium sheet survey (M27) Karyotype analysis: Chromosome number, size and form (M33) Nuclear and organelle markers combined (M42) Organelle DNA analysis (M43) Paternity analysis (M45) Plant material preservation: Cryopreservation and freeze-drying (M47) Plant material preservation: Rapid drying (M48) Ribosomal DNA (rDNA) analysis (M65) Selfing and outcrossing rate (M69) Spatial autocorrelation analysis and Moran's I (M72)
Genetic drift	Computer programs for analysis of genetic data (M8) Effective population-size N_e (M11) F-statistics (M19) Genetic distance (M22) Genetic neighbourhood-size (M24) Haplotype statistics (M26) Linkage disequilibrium (M34) Selfing and outcrossing rate (M69) Spatial autocorrelation analysis and Moran's I (M72)
Genetic load	Fitness measurement (M14) Inbreeding depression estimation (M32)
Genetic neighbourhood-area	Effective population-size N_e (M11) Genetic neighbourhood-size (M24) Paternity analysis (M45) Pollen carry-over between flowers (M49) Spatial autocorrelation analysis and Moran's I (M72)
Gene-transfer	See Vertical gene-transfer
Genotype	Amplified restriction fragment polymorphism (AFLP) (M1) Bioassay (M3) Diallel cross (M9) DNA sequencing (M10) Gas chromatography (GC) (M20) Genetic map construction (M23)

Genotype, contd.	High performance liquid chromatography (HPLC) (M29)
	Immunological methods (M30)
	In situ hybridization to chromosomes (M31)
	Karyotype analysis: Chromosome number, size and form (M33)
	Meiotic analysis: Chromosome pairing and recombination (M37)
	Microsatellite markers (M38)
	Organelle DNA analysis (M43)
	Polymerase chain reaction (PCR) (M59)
	Protein electrophoresis: Isozyme analysis (M60)
	Random amplified polymorphic DNA (RAPD) (M61)
	Restriction fragment length polymorphism (RFLP) (M64)
	Ribosomal DNA (rDNA) analysis (M65)
	Southern and Northern hybridization (blotting) (M71)
	Thin layer chromatography (TLC) (M74)
	Two-dimensional paper chromatography (M77)

Genotype, contd.

High performance liquid chromatography (HPLC) (M29)
Immunological methods (M30)
In situ hybridization to chromosomes (M31)
Karyotype analysis: Chromosome number, size and form (M33)
Meiotic analysis: Chromosome pairing and recombination (M37)
Microsatellite markers (M38)
Organelle DNA analysis (M43)
Polymerase chain reaction (PCR) (M59)
Protein electrophoresis: Isozyme analysis (M60)
Random amplified polymorphic DNA (RAPD) (M61)
Restriction fragment length polymorphism (RFLP) (M64)
Ribosomal DNA (rDNA) analysis (M65)
Southern and Northern hybridization (blotting) (M71)
Thin layer chromatography (TLC) (M74)
Two-dimensional paper chromatography (M77)

Heterosis

Diallel cross (M9)
Fitness measurement (M14)
Heterosis analysis: Over mid-parent (HMP) and better parent (HBP) (M28)
Morphological character analysis (M39)
Photosynthetic efficiency (M46)

Heterostyly

Fluorescent dyes and marking of pollen (M17)
Pollinator preference experiments (M58)

Horizontal gene-transfer

See Marker
See also Chapter 8

Hybrid

Amplified restriction fragment polymorphism (AFLP) (M1)
Bagging of flowers (M2)
Chromosome painting (M5)
Computer programs for analysis of genetic data (M8)
Diallel cross (M9)
DNA sequencing (M10)
Emasculation of flowers (M12)
Flow cytometry (M15)
Fluorescent dyes and marking of pollen (M17)
Gas chromatography (GC) (M20)
Herbarium sheet survey (M27)
High performance liquid chromatography (HPLC) (M29)
Immunological methods (M30)
In situ hybridization to chromosomes (M31)
Karyotype analysis: Chromosome number, size and form (M33)
Meiotic analysis: Chromosome pairing and recombination (M37)
Microsatellite markers (M38)
Morphological character analysis (M39)
Nuclear and organelle markers combined (M42)
Organelle DNA analysis (M43)
Plant material preservation: Cryopreservation and freeze-drying (M47)

Hybrid, contd.	Plant material preservation: Rapid drying (M48)
	Pollen counts (M50)
	Pollen deposition on stigmas: Pollen numbers and fertility (M51)
	Pollen germination tests (M52)Pollen viability tests (M54)
	Polymerase chain reaction (PCR) (M59)
	Protein electrophoresis: Isozyme analysis (M60)
	Random amplified polymorphic DNA (RAPD) (M61)
	Restriction fragment length polymorphism (RFLP) (M64)
	Ribosomal DNA (rDNA) analysis (M65)
	Seed germination tests (M67)
	Seed viability tests (M68)
	Southern and Northern hybridization (blotting) (M71)
	Thin layer chromatography (TLC) (M74)
	Transplantation experiment (M76)
	Two-dimensional paper chromatography (M77)
Hybrid depression	Diallel cross (M9)
	Fitness measurement (M14)
	Heterosis analysis: Over mid-parent (HMP) and better parent (HBP) (M28)
	Morphological character analysis (M39)
	Photosynthetic efficiency (M46)
Hybrid vigour	Diallel cross (M9)
	Fitness measurement (M14)
	Heterosis analysis: Over mid-parent (HMP) and better parent (HBP) (M28)
	Morphological character analysis (M39)
	Photosynthetic efficiency (M46)
Incompatibility	Bagging of flowers (M2)
	Classification of fruits, seeds and ovules (M6)
	Diallel cross (M9)
	Experimental pollination (M13)
	Forced fertilization (M18)
	High performance liquid chromatography (HPLC) (M29)
	Marking plants and flowers: Flowering phenology analysis (M36)
	Pollen counts (M50)
	Pollen deposition on stigmas: Pollen numbers and fertility (M51)
	Pollen germination tests (M52)
	Seed development analysis (M66)
	Stigma receptivity test (M73)
Insect-pollination	Pollen carry-over between flowers (M49)
	Pollen deposition on stigmas: Pollen numbers and fertility (M51)
	Pollinator attraction: Chemical cues (M55)
	Pollinator attraction: Visual cues (M56)
	Pollinator foraging behaviour (M57)
	Pollinator preference experiments (M58)
Insert	See Marker

Insertion	See Marker

Introgression
Amplified restriction fragment polymorphism (AFLP) (M1)
Biogeographical assay and monitoring (M4)
Chromosome painting (M5)
Composite character index (M7)
Computer programs for analysis of genetic data (M8)
DNA sequencing (M10)
Experimental pollination (M13)
Fitness measurement (M14)
Flow cytometry (M15)
Genetic distance (M22)
Haplotype statistics (M26)
Herbarium sheet survey (M27)
Immunological methods (M30)
In situ hybridization to chromosomes (M31)
Karyotype analysis: Chromosome number, size and form (M33)
Meiotic analysis: Chromosome pairing and recombination (M37)
Microsatellite markers (M38)
Morphological character analysis (M39)
Nuclear and organelle markers combined (M42)
Plant material preservation: Cryopreservation and freeze-drying (M47)
Plant material preservation: Rapid drying (M48)
Protein electrophoresis: Isozyme analysis (M60)
Random amplified polymorphic DNA (RAPD) (M61)
Restriction fragment length polymorphism (RFLP) (M64)
Ribosomal DNA (rDNA) analysis (M65)
Southern and Northern hybridization (blotting) (M71)

Mammal-pollination
Pollen carry-over between flowers (M49)
Pollen deposition on stigmas: Pollen numbers and fertility (M51)
Pollinator attraction: Chemical cues (M55)
Pollinator attraction: Visual cues (M56)
Pollinator foraging behaviour (M57)
Pollinator preference experiments (M58)

Marker
Amplified restriction fragment polymorphism (AFLP) (M1)
Gas chromatography (GC) (M20)
High performance liquid chromatography (HPLC) (M29)
Immunological methods (M30)
Microsatellite markers (M38)
Polymerase chain reaction (PCR) (M59)
Protein electrophoresis: Isozyme analysis (M60)
Random amplified polymorphic DNA (RAPD) (M61)
Restriction fragment length polymorphism (RFLP) (M64)
Southern and Northern hybridization (blotting) (M71)
Thin layer chromatography (TLC) (M74)
Two-dimensional paper chromatography (M77)

Metapopulation

Computer programs for analysis of genetic data (M8)
Effective population-size N_e (M11)
F-statistics (M19)
Gene-flow estimation with private alleles (M21)
Genetic distance (M22)
Genetic neighbourhood-size (M24)
Genet identification (M25)
Haplotype statistics (M26)
Linkage disequilibrium (M34)
Morphological character analysis (M39)
Nuclear and organelle markers combined (M42)
Paternity analysis (M45)
Ribosomal DNA (rDNA) analysis (M65)
Spatial autocorrelation analysis and Moran's I (M72)

Natural hybrid

Biogeographical assay and monitoring (M4)
Composite character index (M7)
See also Hybrid

Nectar production

Bagging of flowers (M2)
Marking plants and flowers: Flowering phenology analysis (M36)
Nectar production: Chemical composition (M40)
Nectar production: Volume and rate (M41)
Pollinator attraction: Chemical cues (M55)
Pollinator foraging behaviour (M57)
Reproductive allocation measures (M63)

Outcrossing

Bagging of flowers (M2)
Computer programs for analysis of genetic data (M8)
Emasculation of flowers (M12)
Experimental pollination (M13)
Fluorescent dyes and marking of pollen (M17)
Forced fertilization (M18)
F-statistics (M19)
Gene-flow estimation with private alleles (M21)
Genetic distance (M22)
Genetic neighbourhood-size (M24)
Haplotype statistics (M26)
Inbreeding depression estimation (M32)
Linkage disequilibrium (M34)
Marking plants and flowers: Flowering phenology analysis (M36)
Morphological character analysis (M39)
Nuclear and organelle markers combined (M42)
Paternity analysis (M45)
Pollen carry-over between flowers (M49)
Pollen counts (M50)
Selfing and outcrossing rate (M69)
Spatial autocorrelation analysis and Moran's I (M72)

Ovule development

Emasculation of flowers (M12)
Experimental pollination (M13)
Marking plants and flowers: Flowering phenology analysis (M36)
Ovule counts (M44)

Ovule development, contd.	Seed development analysis (M66) Seed germination tests (M67) Seed viability tests (M68)
Phenology	Bagging of flowers (M2) Classification of fruits, seeds and ovules (M6) Emasculation of flowers (M12) Experimental pollination (M13) Fluorescent dyes and marking of pollen (M17) Marking plants and flowers: Flowering phenology analysis (M36) Ovule counts (M44) Pollen counts (M50) Pollen deposition on stigmas: Pollen numbers and fertility (M51) Pollen germination tests (M52) Pollen traps (M53) Pollen viability tests (M54) Pollinator attraction: Chemical cues (M55) Pollinator attraction: Visual cues (M56) Pollinator preference experiments (M58) Reproduction in alien plants: Community and invader analysis (M62) Seed development analysis (M66) Seed germination tests (M67) Seed viability tests (M68) Sexual morph distribution (M70) Stigma receptivity test (M73) Thinning of flowers (M75) Transplantation experiment (M76)
Plant competition	Photosynthetic efficiency (M46) See also vol. 1.
Pleiotropy	Diallel cross (M9)
Pollen competition	Bagging of flowers (M2) Experimental pollination (M13) Pollen germination tests (M52) Seed germination tests (M67)
Pollen dispersal	Bagging of flowers (M2) Experimental pollination (M13) Fluorescent dyes and marking of pollen (M17) Genetic neighbourhood-size (M24) Marking plants and flowers: Flowering phenology analysis (M36) Paternity analysis (M45) Pollen carry-over between flowers (M49) Pollen deposition on stigmas: Pollen numbers and fertility (M51) Pollen traps (M53) Pollen viability tests (M54) Pollinator foraging behaviour (M57) Spatial autocorrelation analysis and Moran's I (M72)

Pollen germination	Bagging of flowers (M2) Experimental pollination (M13) Pollen counts (M50) Pollen deposition on stigmas: Pollen numbers and fertility (M51) Pollen germination tests (M52) Pollen viability tests (M54) Stigma receptivity test (M73) Thinning of flowers (M75)
Pollen production	Bagging of flowers (M2) Marking plants and flowers: Flowering phenology analysis (M36) Pollen counts (M50) Pollen deposition on stigmas: Pollen numbers and fertility (M51) Pollen traps (M53) Reproductive allocation measures (M63) Thinning of flowers (M75)
Pollen viability	Experimental pollination (M13) High performance liquid chromatography (HPLC) (M29) Marking plants and flowers: Flowering phenology analysis (M36) Pollen counts (M50) Pollen germination tests (M52) Pollen viability tests (M54)
Pollinator activity	Fluorescent dyes and marking of pollen (M17) Marking plants and flowers: Flowering phenology analysis (M36) Pollen deposition on stigmas: Pollen numbers and fertility (M51) Pollinator attraction: Chemical cues (M55) Pollinator attraction: Visual cues (M56) Pollinator foraging behaviour (M57) Reproduction in alien plants: Community and invader analysis (M62)
Pollinator foraging behaviour	Fluctuating asymmetry of vegetative and floral traits (FA) (M16) Gas chromatography (GC) (M20) Nectar production: Chemical composition (M40) Nectar production: Volume and rate (M41) Pollen carry-over between flowers (M49) Pollinator attraction: Chemical cues (M55) Pollinator attraction: Visual cues (M56) Pollinator foraging behaviour (M57) Pollinator preference experiments (M58) Reproduction in alien plants: Community and invader analysis (M62)
Polymorphism	Amplified restriction fragment polymorphism (AFLP) (M1) Bioassay (M3) DNA sequencing (M10) Gas chromatography (GC) (M20) Herbarium sheet survey (M27) High performance liquid chromatography (HPLC) (M29)

Polymorphism, contd.   Immunological methods (M30)
                         Microsatellite markers (M38)
                         Polymerase chain reaction (PCR) (M59)
                         Protein electrophoresis: Isozyme analysis (M60)
                         Random amplified polymorphic DNA (RAPD) (M61)
                         Restriction fragment length polymorphism (RFLP) (M64)
                         Southern and Northern hybridization (blotting) (M71)
                         Thin layer chromatography (TLC) (M74)
                         Two-dimensional paper chromatography (M77)

Recombination   Genetic map construction (M23)
                         Meiotic analysis: Chromosome pairing and recombination (M37)

Reproductive allocation Bagging of flowers (M2)
                         Classification of fruits, seeds and ovules (M6)
                         Experimental pollination (M13)
                         Marking plants and flowers: Flowering phenology analysis (M36)
                         Nectar production: Chemical composition (M40)
                         Nectar production: Volume and rate (M41)
                         Ovule counts (M44)
                         Pollen counts (M50)
                         Pollinator attraction: Chemical cues (M55)
                         Pollinator attraction: Visual cues (M56)
                         Reproduction in alien plants: Community and invader analysis
                          (M62)
                         Reproductive allocation measures (M63)
                         Seed development analysis (M66)
                         Seed germination tests (M67)
                         Seed viability tests (M68)
                         Sexual morph distribution (M70)
                         Thinning of flowers (M75)
                         Transplantation experiment (M76)

Seed abortion   Classification of fruits, seeds and ovules (M6)
                         Marking plants and flowers: Flowering phenology analysis (M36)
                         Ovule counts (M44)
                         Reproductive allocation measures (M63)
                         Seed development analysis (M66)
                         Seed germination tests (M67)
                         Seed viability tests (M68)
                         Thinning of flowers (M75)

Seed germination   Bagging of flowers (M2)
                         Experimental pollination (M13)
                         Marking plants and flowers: Flowering phenology analysis (M36)
                         Seed development analysis (M66)
                         Seed germination tests (M67)
                         Seed viability tests (M68)
                         Thinning of flowers (M75)
                         See also vol. 1

Seed production	Bagging of flowers (M2) Emasculation of flowers (M12) Experimental pollination (M13) High performance liquid chromatography (HPLC) (M29) Marking plants and flowers: Flowering phenology analysis (M36) Ovule counts (M44) Reproductive allocation measures (M63) Seed development analysis (M66) Seed germination tests (M67) Seed viability tests (M68) Stigma receptivity test (M73) Thinning of flowers (M75) Transplantation experiment (M76) See also vol. 1
Self-pollination	Bagging of flowers (M2) Emasculation of flowers (M12) Experimental pollination (M13) Inbreeding depression estimation (M32) Marking plants and flowers: Flowering phenology analysis (M36) Ovule counts (M44) Paternity analysis (M45) Pollen carry-over between flowers (M49) Pollen counts (M50) Pollen deposition on stigmas: Pollen numbers and fertility (M51) Pollen germination tests (M52) Pollinator foraging behaviour (M57) Seed development analysis (M66) Seed germination tests (M67) Seed viability tests (M68) Spatial autocorrelation analysis and Moran's I (M72) Transplantation experiment (M76)
Sex distribution	Marking plants and flowers: Flowering phenology analysis (M36) Ovule counts (M44) Sexual morph distribution (M70)
Stress tolerance	Bioassay (M3) Fluctuating asymmetry of vegetative and floral traits (FA) (M16) Gas chromatography (GC) (M20) Photosynthetic efficiency (M46)
Structural chromosome changes	Chromosome painting (M5) Flow cytometry (M15) Genetic map construction (M23) *In situ* hybridization to chromosomes (M31) Karyotype analysis: Chromosome number, size and form (M33) Meiotic analysis: Chromosome pairing and recombination (M37)

Vertical gene-transfer   Amplified restriction fragment polymorphism (AFLP) (M1)
                                Biogeographical assay and monitoring (M4)
                                Chromosome painting (M5)
                                Composite character index (M7)
                                Computer programs for analysis of genetic data (M8)
                                DNA sequencing (M10)
                                Experimental pollination (M13)
                                Fitness measurement (M14)
                                Flow cytometry (M15)
                                Gene-flow estimation with private alleles (M21)
                                Genetic distance (M22)
                                Haplotype statistics (M26)
                                Herbarium sheet survey (M27)
                                Immunological methods (M30)
                                In situ hybridization to chromosomes (M31)
                                Karyotype analysis: Chromosome number, size and form (M33)
                                Microsatellite markers (M38)
                                Morphological character analysis (M39)
                                Nuclear and organelle markers combined (M42)
                                Organelle DNA analysis (M43)
                                Plant material preservation: Cryopreservation and freeze-drying
                                  (M47)
                                Plant material preservation: Rapid drying (M48)
                                Protein electrophoresis: Isozyme analysis (M60)
                                Random amplified polymorphic DNA (RAPD) (M61)
                                Restriction fragment length polymorphism (RFLP) (M64)
                                Ribosomal DNA (rDNA) analysis (M65)
                                Southern and Northern hybridization (blotting) (M71)

Visual attractant   Pollinator attraction: Visual cues (M56)
                                Reproduction in alien plants: Community and invader analysis
                                  (M62)
                                Reproductive allocation measures (M63)

Water-pollination   Pollen deposition on stigmas: Pollen numbers and fertility (M51)

Wind-pollination   Pollen traps (M53)

5. List of methods with corresponding subcategories

The potential applicability of the methods described in Chapter 7 for studying various subcategories are listed below.

Method		Subcategory
M1.	Amplified restriction fragment polymorphism (AFLP)	introgression marker vertical gene-transfer hybrid genotype polymorphism
M2.	Bagging of flowers	pollen competition pollen germination pollen production agamospermy seed germination seed production nectar production pollen dispersal cross-pollination incompatibility phenology reproductive allocation self-pollination outcrossing hybrid
M3.	Bioassay	genotype polymorphism stress tolerance
M4.	Biogeographical assay and monitoring	introgression vertical gene-transfer natural hybrid
M5.	Chromosome painting	introgression vertical gene-transfer hybrid structural chromosome changes
M6.	Classification of fruits, seeds and ovules	fertilization fruit abortion seed abortion incompatibility phenology reproductive allocation

M7.	Composite character index	introgression vertical gene-transfer natural hybrid
M8.	Computer programmes for analysis of genetic data	introgression outcrossing vertical gene-transfer hybrid genetic diversity genetic drift metapopulation
M9.	Diallel cross	incompatibility heterosis hybrid hybrid depression hybrid vigour epistasis genotype Pleiotropy
M10.	DNA sequencing	introgression vertical gene-transfer hybrid genotype genetic diversity polymorphism
M11.	Effective population-size N_e	genetic drift genetic neighbourhood-area metapopulation
M12.	Emasculation of flowers	agamospermy ovule development seed production cross-pollination phenology self-pollination outcrossing hybrid

M13.	Experimental pollination	pollen competition
		pollen germination
		pollen viability
		agamospermy
		fertilization
		ovule development
		seed germination
		seed production
		pollen dispersal
		cross-pollination
		incompatibility
		phenology
		reproductive allocation
		self-pollination
		introgression
		outcrossing
		vertical gene-transfer
		artificial hybrid
M14.	Fitness measurement	introgression
		vertical gene-transfer
		heterosis
		hybrid depression
		hybrid vigour
		gene pool
		genetic diversity
		genetic load
M15.	Flow cytometry	introgression
		vertical gene-transfer
		hybrid
		structural chromosome changes
M16.	Fluctuating asymmetry of vegetative and floral traits (FA)	pollinator foraging behaviour stress tolerance
M17.	Fluorescent dyes and marking of pollen	pollen dispersal
		pollinator activity
		cross-pollination
		heterostyly
		phenology
		outcrossing
		hybrid
M18.	Forced fertilization	fertilization
		incompatibility
		outcrossing
		artificial hybrid

M19.	F-statistics	outcrossing genetic drift genetic diversity metapopulation
M20.	Gas chromatography (GC)	chemical attractant pollinator foraging behaviour marker hybrid genotype polymorphism stress tolerance
M21.	Gene-flow estimation with private alleles	outcrossing vertical gene-transfer genetic diversity metapopulation
M22.	Genetic distance	introgression outcrossing vertical gene-transfer genetic diversity genetic drift metapopulation
M23.	Genetic map construction	genotype recombination structural chromosome changes
M24.	Genetic neighbourhood-size	pollen dispersal outcrossing genetic drift genetic neighbourhood-area metapopulation
M25.	Genet identification	gene pool genetic diversity metapopulation
M26.	Haplotype statistics	introgression outcrossing vertical gene-transfer genetic diversity genetic drift metapopulation
M27.	Herbarium sheet survey	introgression vertical gene-transfer hybrid genetic diversity polymorphism

M28.	Heterosis analysis: Over mid-parent (HMP) and better parent (HBP)	heterosis hybrid depression hybrid vigour
M29.	High performance liquid chromatography (HPLC)	pollen viability seed production incompatibility marker hybrid genotype polymorphism
M30.	Immunological methods	introgression marker vertical gene-transfer hybrid gene expression gene stability genotype polymorphism
M31.	*In situ* hybridization to chromosomes	introgression vertical gene-transfer hybrid genotype structural chromosome changes
M32.	Inbreeding depression estimation	self-pollination outcrossing genetic load
M33.	Karyotype analysis: Chromosome number, size and form	introgression vertical gene-transfer hybrid genotype structural chromosome changes genetic diversity
M34.	Linkage disequilibrium	outcrossing genetic drift metapopulation
M35.	Local adaptation analysis	cline

M36. Marking plants   pollen production
      and flowers: Flowering   pollen viability
      phenology analysis   agamospermy
                                  fruit abortion
                                  ovule development
                                  seed abortion
                                  seed germination
                                  seed production
                                  nectar production
                                  pollen dispersal
                                  pollinator activity
                                  cross-pollination
                                  incompatibility
                                  phenology
                                  reproductive allocation
                                  self-pollination
                                  sex distribution
                                  outcrossing

M37. Meiotic analysis: introgression
      Chromosome pairing   hybrid
      and recombination   genotype
                                  recombination
                                  structural chromosome changes

M38. Microsatellite   introgression
      markers   vertical gene-transfer
                                  marker
                                  hybrid
                                  genotype
                                  polymorphism

M39. Morphological character   introgression
      analysis   outcrossing
                                  vertical gene-transfer
                                  heterosis
                                  hybrid
                                  hybrid depression
                                  hybrid vigour
                                  metapopulation

M40. Nectar production: chemical attractant
      Chemical composition   nectar production
                                  pollinator foraging behaviour
                                  reproductive allocation

M41. Nectar production: nectar production
      Volume and rate   pollinator foraging behaviour
                                  reproductive allocation

M42.	Nuclear and organelle markers combined	introgression outcrossing vertical gene-transfer hybrid genetic diversity metapopulation
M43.	Organelle DNA analysis	vertical gene-transfer hybrid genotype genetic diversity
M44.	Ovule counts	agamospermy fruit abortion ovule development seed abortion seed production cross-pollination phenology reproductive allocation self-pollination sex distribution
M45.	Paternity analysis	pollen dispersal cross-pollination self-pollination outcrossing genetic diversity genetic neighbourhood-area metapopulation
M46.	Photosynthetic efficiency	heterosis hybrid depression hybrid vigour stress tolerance plant competition (see vol. I)
M47.	Plant material preservation: Cryopreservation and freeze-drying	introgression vertical gene-transfer hybrid gene pool genetic diversity
M48.	Plant material preservation: Rapid drying	introgression vertical gene-transfer hybrid gene pool genetic diversity

M49. Pollen carry-over   pollen dispersal
      between flowers   pollinator foraging behaviour
                                   bird-pollination
                                   insect-pollination
                                   mammal-pollination
                                   cross-pollination
                                   self-pollination
                                   outcrossing
                                   genetic neighbourhood-area

M50. Pollen counts   pollen germination
                                   pollen production
                                   pollen viability
                                   cross-pollination
                                   incompatibility
                                   phenology
                                   reproductive allocation
                                   self-pollination
                                   outcrossing
                                   hybrid

M51. Pollen deposition   pollen germination
      on stigmas: Pollen   pollen production
      numbers and fertility   pollen dispersal
                                   pollinator activity
                                   bird-pollination
                                   insect-pollination
                                   mammal-pollination
                                   water-pollination
                                   cross-pollination
                                   incompatibility
                                   phenology
                                   self-pollination
                                   hybrid

M52. Pollen germination tests   pollen competition
                                   pollen germination
                                   pollen viability
                                   fertilization
                                   cross-pollination
                                   incompatibility
                                   phenology
                                   self-pollination
                                   hybrid

M53. Pollen traps   pollen production
                                   pollen dispersal
                                   wind-pollination
                                   cross-pollination
                                   phenology

M54. Pollen viability tests

pollen germination
pollen viability
agamospermy
pollen dispersal
phenology
hybrid

M55. Pollinator attraction:
      Chemical cues

chemical attractant
nectar production
pollinator activity
pollinator foraging behaviour
insect-pollination
mammal-pollination
phenology
reproductive allocation

M56. Pollinator attraction:
      Visual cues

pollinator activity
pollinator foraging behaviour
visual attractant
bird-pollination
insect-pollination
mammal-pollination
phenology
reproductive allocation

M57. Pollinator foraging
      behaviour

nectar production
pollen dispersal
pollinator activity
pollinator foraging behaviour
bird-pollination
insect-pollination
mammal-pollination
cross-pollination
self-pollination

M58. Pollinator preference
      experiments

pollinator foraging behaviour
bird-pollination
insect-pollination
mammal-pollination
heterostyly
phenology

M59. Polymerase chain
      reaction (PCR)

marker
hybrid
genotype
polymorphism

M60. Protein electrophoresis: introgression
      Isozyme analysis   marker
                                   vertical gene-transfer
                                   hybrid
                                   gene expression
                                   gene stability
                                   genotype
                                   polymorphism

M61. Random amplified   introgression
      polymorphic DNA   marker
      (RAPD)   vertical gene-transfer
                                   hybrid
                                   genotype
                                   polymorphism

M62. Reproduction in alien   chemical attractant
      plants: Community and   pollinator activity
      invader analysis   pollinator foraging behaviour
                                   visual attractant
                                   cross-pollination
                                   phenology
                                   reproductive allocation

M63. Reproductive allocation   pollen production
      measures   fruit abortion
                                   seed abortion
                                   seed production
                                   chemical attractant
                                   nectar production
                                   visual attractant
                                   reproductive allocation

M64. Restriction fragment   introgression
      length polymorphism   marker
      (RFLP)   vertical gene-transfer
                                   hybrid
                                   genotype
                                   polymorphism

M65. Ribosomal DNA   introgression
      (rDNA) analysis   vertical gene-transfer
                                   hybrid
                                   genotype
                                   genetic diversity
                                   metapopulation

M66. Seed development analysis

agamospermy
fertilization
fruit abortion
ovule development
seed abortion
seed germination
seed production
cross-pollination
incompatibility
phenology
reproductive allocation
self-pollination

M67. Seed germination tests

pollen competition
fruit abortion
fertilization
ovule development
seed abortion
seed germination
seed production
cross-pollination
phenology
reproductive allocation
self-pollination
hybrid

M68. Seed viability tests

fruit abortion
ovule development
seed abortion
seed germination
seed production
cross-pollination
phenology
reproductive allocation
self-pollination
hybrid

M69. Selfing and outcrossing rate

outcrossing
genetic diversity
genetic drift

M70. Sexual morph distribution

phenology
reproductive allocation
sex distribution

M71. Southern and Northern   introgression
      hybridization (blotting)   marker
                                     vertical gene-transfer
                                     hybrid
                                     gene expression
                                     gene stability
                                     genotype
                                     polymorphism

M72. Spatial autocorrelation   pollen dispersal
      analysis and Moran's I   cross-pollination
                                     self-pollination
                                     outcrossing
                                     genetic diversity
                                     genetic drift
                                     genetic neighbourhood-area
                                     metapopulation

M73. Stigma receptivity test   pollen germination
                                     fertilization
                                     seed production
                                     incompatibility
                                     phenology

M74. Thin layer   chemical attractant
      chromatography (TLC)   marker
                                     hybrid
                                     genotype
                                     polymorphism

M75. Thinning of flowers   pollen germination
                                     pollen production
                                     fruit abortion
                                     seed abortion
                                     seed germination
                                     seed production
                                     phenology
                                     reproductive allocation

M76. Transplantation   seed production
      experiment   cross-pollination
                                     phenology
                                     reproductive allocation
                                     self-pollination
                                     hybrid
                                     cline

M77. Two-dimensional paper   marker
      chromatography   hybrid
                                     genotype
                                     polymorphism

6. Synopsis of subcategories and recommended methods

By: Klaus Ammann, Francois Felber, Yolande Jacot, Rikke Bagger Jørgensen, Gösta Kjellsson, Jens Mogens Olesen, Marianne Philipp, Pia Rufener Al Mazyad, Mikkel Heide Schierup and Vibeke Simonsen

The different subcategories listed in Chapter 3 are examined in relation to the risk assessment of GMPs in natural and agricultural ecosystems. Methods which are suitable as test procedures for specific items are suggested. Subcategories lacking relevant methods are indicated in the margin as "need for new methods" and commented on, while interesting research subjects which have had little investigation are marked "research need". For the discussion of subjects under the categories "Genetic engineering techniques", "Inserted trait" and "Test procedures", see Chapters 8, 9 and 10, respectively. A list of the references for each subcategory is shown in Section 6.1 and a list of reviews for each category is shown in Section 6.2.

Category	**Pollen development and production**
Subcategory	Pollen competition, pollen germination, pollen production, pollen viability.
Pollen development and survival	Pollen grains are produced in anthers. After meiosis the haploid microspore goes through one or two mitotic divisions. Pollen grains thus consist of one vegetative cell and one or two generative cells when they are dispersed. After release from the anthers, pollen normally have a very short life, from about 30 minutes to one day, but much longer time is found in some species. The targets (stigma surfaces) for the dispersal of the pollen grains have a very limited size, and it is crucial for the success of the male function to hit a legitimate stigma within the life time of pollen grains.
Pollen germination	The requirements for pollen germination are complex. Pollen needs water and oxygen to germinate. A solution with the right osmotic potential is necessary for growth, otherwise the pollen grains will burst. A wide range of substances has been shown to have a stimulatory effect on pollen-tube growth *in vitro*. Boron is necessary for successful **pollen germination**. It aids sugar uptake and is involved in pectin synthesis.
Fertilization	Legitimate pollen will most often adhere to the stigma. Pollen grains become hydrated after arriving on a suitable stigma, and pollen tubes emerge through the pore of the pollen grains. In most plants only one pollen tube is produced per pollen. Pollen tubes grow down the style and it requires between 12 and 48 hours from **pollen germination** to **fertilization** in most plants. In some species it proceeds much faster (15 min.) or very slowly (several months). Pollen sizes and amount of resources are not related to pollen tube growth rate. The two male gametes in the pollen grain move to an apical position in the pollen tube and perform the double fertilization.

Pollen competition

In most cases the amount of pollen arriving on a stigma is much larger than the number of ovules in the corresponding ovary. This means that only a part of the pollen tubes will succeed in **fertilization** (1005). Consequently, **pollen tube competition** will take place down the styles. The outcome of this competition depends on both the genotype of the style and the pollen tubes. The interaction between style and pollen tubes may occur through the transfer of resources from the style tissue to the pollen tube. Certain pollen tube genotypes receive more resources than others and thus grow faster. It has been shown that some pollen tube genotypes grow faster on some style geno-types than on others. Furthermore, the variation between offspring becomes smaller after intensive pollen tube competition than after less **pollen competition**.

Pollen production

The size of the **pollen production** per flower is often related to the number of ovules (P/O-ratio) and is characteristic for the breeding system. It is further related to the way of dispersal. Plants with wind dispersed pollen most often produce a large amount of pollen, whereas plants with insect dispersed pollen produce less, and plants which are mainly selfing produce the smallest amount of pollen per ovule. The size of the **pollen production** influences the intensity and distance of **pollen dispersal** and establishment from, e.g., GMP to related populations in the surroundings.

Research need

Little is known about the exact way that interaction and communication between pollen tubes and style take place. **Pollen development** and the derived **pollen viability** and possibility for germination are highly influenced by environmental factors such as temperature and nutritional status. **Pollen production** and the **viability** might reflect the health (vigour) of the plants. **Pollen viability** and **pollen germination** together with **pollen production** are influencing the male function of the plants. The effect of environments on **pollen development and production** and its consequences for the male function (e.g., male sterility) need more research.

Recommended methods

Pollen production and **viability** are determined by **pollen counts** (M50) and by **pollen viability tests** (M54). **Pollen viability** and **pollen germination** (M52) are determined by different staining procedures and by germinating pollen on stigmas or on artificial media. The range of tests on **pollen viability** and **germination** from staining over germination on artificial media to germination on stigmas are increasingly valuable, but also increasingly demanding in terms of equipment. The result of **pollen competition** through the styles can be detected by different fitness components determined by **fitness measurements** (M14).

Reviews

758, 1005.

Category	**Seed development and production**
Subcategory	Agamospermy, fertilization, fruit abortion, ovule development, seed abortion, seed production.
Ovule development and fertilization	Ovules consist of a female gametophyte: The embryo sac surrounded by nucellus and one or two integuments through which there is an opening (micropyle). Most angiosperms (90%) possess haploid embryo sacs consisting of seven cells: One egg cell, two synergid cells, three antipodal cells and one central cell (with two nuclei). Stigmas are usually receptive to pollen grains over a relatively short period while the flower is open. Receptivity can often be seen from the generally marked turgidity of the stigmatic papillae and concluded by the necrosis of these cells. After pollen germination on stigmas and pollen tube growth down the styles, pollen tubes most often enter the ovule through the micropyle and one of the synergid cells. One of the male gametes will enter the egg cell and the other the primary endosperm cell. Hereby the diploid zygote and the primary triploid endosperm cell is produced. This double **fertilization** is very special to the angiosperms (454). **Ovule development** depends in some species on the position of the ovule in the fruit and on differences in resorce supply and pollen competition. Ovules positioned closest to the fruit stalk will in some cases receive most resources and attain fastest development. Ovules positioned closest to the style will in some cases be fertilized by the fastest growing pollen tubes resulting in fast developing ovules.
Seed development	During **seed development**, embryos with cotyledons and primary root develop from the zygotes by a number of mitotic cell divisions. The endosperm may develop into either many or few cells. The fertilized ovule thus consists of four independent and genetically different generations:

a) The sporophyte (diploid) integument of the ovule forming seed coat.
b) The female gametophyte (haploid) embryo sac wall and antipodals.
c) The offspring sporophyte (diploid) embryo.
d) The endosperm (triploid) product of an independent fusion between a sperm cell and diploid primary endosperm nucleus.

Agamospermy	Species with **agamospermy** do not need **fertilization** to develop mature seeds. There are three types of agamospermy:

a) Diplospory, where the embryo sac develops from the archesporium which has not gone through the regular meiosis and is still diploid.
b) Apospory, where the embryo sac develops from nucellus.
c) Adventitious embryony, where the embryo develops directly from some other cells in the ovule.

Some agamospermous species need pollination to stimulate development of embryo sacs into embryos (pseudogamy).

Seed production

Seed production is highly different among plant species. Number of ovules is inversely related to the mature seed weight. Number of ovules may also depend on the strategy of the plant: Ruderal plants (r-strategists) often possess seeds which are smaller than the seeds of competitive species (C-strategists), see the CSR-strategy system (320). The seed dispersal system is related to the size and number of ovules and seeds, but in a less predictive manner. Seeds adapted to animal dispersal are generally the largest.

Reproductive allocation, seed and fruit abortion

The proportion of the total resource budget that is devoted to reproduction is called the reproductive effort (RE). **Reproductive allocation** means allocation to acquisition of mates, production of gametes and parental care. Seed maturation demands considerable resource allocation, and in most species a high number of ovules and ovaries are aborted due to resource limitation (116, 663, 897). **Seed abortions** and **fruit abortions** do not take place randomly. Ovaries with few fertilized ovules or with damages from herbivores are primarily aborted. Abortion of developing ovules is also, to some degree, determined by the genotypes of the embryos in relation to the seed parent plant. If resources are added to the plant (water, nutrients) a higher percentage of ovaries and fertilized ovules is likely to mature.

Recommended methods

Ovule production can be estimated by **ovule counts** (M44). Viability of ovules is measured as **seed production** or by **seed viability tests** (M68). The phenology of stigmas is observed by **stigma receptivity test** (M73). Development of ovules from fertilization to mature seeds can be inspected by **seed development analysis** (M66). Ovules can abort at different stages: Non-fertilized ovules, fertilized ovules in the initial stage and in later stage. These stages of abortion can be recorded as described in **classification of fruits, seeds and ovules** (M6), and further investigated by, e.g., **thinning of flowers** (M75). A number of methods are used in order to inspect ovule and seed development. See **seed development analysis** (M66). Viability and germinability of the mature seeds can be measured by different **seed viability tests** (M68) and **seed germination tests** (M67). Germination experiments can be performed on different substrate from soil to moist filter paper and, in addition, light and temperature regimes can be varied.

Reviews

116, 454, 663, 897.

Category	**Pollination**
Subcategory	Chemical attractants, nectar production, pollen dispersal, pollinator activity, pollinator foraging behaviour, visual attractant.

Pollination, the process of transfer of pollen from the anthers to the stigma of a plant, is a prerequisite for fertilization of ovules and subsequent development of seeds. The various **pollination syndromes** (see p. 55) can be divided into the use of biotic vectors (e.g., pollinators) and abiotic vectors (e.g., wind), with biotic vectors being most common.

Pollinator attraction

Biotic vectors are attracted to flowers by **visual attractants** such as colour, size and shape, see **pollinator attraction: Visual cues** (M56), and by **chemical attractants** such as scents and food rewards, see **pollinator attraction: Chemical cues** (M55). Insects are able to discriminate between even minor differences in colour and size, see **fluctuating asymmetry** (M16), which may give differences in pollination success of individual plants. Bumblebees, for example, prefer the most symmetric flowers which also have more nectar (585, 587). Seed abortion rate also increases with increasing asymmetry of both pollen donors and recipients (586).

Scent is an important attractant, especially for pollinators which are active during the night. Specific flower scents may enhance the fidelity of insects to particular plant species by which intraspecific pollen transfer is increased. Bees are, in general, most sensitive to volatile chemicals. In many plant species, the maximum scent production is coordinated with the time when the pollen is ripe and the flower is ready for pollination. Some flowers mimic the smell of decaying protein or faeces to attract carrion and dung insects, or imitate female pheromones to attract male pollinators to the flower.

Pollinator rewards

Once attracted to the flower, pollinators are usually rewarded by nectar or pollen. The nectar itself may function as a **chemical attractant**. If the pollination syndrome is specialised, the **nectar production** and the pollinator can be adapted to one another with respect to volume, concentration and chemical composition of the nectar, e.g., the larger the pollinator the larger the nectar volume per flower or inflorescence. The content of sugars and amino acids in nectar is often also related to the needs of the most common pollinator of the species.

Pollinator activity

Pollinator activity is dependent on climatic conditions. Insect activity is low during cold, rainy or windy periods. The **pollinator foraging behaviour** (M57) is influenced by the amount of rewards. When the nectar amount is large in a given plant or a given patch, the pollinator will stay on the plant or in the patch and will fly short distances only. If, however, the nectar rewards are small, the pollinator will take longer flights. Some pollinators, especially bees, forage at only one species of plants during each nectar or pollen collecting bout. The effect of this behaviour is that pollen is transported between plants belonging to the same species. The probability of hybridization with related species is thus decreased. Other insects, e.g., flies and beetles, are more random in their behaviour, being mainly attracted to plants and patches with many flowers.

The specific **foraging behaviour** influences the **pollen dispersal** distances, which will be short when abundant rewards are close and longer when the rewards are scattered and small. Most pollen obtained in a flower are deposited on the next flower visited, but a fraction remains on the pollinator and can be carried further to flowers visited later, a phenomenon termed **pollen carry-over between flowers** (M49). The amount of pollen carry-over depends on the plant's sex distribution (see **breeding system**, p. 59) and on the insect's ability to clean itself during flights between flowers.

Pollinators and introduced plants

The success of an introduced plant (e.g., a GMP) depends to some extent on the availability of pollinators in the habitat where it invades. Specialised pollinators may not visit the invader, which is initially faced with only those generalist pollinators (and seed dispersers) that are able to shift from resident plants to the new species, see **reproduction in alien plants** (M62). The change in number and strength of interactions caused by these shifts may partially determine the outcome of an immigration event. The pool of potentially interacting generalists already present in the receiver habitat is thus assumed to be very important to the success of an invasion.

Recommended methods

Studies of **pollinator activity** should be done at a locality where the plant species grow naturally. The interaction between flowers and pollinators is observed as described in **pollinator attraction: Chemical cues, visual cues** (M55, M56), **pollinator foraging behaviour** (M57) and **pollinator preference experiments** (M58). The importance of pollinators for the plants can be inferred from **bagging of flowers** (M2), **emasculation of flowers** (M12), **experimental pollination** (M13) and **fitness measurements** (M14). **Nectar production** can be investigated by looking at the **chemical composition** (M40) and **volume and rate** (M41) of nectar. **Pollen dispersal** distances can be measured both directly and indirectly. The direct method is through observation of **pollinator foraging behaviour** (M57), quantification of **pollen carry-over** (M49) or using **fluorescent dyes and marking of pollen** (M17). The indirect methods are through the study of gene-flow from genetic markers (see also **gene-flow**, p. 61) and include both **paternity analysis** (M45) and **spatial autocorrelation** (M72). **Pollen dispersal** estimated from the direct methods provides a snap-shot picture of the *potential* for **gene-flow** through pollen, whereas **pollen dispersal** estimated through the indirect methods yields an average picture over several seasons of the *successful* **pollen dispersal**. By comparing results from these two approaches, information on the effect of **pollen carry-over**, post-pollination discrimination against certain pollen (e.g., inbred pollen), and the relative effect of **pollen dispersal** compared to seed dispersal, may be acquired.

Research need

Very little information exists on the initial reproductive behaviour of intentionally released plants in relation to the characteristic receiver habitat. Experimental introductions may tell us about qualities important to invaders in their initial establishment and features specifically related to the level of invasibility of a natural habitat. Achieving a certain level of reproductive success is a fundamental challenge to any invader in its establishment and is determined both by factors intrinsic to the invader itself and by abiotic and biotic characteristics of the recipient habitat. New methods like **reproduction of alien plants: Community and invader analysis** (M62) may answer some of these questions.

Research need

Pollination biological studies on small, non-specialised insects have been somewhat neglected compared to studies involving bees and butterflies. Studies including both direct observation of pollinators and gene-flow study of the genetic processes following pollination give important information but have rarely been done. Such studies could be applied to investigations of the success of hybridization between wild species and GMPs.

Reviews

126, 183, 227, 342, 431, 587, 727, 748, 758, 962.

Category

Pollination syndrome

Subcategory

Ant-pollination, bat-pollination, bee-pollination, buzz-pollination, beetle-pollination, bird-pollination, butterfly-pollination, fly-pollination, insect-pollination, mammal-pollination, water-pollination, wind-pollination.

Types of syndromes

A plant species usually relies on a specific **pollination syndrome,** although there are large differences in how specialised the flowers are for a specific set of pollinators. Knowledge of the types of pollination interactions (syndromes) is valuable for making sensible decisions before detailed analyses of gene-flow and hybridization are made. The characteristics of the major types of **pollination syndromes** are described below.

Insect-pollination

The pollination of flowers by insects is usually based on food-rewards (i.e., nectar, pollen and oil), but deceit mechanisms also occur (e.g., floral mimicry and pseudocopulation) especially in orchids. The ability of insects to perceive and discriminate between different flower types varies greatly (e.g., high for bees and low for beetles). The major insect pollinator groups (in approx. order of importance) are: Bees, flies, butterflies and moths, beetles and wasps. Many crops largely depend on insects (mainly bees) for pollination, e.g., onion (*Allium*), alfalfa (*Medicago sativa*) and red clover (*Trifolium pratense*).

Ant-pollination

Ants occasionally visit flowers and nectaries for sugar but rarely carry pollen. Thus very few plants (if any) are adapted to **ant-pollination.** A few cases of observed pollination by ants exist, e.g., in *Glaux* and *Scleranthus*. Ants invading plants with extrafloral nectaries may give protection against herbivore attack.

Bee-pollination

Flowers pollinated by bees are often semiclosed and zygomorphic with a lip for landing, yellow and blue colours with nectar guides and moderate nectar in short spur or sac, anthers and stigma are often hidden. Specialized bee-flowers include: Corydalis (*Corydalis*), larkspur (*Delphinium*), dead-nettle (*Lamium*), lousewort (*Pedicularis*), thunbergia (*Thunbergia*) and clover (*Trifolium*). Many actinomorphic flowers are also pollinated by bees (e.g., *Asteraceae, Cistaceae* and *Rosaceae*). Social bees are very important pollinators: Honeybees (*Apis*), bumblebees (*Bombus*) and the genera *Melipona* and *Trigona* in tropical America. Solitary bees (e.g., *Andrena, Anthophora, Megachile, Osmia* and *Xylocopa*) are especially important early in the season in temperate and mediterranean regions. Generally both nectar and pollen are collected.

Buzz-pollination

Bees sometimes vibrate (buzz) flowers with poricidal anthers for collecting pollen. Buzz-pollinated plants require certain bees for pollen transfer. Examples include: *Cassia,* wintergreens (*Pyrolaceae*) and potato and tomato (*Solanaceae*). Some bees are effective buzzers, e.g., bumblebees (*Bombus*), *Anthophora, Melipona* and *Xylocopa*, while others are not, e.g., honeybees (*Apis*) and leafcutter bees (*Megachilidae*).

Beetle-pollination

Flowers pollinated by beetles often have bowl-shaped flowers with dull white to greenish colour, strong fruity scent, pollen and nectar easily accessible. Typical examples are: Carolina allspice (*Calycanthus*) and magnolia (*Magnolia*).

Butterfly-pollination

Butterfly-pollinated flowers are usually pink or red and have narrow, mostly long, tubes or spurs with nectar hidden. Moth-pollinated flowers are mostly white with a heavy scent at night. Typical examples include: *Buddleia,* pink (*Dianthus*), red valerian (*Centranthus*) and lily (*Lilium*) for butterflies and honeysuckle (*Lonicera*), butterfly orchid (*Plantanthera*) and campion (*Silene*) for moths.

Fly-pollination

Flies will occasionally visit flowers which are primarily pollinated by bees (e.g., *Ranunculus, Filipendula* and *Thymus*). However some species are adapted to **fly-pollination** offering nectar, others are pollinated by deceit mechanisms. Syndrome characters for general **fly-pollination** include: Flowers regular, flat, with yellow to greenish colour or sometimes white, shiny structures and exposed anthers and stigma. Nectar easily accessible, often on disk-shaped nectaria, ± nectar guides. Typical plant genera are: Moschatel (*Adoxa*), spurge (*Euphorbia*) and saxifrage (*Saxifraga*). Syndrome characters for **fly-pollination** by deceit (sapromyophily) include: Flowers regular and flat, or intricate deep traps, with dull brown to reddish brown or greenish colours often with spots, hairs and tactile appendages and a rotten or decaying smell. Anthers and stigma often hidden inside the flower. No nectar or food reward. Many intricate flower or inflorescence types exist, trap-flowers (e.g., *Aristolochia, Arum, Stapelia*) catch mechanisms (e.g., *Bulbophyllum*), brood place mimics (e.g., *Stapelia*) and fungus mimics (e.g., *Arisarum* and *Masdevallia*). Pollinators of sapromyophile flowers are primarily carrion and dung-flies and fungus gnats (*Mycetophilidae*), however beetles may also be attracted.

Wasp-pollination

Wasps are generally not important as pollinators but often visit fly-pollinated flowers and others, e.g., maple (*Acer*) and ivy (*Hedera*). In some cases stronger interactions for cup-shaped flowers with brown colour may exist, e.g., between *Vespa* and figwort *(Scrophularia nodosa)* and *Cotoneaster*.

Bird-pollination

Flowers pollinated by birds have a deep tube or spur, brightly coloured (often red with a yellow part), protruding anthers and stigma, ovary protected, abundant nectar, without any scents. Mainly plants in tropical areas, but also occurring in N. America and temperate Australia, e.g., *Strelitzia, Fuchsia, Hibiscus, Bromeliaceae* and *Eucalyptus*. Important bird pollinators include: Sunbirds (*Nectariniidae*), hummingbirds (*Trochilidae*), honey-eaters (*Meliphagidae*) and lorikeets (*Trichoglossidae*).

Mammal-pollination

Some mammals regularly visit plants for nectar or to eat flowers (see also **bat-pollination**). Plants adapted to pollination by mammals are generally open with brush-flowers, and mainly found in Australia, e.g., marsupials: Honey-possum (*Tarsipes*), in S. Africa: Rodents and in Madagascar: Lemurs (*Varecia*). Plants pollinated by mammals include *Banksia* and *Ravenala*.

Bat-pollination

Bat-pollinated plants generally have large, open brush-flowers, placed outside the foliage, with white or drab colours and a strong aromatic smell at night, large nectar and pollen production. Many trees and large herbs in tropical Asia, Africa and America, e.g., baobab (*Adansonia*), sausage tree (*Kigelia*), kapok (*Ceiba*) and banana (*Musa*).

Water-pollination

Pollination may either take place on the water surface or in the water. Pollen grains may be floating or slowly sinking according to their morphology and size. Some aquatic plants (but not the majority) are pollinated by water, e.g., *Ruppia, Valisneria, Ceratophyllum* and *Zostera*.

Wind-pollination

Flowers pollinated by wind are often unisexual with small or lacking perianth and no **chemical attractants**, anthers and stigmas exposed, large production of small pollen grains. Thus the pollination process is much less precise than **insect-pollination**. Dispersal distances depend, e.g., on fall-rate and vegetation structure. Common type of pollination among grasses (*Poaceae*), sedges (*Cyperaceae*), rushes (*Juncaceae*), conifers (*Coniferae*) and Northern temperate trees (e.g., *Quercus, Betula* and *Populus*).

Recommended methods

Interactions between plants and pollinators which determine **pollination syndromes** largely depend upon rewards (nectar and pollen) and specific signals (colour, scent) from flowers to pollinators. How the rewards function can be studied by measuring the **nectar production: Volume and rate** (M56) and its **chemical composition** (M55).

Most flower-visiting animals are able to distinguish between colours (118). The visual spectrum of bumblebees and honeybees ranges at least from 336 to 532 nm, i.e., ultraviolet to yellow-green (408, 520). Butterflies and a few beetles are able to see red (600-700 nm). The colour perception of some flies have an optimum for yellow. Thus, many insect groups seem to have innate preferences for particular colours. Birds are sensitive to red and at least some can see long-waved ultraviolet. Bats mainly use scent as their flower cues. Quantification of colour spectres of flowers may be done by the method, **pollinator attraction: Visual cues** (M56). The attraction of specific pollinators to specific flowers may include tests of **pollinator preference experiments** (M58), **pollinator foraging behaviour** (M57) and **pollinator attraction: Chemical cues** (M55). Methods to study the molecular composition of flower scents include the use of **gas chromatography** (M20).

Wind-dispersed pollen of plants with large pollen production (i.e., grasses and some trees, see above) have an enhanced probability of long-distance dispersal, even if the majority of pollen is dispersed within small distances from the source. In palynology (pollen analysis of fossil and recent pollen depositions) transport over hundreds of kilometres are well studied phenomena, recognised especially when exotic pollen appears in a totally different environment (e.g., Mediterranean and North African pollen have been found in glacier ice and in lake depositions in Switzerland). Regional and long-distance transport complicates the determination of strict safety distances for pollen flow. **Pollen traps** (M53) are useful for studying dispersal patterns and pollen production of wind pollinated plants. Pollen of many genera may be identified from position of pores and the structure of the outer wall (i.e., exine). If pollen are marked (e.g., by radioactive tracers) more specific information on source may be obtained.

Research need

Knowledge on probability of pollen transport over long distances is required, and the viability of pollen in relation to distance and time needs to be tested.

Need for new methods

Specific methods to identify pollen sources for long-distance dispersal need to be developed. This issue could become especially important for risk assessment when transgenics of common forest trees are developed in future.

Reviews

342, 360, 724.

Category	Breeding system
Subcategory	Cross-pollination, flowering phenology, heterostyly, incompatibility, reproductive allocation, self-pollination, sex distribution.
Breeding systems	The study of **breeding systems** includes all events responsible for selection of the gametes which ultimately unite to produce mature seeds. The organisation of the **breeding systems** is fundamental to the genetic structure and variation at species, population and individual level. Studies of **breeding systems** are closely related to studies of outcrossing rate, inbreeding and outbreeding depression, pollination ecology, genetic variation and gene dispersal.
Flowering phenology	Morphologically there is great variation in the way the female and male function are organised in space and time in plants. Studies of **flowering phenology** at flower or inflorescence level (456) are important for understanding the function of the flower and the breeding system.
Sex distribution, self- and cross-pollination	Most plant species have hermaphrodite flowers which are able to self-pollinate, thus resulting in selfing (autogamy). This has been circumvented for those species which have a **self-incompatibility** system preventing self-fertilization (142, 266). The amount of selfing in relation to outcrossing (a mixed mating system) varies between species and with time (374, 514). **Self-pollination** (and self-fertilization) is further limited and consequently **cross-pollination** (allogamy) is promoted in many cases by a temporal **sex distribution** within individual flowers (dichogamy). In many species the male function is expressed first and is succeeded by the female function (protandry). In other, but much fewer species, the female function precedes the male function (protogyny) (83). Some plant species possess flowers in which there are a spatial separation of female and male functions (herkogamy and **heterostyly**). Some species have flowers in which only one sex is represented (dicliny). In some species, female and male flowers are found on the same individuals (monoecy), in others, they are found on different individuals (dioecy). Male and hermaphrodite flowers can also be found on the same individual (andromonoecy) in some species and, in other species, on separate individuals (androdioecy). Likewise, female and hermaphrodite flowers can be found on the same individual (gynomonoecy) or on separate individuals (gynodioecy).
Functional gender	In a population of a species with hermaphroditic flowers, all individuals and flowers may not necessarily be functioning as hermaphrodites. In flowers where ovules or ovaries are aborted their function will be male only. In flowers, which are setting seeds, the pollen may not succeed in fertilization of other flowers and they will thus be functioning as females only. The functional gender (sex) of a flower or an individual can therefore be different from the morphologically observed sexual morph determined by **sexual morph distribution** (M70).

Research need

The results of **breeding system** experiments are highly dependent on local climatical conditions such as temperature, sunshine and wind. Recently, it has been shown that nutrient composition of soils also highly influences the performance of pollen grains. However, more research is needed in this area. Most studies of **breeding systems** have been made at the population level. Studies on interactions of **breeding systems** among species within a community are rare and more information is urgently required.

Recommended methods

Observations on **breeding systems** should be performed in the field in natural populations, but additional pollination experiments and observations in greenhouses and on cultivated plants on experimental fields are most useful. The functioning of the **breeding system** is studied by observation of flowering **phenology** in single flowers and inflorescences through **marking of plants and flowers** (M36). Furthermore, observation of pollination is done by looking at **pollinator preference experiments** (M58) and **pollinator foraging behaviour** (M57), combined with **bagging of flowers** (M2), **emasculation of flowers** (M12) and **experimental pollination** (M13). Inclusion of **diallel cross** (M9) may reveal additional information on the **breeding system**. The results of these manipulations are measured by either **pollen deposition on stigmas** (M51) or by seed set and seed performance. Seed performance is measured by **seed development analysis** (M66), **seed viability tests** (M68) and **seed germination tests** (M67, see also M18 and M19 in vol. 1). Fitness of seedlings and developing individuals may be determined by **fitness measurements** (M14).

Allocation to different flower functions is an important subject in the evaluation of **breeding systems**. **Reproductive allocation analysis** (M63) can be used for quantification. A problem is to delimit which structures belong to female and male function. Resources used on advertising structures are now believed mostly to benefit male function.

Reviews

83, 142, 266, 374, 456, 514.

Category	**Gene-flow**
Subcategory	Gene-transfer.
Gene-transfer	**Gene-transfer** is the transfer of genes from one individual to another. **Gene-transfer** occurs naturally as transfer of genes from one gene-pool to another, i.e., at the population or the species level. In this case it is synonymous with **gene-flow**. The term **gene-transfer** also denotes the transfer of alien genes into plant species in genetic engineering (see Chapter 8). **Gene-transfer** may be divided into **vertical** and **horizontal gene-transfer**, see below. The integration of the alien genes into the genome is known as **introgression** and is the endpoint of **gene-transfer**.

Gene-transfer or **gene-flow** at the population level is a fundamental issue to risk assessment, because it represents an avenue by which engineered genes may escape from cultivated fields. |
| *Subcategory* | Vertical gene-transfer.

Vertical gene-transfer refers to sexual transfer of genes between two genetic different entities, e.g., between two distinct populations or two species. **Hybridization** through **outcrossing** is needed for a successful **vertical gene-transfer**. |
Recommended methods	Direct measurements of **gene-transfer** at the population or the species level are based on seed dispersal (see vol. 1) and pollen dispersal (see **pollination**, p. 53). Indirect quantification of **gene-transfer** by **markers** have been extensively done, e.g., by **paternity analysis** (M45) or **spatial autocorrelation** (M72), see subcategory **marker**, p. 64. In this context, **F-statistics** (M19) and **genetic distance** (M22) may also be employed.
Research need	There is a lack of empirical data on the relative importance of seeds as compared to pollen for **gene-flow**. Furthermore, seeds, unlike pollen, have the ability to colonize new areas and the number of seeds involved in colonization has rarely been estimated. Seed dispersal can be difficult to monitor and quantify by direct methods. This may be somewhat rectified by using multilocus techniques for the study of **gene-flow** by seeds. The comparisons of values of **gene-flow**, based on biparental and maternal inherited genetic markers by **nuclear and organelle markers combined** (M42), should aid understanding of the relative contributions of seed and pollen to **gene-flow** processes. Measurements of **pollen dispersal** and realized **vertical gene-transfer** may give different results due to factors which influence **pollen dispersal** and **gene-transfer** differently (e.g., mode of dispersal versus homology between the genomes of donor and recipient). Comparison of the two types of measurements may therefore quantify the importance of these factors.
Reviews	17, 223, 313, 314, 315, 340, 499, 504, 557, 764, 852, 853, 854, 858.

Subcategory   Introgression.

When crosses between plants result in a stable incorporation of genes from one gene pool into another, differently composed, gene pool, the process is called **introgression** or introgressive hybridization (27, 765). **Introgression** is often difficult to prove with certainty because shared traits may also be the result of common ancestors or convergent evolution (evolution of a similar trait in more than one taxon). Tests and monitoring of **introgression** are important when possible environmental effects of a transgenic plant are analyzed. The transfer of transgenes to a temporary hybrid (most hybrids are temporary, at least if they are interspecific) is less likely to have an impact on the environment, compared to stable **introgression** of transgenes in a population. **Gene-flow** from a transgenic, herbicide-resistant rape (*Brassica napus*) to the weedy relative *B. rapa ssp. campestris* (Figure 6.1), often abbreviated *B. campestris*, has been detected in field trials (578).

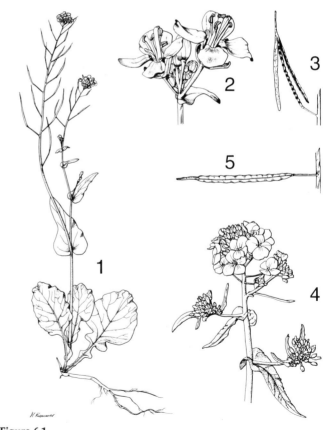

Figure 6.1.
1) *Brassica campestris*, 2) inflorescence of *B. campestris*, 3) open fruit of *B. campestris*, 4) inflorescence of rape, *Brassica napus*, 5) closed fruit of *B. napus*.

Recommended methods

The most powerful way to detect **introgression** is by tracking linked markers, i.e., markers located closely to each other on a chromosome. If a putative introgressive plant reveals multiple linked markers arranged in the same way as in the potential donor plant, mutual ancestors or convergent evolution can be excluded. An excellent tool for this approach is comparative linkage mapping which requires **genetic map construction** (M23) of the donor and recipient species. Saturated linkage maps can most easily be generated by the use of molecular **markers** revealed by, e.g., **AFLP** (M1), **microsatellites** (M38), **RFLP** (M64), **RAPD** (M61) and **isozyme analysis** (M60) which are methods that provide many markers in a short time. Reliable methods providing a high extent of **polymorphism** are preferred, e.g., **PCR** (M59) based methods such as **AFLP** (M1). DNA-fragments that are assumed to be the product of recent **introgression** should reveal total DNA homology when compared to the parental sequences by **DNA sequencing** (M10).

When donors with novel genes are introduced to the environment, introgressive **hybridization** cannot be mimicked by convergent evolution or explained by mutual ancestors. Thus, novel genes (see Chapter 9) may serve as ideal markers of **introgression** (403). Engineered genes can be revealed by their phenotype, e.g., their **gene expression** detected by **AFLP** (M1) and RT-PCR (M59) or by confirming the presence of the genes in the genome by methods such as **PCR** (M59) and **Southern hybridization** (M71).

Reviews

5, 37, 187, 403, 762, 763, 765.

Subcategory

Outcrossing.

Outcrossing, the successful fertilization of an ovule by foreign pollen, is a process by which engineered genes or genes different to the gene pool studied may escape to the wild flora. **Outcrossing** is thus a prerequisite for **gene-transfer**. The **outcrossing** rate, the relative fraction of cross-fertilizations to the total number of matings, is therefore of major importance for risk assessment, as it indicates the potential for **gene-transfer**. The **outcrossing** rate depends on the type of **pollination** and the **breeding system**. Moreover, it greatly influences the genetic structure of populations.

Recommended methods

Two categories of estimations of outcrossing rates have been developed. The first category consists of empirical and experimental investigations on **pollination** and **breeding system** which are connected to **outcrossing** and are described in the respective sections. The second category relates to the genetic structure of natural populations and the analysis of natural progeny. **Outcrossing** rate may be inferred from the analysis of the genetic structure of the population by **F-statistics** (M19). The methods, **selfing and outcrossing rates** (M69) and also **paternity analysis** (M45), are useful for evaluation of **outcrossing** in a population when analysing maternal progeny. Both approaches are necessary and produce complementary information on the ability of the species for **outcrossing** and hence the potential for **gene-transfer**.

Reviews   58, 980.

Subcategory   Horizontal gene-transfer.

Horizontal gene-transfer refers to non-sexual gene-transfer among organisms, which may belong to unrelated systematic groups.

Horizontal gene-transfer of transgenes into natural or semi-natural habitats has not been detected up to now. All proven cases of **horizontal gene-transfer** have been provoked in artificial systems or small scale field experiments, where a high **gene-flow** pressure has been induced by favourable environmental conditions. **Horizontal gene-transfer** between higher plants and soil microbes and aphids thus remain hypothetical, despite some seemingly positive reports (723). Consequently, negative effects based on horizontal gene-transfer are hypothetical as well (810).

Reviews   22, 47, 723, 810.

Subcategory   Marker.

A **marker** is a recognizable trait, found in an individual plant. **Markers** are useful for determination of **gene-flow**, **hybridization**, **introgression** etc., when the marker is **polymorphic**, i.e., recognized in more than one form among the individuals studied. A **marker** can be a morphological, cytological or molecular trait, primarily determined by the genotype of the plant, but also influenced by the enviroment to which a specific individual is exposed. Low or no effect of the environment on the **marker** is usually desirable as this will increase the applicability of the **marker** for genetic studies.

A brief review on different **markers** (see Table 6.1), their applicability and the methods used for detection, are given below.

Morphological markers   Qualitative morphological **markers** are traits which can be directly recognised. They include: Flower colour, leaf pattern, etc. (see **morphological markers** M46 in vol. 1). Quantitative **markers** are traits which have to be measured, e.g., by length or weight. However, more advanced methods may be applied such as, e.g., determination of leaf area (see M35 and M36 in vol. 1).

Heritability   The extent to which the variance in a given measured trait is determined by genetic differences is denoted the heritability. The heritability is dependent on the specific environment and is often maximized when the environment is homogeneous, e.g., a green-house or a growth chamber. Many quantitative traits have proved to have heritabilities of the order of 50%, i.e., half the variation is genetic. However, large diferences between traits exist.

Recommended methods	Morphological markers may be analysed by **morphological character analysis** (M39) followed by a relevant statistical method, e.g., **principal component analysis (PCA)** (see M54 in vol. 1).
Cytological markers	Cytological **markers** encompass number and form of chromosomes which may be caused by **structural chromosome changes**, see Chapter 2. The shape of a chromosome may be changed by the insertion of a gene, e.g., in a transgenic plant.
Recommended methods	The chromosome number may be determined by **karyotype analysis: Chromosome number, size and form** (M33) or by **flow cytometry** (M15). The changed form of a chromosome may be detected by **chromosome painting** (M5), *in situ* **hybridization to chromosomes** (M31), **karyotype analysis: Chromosome number, size and form** (M33) and **meiotic analysis: Chromosome pairing and recombination** (M37). **Polymerase chain reaction (PCR)** (M59) is a versatile method for detection of inserts, e.g., caused by insertion of alien genes in plants.
Metabolic marker	Metabolic **markers** include the production of chemical compounds such as: Ascorbic acid, proline, hexoses or other secondary metabolites. Furthermore, a quantification of oxygen "evolution" (formation) during photosynthesis, is included. Metabolic **markers** are often highly influenced by the environment, e.g., an increased production of proline is initiated by water stress.
Recommended methods	Methods such as **gas chromatography (GC)** (M20), **high performance liquid chromatography (HPLC)** (M29), **thin layer chromatography (TLC)** (M74) and **two dimensional paper chromatography (M77)** are valuable for determining chemical compounds. The formation of oxygen may be determined by **photosynthetic efficiency** (M46). Metabolic **markers** may be applied either as qualitative **markers** by presence or absence in the plant or as quantitative **markers** by, e.g., determination of the concentration of the compound.
Protein markers	Proteins cover a wide range of functions in the organisms such as constituents of the cell wall, transporting agents to catalysts (enzymes) and so forth. Thus proteins are essential to sustain life processes. The life stage of an individual may have an effect on the expression of a particular protein. Environmental factors may also have an influence on the production of proteins, e.g., the enzyme alcohol dehydrogenase is expressed in barley roots only when the plants are grown in water culture.
Recommended methods	Protein **markers** can be detected by **immunological methods** (M30) such as Western blotting or by **protein electrophoresis** (M60). Both methods are able to recognise codominant markers. Detection of a specific protein is the ultimate proof for expression of the determining gene, which is of great importance in the study of transgenic plants.

Nucleic acid markers

Nucleic acids in plants comprise of DNA, which contains the genetic information and of RNA, which is needed for production of proteins, i.e., translation of the genetic information. DNA is located in the nucleus, in the plastids and the mitochondria, whereas RNA is found throughout the cell. The effect of the environment on **markers** based on nucleic acids is presumably very low, though specific environments may increase the mutation rate.

Recommended methods

Restriction fragment length polymorphism (RFLP) (M64) is a method revealing nucleic acid **markers**. Correlations between an agronomic quantitative morphological trait and a particular **RFLP-marker** are found in several cases and the **RFLP-marker** is then termed a quantitative trait locus (QTL). A QTL is thus either directly involved in the determination of the trait or it is closely linked to it. **RFLP** (M64) combined with **Southern and Northern hybridization (blotting)** (M71) is a powerful tool for revealing the presence and the expression of specific genes in the genome, e.g., insertions of alien genes. **Polymerase chain reaction (PCR)** (M59) has improved the applicability of nucleic acid **markers**, as the amount of nucleic acid needed is heavily reduced. **Amplified restriction fragment polymorphism (AFLP)** (M1), **microsatellite markers** (M38) and **random amplified polymorphic DNA (RAPD)** (M61) are all based on **polymerase chain reaction (PCR)** (M59). Depending on the method, the **marker** may be either dominant or codominant.

Markers available and influence of genes and enviroment

Table 6.1. List of markers and evaluation of the influence by genes and by environment		
Type of marker	Degree of influence	
	by genes	by environment
Morphological	Partial	Partial to large
Cytological	Large	Small
Metabolic	Partial	Partial to large
Protein	Large	Small
Nucleic acid	*	None?

*: Genes are composed by nucleic acids, which may or may not result in gene products

Important for markers

The techniques mentioned above will only detect dissimilarities among individuals. Even, if a **marker** is found by a specific method to be identical for two plants, another method may still detect differences. Consequently, for protein and nucleic acid **markers**, the identity has to be proved by sequencing, either amino acid sequencing (see standard biochemistry books) or **DNA sequencing** (M10).

Reviews

47, 50, 76, 238, 243, 383, 395, 576, 680, 736, 761, 781, 809, 824, 869, 942, 971, 985, 992.

Category	**Hybridization**
Subcategory	Artificial hybrid, hybrid, natural hybrid.

Hybrids are the result of a cross between two genetically distantly related plants. **Hybrids** may be produced either through natural **outcrossing** or through human impact. When hybrids are produced without human interference (natural hybrids), the parents are not known, whereas the parentage of artificial hybrids, e.g., produced by **experimental pollination** (M13) or **forced fertilization** (M18), usually can be verified. The success of a **hybrid** depends on its fitness compared to its parents. Differences between parental genomes of a **hybrid** often cause segregation disorders, leading to a much reduced fitness. However, the hybrid can avoid the segregation disorder by polyploidization. Possible backcrosses or crosses with other **hybrids** (hybrid swarms) are factors which will also increase the success of a **hybrid**. In the context of transgenic plants, the incidence of **hybridization** between species is of great interest, as it may enhance the risk for dispersal of the transgenes.

Recommended methods

Various **markers** may be used for the identification of **hybrids**. The inheritance of the **marker** has to be verified by crossing experiments, e.g., **diallel cross** (M9). When analysing a population which consists of **hybrids** and backcrossed individuals, **composite character index** (M7) may be useful for identification of the **hybrids**. Depending of the choice of **marker**, different treatments of the plant material are needed before **marker** determination. This may be done either by **plant material preservation: Cryopreservation and freeze-drying** (M47) or **plant material preservation: Rapid drying** (M48).

Hybrids may be identified by **herbarium sheet survey** (M22) applying **morphological character analysis** (M39), followed by relevant statistical methods such as **principal component analysis (PCA)** (M54 in vol. 1). **Herbarium sheet survey** (M22) allows identification of present-day hybrids as well as hybrids from the past. This may add more information on which species that have the ability for **hybridization**.

Karyotypic analysis: Chromosome number, size and form (M33) is used for determining the chromosome number of a putative **hybrid**. **Flow cytometry** (M15) can be a rapid and reliable method for determining nuclear DNA content of plants, if the parental species have significant differences in DNA content (576). **Meiotic analysis: Chromosome pairing and recombination** (M37) can confirm the **hybrid** nature of a plant (238, 700), as non-pairing chromosomes may occur. *In situ* **hybridization** (M31) offers possibilities for detection of either chromosome or gene **introgression** within a taxon or among taxa (395).

Metabolic **markers** revealed through methods such as **GC** (M20), **HPLC** (M29), **TLC** (M74) and **two-dimensional paper chromatography** (M77) may be applied depending on the difference between the parental taxa. The greater the difference, the more a reliable identification of **hybrids** is possible. A prerequisite is that profiles of the possible parental species are available. Metabolic **markers** may be less relevant to use for identification of **hybrids**, compared to morphological, protein and nucleic acid **markers**.

Protein **markers** determined either by **protein electrophoresis** (M60) or by **immunological methods** (M30), such as Western blotting, have shown to be well-suited for identification of **hybrids**.

Nucleic acid **markers** determined either by **AFLP** (M1), by **RAPD** (M61) or by **RFLP** (M64) are more efficient tools for revealing **hybrids** than protein **markers** due to the greater number of bands in the DNA-profiles.

Ideally, a combination of morphological, chemical, molecular and cytological **markers**, applied to plants collected in nature or from experiments should be used (e.g., 974). Such a combination of different methods provides a much more precise and reliable analysis than when a single **marker** is used (910).

Research need
It has been emphasized, that sampling for studies of **hybridization** and introgression using molecular **markers** should be extended to cover a large area due to the potential dispersal of pollen over long distances (759). Hence, adequate sampling strategies have to be elucidated.

Need for new methods
Hybridization is difficult to study without a large number of genetic **markers** and a detailed **genetic map** (M23). Advances in molecular techniques, especially **PCR** analysis (M59), have made this a much less expensive task than previously. A great number of **markers** will be added to **genetic map** (M23) for crop species (i.e., the potential GMPs) in the near future.

Reviews
66, 82, 414, 680, 756, 759, 761, 807.

Subcategory
Heterosis, hybrid depression, hybrid vigour.

Heterosis or **hybrid vigour** is an increased performance, i.e., fitness, of the hybrid compared to the parents, whereas a decrease in fitness is called **hybrid depression**. **Heterosis** is often found for natural hybrids (37).

Recommended methods
Hybrid vigour or **hybrid depression** may be detected by applying **fitness measurement** (M14). For many crop species, the occurrence of **heterosis** is an important factor, which can be measured in field trials with randomised complete block design (96, 843). The **heterosis** or hybrid performance is then calculated as **heterosis analysis: Over mid parent and better parent** (M28). **Hybrid depression** may be detected in similar manner.

Reviews
37, 246, 940.

Category	**Genome structure**
Subcategory	Epistasis, gene expression, gene stability, genotype, insert, pleiotropy, recombination.

In the process of introgressive hybridization, genes from parental taxa are brought together in new combinations. A new genotype-background may result in altered **interaction** (**epistasis** and **pleiotropy**), **expression** and **stability** of the genes. Both endogenous (natural) and inserted genes (transgenes) are the target of such changes. The best basis for studying alterations at the gene level is knowledge on the genes involved and their inherent **genotype**. The sequence and position of the genes in the genome are important parameters in such studies. The plant environment is a key factor modulating gene expression and consequently the phenotype.

Pleiotrophic effects of a particular gene may be of great importance in the context of transgenic plants, because it may add unexpected new traits to the plant. The new traits may only turn up under very specific conditions, e.g., in a harsh environment.

Recommended methods	Provided the sequence or part of the sequence of a gene is known, its presence and inheritance can be analyzed by **PCR** (M59) with a specific primer (**Southern and Northern hybridization (blotting)** (M71)) or by *in situ* hybridization (M31). Analysis of expression, which may be performed by applying **AFLP** (M1) or **Southern and Northern hybridization** (M71) and inheritance (stability) of genes is made easy if the genes result in distinct phenotypes. The position of a gene in the genome seems to be of extreme importance to both **expression, stability** and genetic **recombination** of the gene. The genomic position of the gene is commonly determined by linkage analysis (M40 in vol. 1) based on **genetic map construction** (M23) of segregating offspring. The genetic map is an illustration on how the genes are organized in the genome. Disclosing the neighbouring sequences of a specific gene is made possible by the **inverse PCR** technique (M59).
Research need	Presently, very few studies exist on changes of the behaviour of genes, inserted into another genetic background after hybridization. A review provides information on possible mechanisms responsible for changes in **expression** and **stability** of transgenes following introgression (184).
Need for new methods	Test procedures for revealing **pleiotropy** are highly needed for risk assessment of transgenic plants.
Reviews	184.
Subcategory	Structural chromosome changes.

Genome structural changes or karyotype changes are often experienced in offspring from crosses between taxa that are separated by rather large genomic differences. Such changes of the genome may affect the success of meiotic recombination and thus the probability that stable introgression of genetic material from one taxon to another will take place.

Recommended methods

Structural changes can be detected by the different methods of **karyotype analysis** (M33). **Chromosome number, size and form** can be revealed in mitotic or meiotic metaphase cells. In a **meiotic analysis** (M37), the chromosomal configurations formed can reveal the nature of both structural and numerical chromosome changes. **Chromosome painting techniques** (M5) and *in situ* **hybridization** (M31), may help in detecting genome changes. Preferably, the combination of **meiotic analysis** with **chromosome painting** should be used. Linkage maps of the taxa involved in an introgressive hybridization determined by **genetic map construction** (M23) can also be of value in the identification of additions, deletions, substitutions, amphidiploidy and other types of genome changes. It is required that the maps are saturated with taxon specific markers and that segregating offspring plants are available for analysis.

Research need

Development of more reproducible and easy methods for **chromosome painting** (M5) would be an advantage in the analysis of genome and chromosomal aberrations. The detection of genomic alterations are now being facilitated by molecular methods providing detailed linkage maps for an increasing number of crop species.

Reviews

27, 37, 403, 765, 1002.

Category

Population

A **population** may be defined as a group of plants of the same species which are separated from other such groups in space and in time. It is often difficult to delimit the **population**, because plants may occur almost continuously over large areas. Therefore, the definition of a **population** in an ecological sense may often be a practical definition, i.e., as the group of plants currently under study. The **population** is often regarded as the basic unit of evolution. The impact of a released GMP on other plant species will partly be through its effect on other species at the **population** level.

In studies of the genetic processes occurring in **populations** (i.e., population genetics) more strict definitions are needed. Hence, the concept of the "panmictic unit" is used. In some cases the "panmictic unit" may be equal to the ecological **population**. However, if gene-flow through seeds or pollen is restricted, matings between widely separated plants in the **population** may be very rare and the **population** is said to be subdivided. In a patchy subdivided **population** the patches are termed subpopulations. If the plants are continuously distributed in space, plants are said to be genetically "isolated by distance". In a subdivided **population** it is possible to define a migration rate between the different subpopulations, whereas in a continuous **population**, an average dispersal distance can be defined. However, in most plant **populations**, the pattern is a mixture of both subdivided and continuous **population** structures.

In cases where gene-flow is unrestricted over large distances, a given ecologically defined **population** may only be part of the "panmictic unit".

Subcategory   Cline.

Cline   A **cline** is a graded series of a genetically determined trait along a spatial dimension. A **cline** can usually be observed morphologically in a hybrid zone between two subspecies or species, e.g., a GMP and a related wild plant species.

Recommended methods   **Clines** can be investigated genetically by using any of the molecular methods. By comparing **nuclear and organelle DNA** (M42) information on the direction of **introgression** in the **hybridization** process can be investigated. **Genetic map construction** (M22) of involved species makes it possible to investigate which genes introgress easily. **Hybridization** usually has fitness effects, so therefore **fitness measurement** (M14) along the **cline** may also aid to the understanding of the frequency of successful **introgression.** **Local adaptation** (M35) may be important in maintaining a **cline** and should be estimated with a proper **fitness measure** (M14).

Research need   The understanding of the genetic processes in populations and how **clines** are maintained is still limited. Theoretically, a number of models have been investigated, but experimental studies are still few and with uncertain conclusions.

Reviews   63, 230.

Subcategory   Gene pool, genetic diversity, polymorphism.

Polymorphism is the existence of more than one form of the same gene or DNA-sequence in the population. **Polymorphism** determines the amount of **genetic diversity** in the species, whereas the **gene pool** refers to all genes in the species, whether they are polymorphic or not. The **gene pool** interacting with the environment determines the phenotypes of the individuals in the population. Differences between phenotypes of individuals are determined by the **genetic diversity**, environmental factors and interactions between the genotypes and the environment.

Recommended methods   **Polymorphism** can, in some cases, be determined from the phenotype alone (e.g., colour polymorphism), but is usually measured by genetic markers in laboratory experiments. These markers may be genes determined by **protein electrophoresis** (M60) and **RFLP analysis** (M64), some **DNA-sequences** (M10) or non-coding stretches of DNA determined by **RAPD** (M61), **microsatellite markers** (M38) or **AFLP** (M1). **Polymorphism** is a prerequisite for application of any statistical analysis of genetic data. When **polymorphism** is determined from a number of markers, it may be used to estimate the **genetic diversity**. The multiple-locus markers such as **AFLP** (M1) or **RAPD-analysis** (M61) may be most powerful in this respect.

A number of methods investigate the distribution of the **genetic diversity**. The breeding system can be analyzed through **selfing and outcrossing rate** (M69). In a continuous population the **genetic neighbourhood-size** (M24) is important and can be further investigated through **spatial autocorrelation** (M72), which yields information on the distribution of dispersal distances within the population. Furthermore, **paternity analysis** (M45) can be used to infer precisely between which individuals mating occurs. The distribution of **genetic diversity** between populations can be investigated by **F-statistics** (M19) or analysis of **genetic distance** (M22). The **gene pool** can be studied through **genetic map construction** (M23) of the species in focus.

The location of the transgene in a GMP can be determined by **genetic map construction** (M23) of the species. By **Southern hybridization** (M71) it is possible to investigate whether the transgene has introgressed into a wild species and, if so, the location can be determined after **genetic map construction** (M23) of the wild species.

Research need

Knowledge on the **hybridization** process requires a large number of genetic markers and a detailed genetic map. Advances in molecular techniques, especially the **PCR** (M59), have made this a much less expensive task than previously and more markers will be added to crop species (incl. GMPs) in the near future. **Genetic diversity** can thus be determined accurately, but the adaptive value of this diversity is still being debated. More studies linking variation at the DNA-level or protein-level to variation in plant fitness are needed.

Reviews

337, 338, 346, 869, 988.

Subcategory

Genetic drift, genetic neighbourhood-area, metapopulation.

Genetic drift is random fluctuations in gene frequencies caused by a finite population-size as opposed to deterministic changes caused by selection. **Genetic drift** causes genetic differentiation, which may lead to independent evolution of different populations.

Recommended methods

The amount of **genetic drift** is determined by the **effective population size** (M11) of the population under study. If a population is subdivided, either in discrete patches (a **metapopulation**) or continuously (in **genetic neighbourhoods**), the size of the subpopulation or **genetic neighbourhood-size** (M24), respectively, in combination with the migration rate, determine the amount of **genetic drift**.

The amount of migration between subpopulations can be determined by **F-statistics** (M19) or from **gene-flow estimation with private alleles** (M21). Migration between subpopulations may be dependent on the distance between the subpopulations. This can be investigated by applying a **genetic distance** (M22) measurement, or by using several hierarchies in an analysis of **F-statistics** (M19). Migration in plants can be through either seeds or pollen. The method **nuclear and organelle markers combined** (M42) can separate these two modes of migration, and **paternity analysis** (M45) can be used to study gene-flow through pollen only. **Computer programs for analysis of genetic data** (M8) facilitate the use of these methods. Investigations of all the above factors require different types of genetic markers: 1. Single-locus codominant markers, such as isozymes, determined by **protein electrophoresis** (M60, see also M34 in vol. 1) and **microsatellite markers** (M38), and 2. Dominant multiple-locus markers, determined by **RAPD-analysis** (M61), **AFLP** (M1) or by **DNA sequencing** (M10).

Research need	The migration rate of pollen or seeds into small populations from larger introduced GMP populations is the major factor determining the opportunity for **hybridization** followed by **introgression.** Though a number of studies have estimated migration rates, the maximum possible migration distance is usually not known. The dependency of migration rate on the size of the receiving population has also been poorly investigated. The power of the traditional markers such as isozymes to detect migration is considerably lower than when using microsatellites. **Microsatellite markers** (M38) are under development for a large number of species including most crops (incl. GMPs) and may be very useful in future. Statistical models integrating different kinds of genetic markers in the analysis may also yield new insight into migration patterns, as in the method **nuclear and organelle markers combined** (M42). Furthermore, additional studies are needed to determine to which extent direct observation of pollen dispersal is a good estimate of gene-flow through pollen.
Reviews	Metapopulation and genetic neighbourhood-area: 60, 337, 338, 346, 501, 620, 988, 1018, 1019. Molecular techniques: 47, 50, 89, 869, 985.
Subcategory	Genetic load.
	Genetic load is a measure of the fitness effects of segregating deleterious alleles, i.e., protein variants which decrease the fitness of an individual in a specific environment.
Recommended methods	**Genetic load** can be estimated through **inbreeding depression estimation** (M32). Inbreeding increases homozygosity at polymorphic loci, and more recessive deleterious alleles will be exposed for selection. An accurate estimation of **genetic load** requires a reliable **fitness measurement** (M14). The amount of **genetic load** is a major determinant for a successful **introgression** of a GMP into a wild species. For the **introgression** to be successful, it has to be advantageous to the population despite the effects **genetic load** may have on fitness.
Need for new methods *Research need*	Methods are needed for establishing a link between genetic variation and **genetic load**, since knowledge on the identity of genes causing **genetic load** and their allele action is limited.
Reviews	143, 923, 951.

6.1. List of subcategories with corresponding references

Agamospermy	10, 120, 428, 615, 636, 637, 731, 785, 812, 837, 967
Altered flower colour	263, 292, 574, 591, 672
Altered metabolic content	185, 202, 245, 279, 302, 458, 473, 474, 523, 709, 915
Ant-pollination	See Insect-pollination
Artificial hybrid	76, 439, 440, 478, 492, 602, 673, 700, 717, 736, 743, 781, 903
Bacteria tolerance	248, 371, 631, 816
Bat-pollination	See Mammal-pollination
Bee-pollination	See Insect-pollination
Beetle-pollination	See Insect-pollination
Biological vector	13, 20, 32, 46, 48, 57, 72, 77, 92, 93, 100, 101, 106, 121, 136, 148, 166, 190, 207, 220, 258, 279, 281, 318, 326, 327, 334, 367, 368, 370, 373, 422, 449, 451, 452, 458, 482, 506, 525, 538, 544, 546, 548, 551, 581, 598, 623, 629, 632, 665, 694, 710, 712, 794, 800, 821, 822, 825, 829, 838, 841, 862, 883, 908, 909, 915, 939, 960, 969, 970, 1026, 1036, 1038
Bird-pollination	242, 360, 724
Butterfly-pollination	See Insect-pollination
Chemical attractant	51, 54, 55, 81, 90, 95, 97, 129, 169, 181, 183, 219, 242, 274, 297, 298, 311, 333, 342, 360, 377, 389, 426, 481, 517, 599, 691, 692, 693, 706, 708, 724, 726, 748, 806, 842, 873, 874, 928, 962, 986, 1032
Chemical poration	48, 220, 253, 272, 282, 460, 507, 547, 571, 611, 642, 661, 712, 741, 813, 840, 877, 970
Cline	33, 230, 231, 278, 470, 718, 757, 814, 978, 1024
Cross-pollination	12, 43, 44, 85, 105, 125, 134, 174, 183, 188, 251, 277, 299, 351, 353, 361, 374, 388, 406, 411, 413, 415, 426, 514, 522, 636, 637, 639, 650, 653, 657, 658, 662, 675, 701, 719, 726, 727, 729, 739, 774, 921, 954, 965, 966, 982, 990, 997
Deliberate release	22, 80, 185, 438, 634, 925, 1002
Drought tolerance	441, 442, 640, 705
Electroporation	9, 14, 34, 40, 41, 48, 67, 91, 92, 130, 146, 147, 155, 220, 258, 269, 404, 505, 571, 712, 713, 741, 840, 970, 994
Epistasis	213, 376, 554

Experimental design	673, 716, 743
Fertilization	132, 182, 214, 267, 330, 348, 454, 466, 704, 902, 911, 943, 954, 1005
Field experiment	1, 8, 15, 25, 27, 33, 37, 59, 71, 85, 96, 98, 99, 113, 118, 126, 128, 143, 154, 160, 161, 176, 183, 196, 197, 198, 201, 208, 214, 215, 218, 222, 224, 225, 226, 227, 230, 235, 241, 242, 254, 265, 275, 277, 278, 299, 336, 342, 349, 350, 351, 353, 362, 363, 377, 387, 388, 390, 396, 407, 408, 415, 426, 429, 431, 432, 446, 453, 457, 465, 467, 470, 478, 480, 492, 500, 501, 503, 517, 520, 532, 541, 553, 554, 558, 566, 567, 569, 583, 584, 585, 586, 587, 588, 590, 594, 598, 614, 633, 634, 653, 655, 656, 657, 658, 659, 668, 669, 670, 671, 673, 678, 679, 685, 688, 695, 716, 718, 721, 724, 726, 727, 729, 743, 748, 754, 757, 758, 782, 801, 806, 814, 833, 843, 849, 857, 861, 864, 865, 866, 890, 898, 935, 950, 965, 966, 978, 979, 983, 999, 1004, 1013, 1016, 1017, 1023, 1024, 1033, 1034
Flowering phenology	See Phenology
Fly-pollination	See Insect-pollination
Frost tolerance	605, 796
Fruit abortion	612, 766, 886
Fungal tolerance	106, 171, 228, 332, 356, 393, 516, 995
Gene expression	9, 32, 56, 57, 73, 78, 86, 121, 131, 135, 148, 155, 161, 167, 186, 190, 200, 202, 203, 212, 220, 272, 306, 327, 334, 368, 370, 373, 379, 392, 393, 405, 410, 422, 423, 435, 451, 452, 460, 462, 472, 475, 482, 506, 507, 535, 538, 539, 544, 546, 548, 561, 565, 571, 573, 575, 581, 582, 625, 627, 629, 632, 642, 665, 687, 694, 741, 750, 767, 800, 813, 822, 825, 829, 838, 841, 876, 885, 908, 909, 915, 939, 955, 1036, 1038
Gene pool	18, 29, 122, 141, 236, 280, 343, 421, 527, 624, 953, 1028
Gene stability	167, 379, 435, 475, 565, 885
Genetic diversity	4, 15, 18, 29, 33, 35, 47, 75, 79, 103, 109, 111, 113, 114, 122, 123, 128, 141, 154, 162, 163, 164, 197, 198, 208, 222, 224, 225, 226, 234, 235, 236, 239, 240, 252, 255, 256, 275, 280, 284, 291, 303, 304, 307, 312, 329, 337, 339, 340, 343, 349, 350, 362, 375, 385, 396, 421, 457, 469, 487, 518, 527, 532, 545, 557, 566, 567, 580, 604, 617, 618, 619, 620, 621, 624, 633, 648, 671, 676, 688, 695, 732, 742, 746, 764, 769, 770, 771, 772, 773, 783, 808, 809, 817, 818, 819, 828, 833, 835, 836, 845, 846, 851, 855, 856, 857, 861, 864, 865, 866, 869, 870, 890, 913, 926, 929, 934, 953, 975, 979, 988, 989, 999, 1000, 1004, 1009, 1015, 1016, 1023, 1028
Genetic drift	4, 35, 47, 75, 98, 99, 108, 109, 110, 111, 123, 124, 128, 138, 163, 176, 218, 234, 235, 239, 240, 250, 252, 254, 284, 303, 304, 305, 307, 312, 346, 362, 369, 375, 386, 421, 424, 465, 469, 479, 487, 501, 503, 532, 545, 580, 590, 604, 618, 619, 620, 635, 648, 676, 685, 688, 695, 732, 746, 769, 770, 771, 772, 773, 783, 809, 818, 819, 833, 835, 836, 846, 855, 857, 865, 866, 934, 979, 988, 989, 999, 1015, 1016, 1017, 1023, 1029, 1033

Genetic load	1, 143, 154, 201, 214, 268, 336, 349, 350, 363, 387, 407, 467, 480, 500, 536, 553, 554, 594, 614, 633, 670, 671, 721, 890, 935, 983, 1004, 1013
Genetic neighbour-hood-area	47, 98, 99, 113, 124, 126, 128, 138, 176, 197, 198, 208, 218, 222, 224, 225, 226, 235, 254, 275, 362, 386, 396, 424, 457, 465, 479, 501, 503, 532, 566, 567, 590, 620, 635, 685, 688, 695, 758, 818, 833, 857, 864, 865, 866, 979, 999, 1015, 1016, 1017, 1023, 1033
Gene-transfer	313, 341, 403, 417, 418, 499, 504, 634, 723, 810, 852, see also Vertical gene-transfer and Horizontal gene-transfer
Genotype	3, 6, 7, 49, 74, 77, 88, 91, 93, 100, 101, 145, 149, 150, 152, 153, 159, 167, 175, 179, 189, 199, 205, 211, 213, 221, 237, 238, 260, 261, 271, 273, 283, 285, 286, 288, 289, 293, 301, 308, 309, 316, 318, 325, 367, 376, 380, 381, 394, 401, 402, 412, 423, 430, 435, 444, 449, 455, 459, 475, 478, 485, 488, 489, 506, 510, 512, 513, 521, 524, 526, 531, 546, 550, 565, 575, 578, 579, 595, 597, 601, 613, 641, 660, 687, 689, 690, 715, 720, 734, 745, 753, 760, 761, 762, 763, 765, 767, 776, 784, 785, 791, 797, 798, 811, 821, 828, 830, 871, 892, 894, 900, 906, 907, 909, 916, 922, 930, 932, 936, 946, 960, 961, 972, 998, 1001, 1002, 1014, 1025, 1026, 1035, 1038, 1039
Greenhouse experiment	1, 6, 24, 27, 85, 88, 118, 143, 154, 160, 172, 183, 201, 214, 215, 242, 277, 299, 331, 336, 342, 349, 350, 351, 353, 363, 387, 388, 394, 408, 415, 426, 431, 432, 444, 446, 453, 467, 480, 500, 512, 517, 520, 541, 553, 554, 558, 583, 584, 585, 586, 587, 594, 614, 633, 653, 657, 658, 659, 668, 670, 671, 678, 679, 721, 726, 727, 729, 748, 752, 760, 782, 806, 890, 898, 935, 950, 965, 966, 983, 1004, 1013, 1025, 1030
Growth chamber ex-periment	154, 349, 350, 633, 671, 717, 890, 903, 1004
Herbicide tolerance	11, 25, 26, 61, 70, 93, 100, 166, 173, 185, 207, 209, 247, 268, 318, 322, 443, 528, 529, 530, 533, 789, 790, 831, 832, 878, 879, 880, 881, 883, 891, 892, 901, 919
Heterosis	15, 96, 246, 478, 492, 569, 588, 673, 716, 743, 780, 801, 861, 940, 1034
Heterostyly	39, 112, 276, 425, 589, 630, 651, 654, 677, 774, 826, 882, 927, 981, 984
Horizontal gene-transfer	723, 810
Hybrid	8, 19, 27, 38, 45, 63, 66, 96, 151, 153, 167, 172, 175, 196, 215, 233, 243, 246, 260, 331, 345, 378, 395, 402, 414, 429, 433, 436, 437, 453, 498, 549, 550, 556, 565, 569, 576, 579, 588, 601, 649, 668, 680, 716, 717, 734, 744, 752, 756, 759, 761, 762, 763, 765, 784, 795, 801, 807, 815, 824, 863, 867, 906, 910, 916, 931, 940, 942, 946, 971, 973, 974, 976, 991, 992, 998, 1007, 1030, 1034, 1035
Hybrid depression	246, 849
Hybrid vigour	15, 96, 246, 492, 673, 843, 849, 861
Incompatibility	142, 266, 502, 555, 568, 606, 612, 698, 820, 839, 943

Insect-pollination	95, 125, 129, 169, 170, 174, 183, 242, 262, 352, 354, 360, 361, 426, 652, 658, 693, 699, 706, 707, 724, 726, 728, 873, 874, 921
Insect tolerance	20, 25, 68, 77, 101, 156, 186, 260, 271, 283, 285, 286, 287, 288, 289, 316, 319, 365, 366, 367, 382, 391, 405, 412, 416, 449, 494, 506, 546, 788, 821, 830, 899, 900, 917, 952
Insert	See Marker
Insertion	See Marker
Introgression	5, 27, 36, 37, 74, 91, 144, 145, 151, 153, 167, 172, 175, 194, 199, 205, 206, 215, 238, 293, 344, 345, 381, 401, 402, 403, 429, 430, 455, 488, 489, 515, 519, 550, 565, 578, 579, 601, 602, 641, 668, 720, 730, 734, 752, 756, 759, 761, 762, 763, 765, 784, 792, 795, 804, 867, 870, 906, 907, 912, 920, 930, 972, 974, 991, 1007, 1022, 1030, 1035, 1039
Laboratory experiment	4, 6, 18, 19, 29, 66, 76, 88, 103, 109, 111, 113, 114, 122, 126, 141, 151, 160, 162, 163, 164, 195, 197, 198, 204, 208, 215, 222, 224, 225, 226, 234, 236, 238, 243, 255, 256, 275, 280, 291, 303, 304, 329, 335, 337, 339, 343, 378, 385, 394, 395, 396, 409, 414, 421, 433, 439, 440, 444, 457, 469, 487, 496, 497, 498, 512, 518, 527, 541, 549, 566, 567, 576, 583, 584, 585, 586, 587, 596, 604, 617, 619, 621, 624, 676, 678, 679, 680, 700, 717, 736, 737, 752, 758, 759, 760, 761, 769, 770, 772, 773, 781, 795, 807, 808, 819, 824, 834, 835, 836, 845, 856, 859, 864, 869, 898, 910, 913, 922, 926, 929, 931, 934, 942, 953, 971, 973, 992, 1000, 1009, 1011, 1025, 1028, 1034, 1035
Long-term monitoring	25, 27, 37, 144, 206, 344, 383, 390, 409, 429, 496, 497, 519, 730, 759, 804, 912, 920, 964, 996, 1007, 1011, 1022
Male sterility	268, 539, 540, 941, 1034
Mammal-pollination	242, 360, 724
Marker	56, 73, 82, 121, 131, 135, 137, 148, 155, 189, 190, 195, 204, 207, 212, 220, 272, 273, 281, 296, 297, 317, 327, 328, 368, 373, 380, 392, 410, 422, 452, 460, 462, 472, 535, 561, 571, 582, 596, 600, 625, 627, 629, 642, 661, 665, 741, 750, 780, 800, 813, 822, 825, 834, 841, 876, 903, 922, 939, 955, 975
Marsupial-pollination	See Mammal-pollination
Metapopulation	35, 38, 45, 47, 75, 98, 99, 103, 108, 110, 113, 114, 122, 123, 124, 128, 138, 162, 164, 176, 197, 198, 208, 218, 222, 224, 225, 226, 233, 235, 239, 240, 250, 252, 254, 255, 256, 275, 284, 291, 304, 305, 307, 312, 329, 337, 339, 343, 346, 362, 369, 375, 385, 386, 396, 424, 457, 465, 479, 501, 503, 518, 532, 545, 556, 566, 567, 580, 590, 617, 618, 619, 620, 621, 635, 648, 676, 685, 688, 695, 732, 746, 769, 771, 783, 791, 808, 809, 815, 818, 833, 845, 846, 855, 856, 857, 864, 865, 866, 869, 913, 926, 929, 953, 979, 988, 989, 999, 1000, 1009, 1015, 1016, 1017, 1023, 1028, 1029, 1033
Microinjection	23, 94, 410, 462, 521, 625, 626, 627, 664, 666, 710, 712, 750, 751, 850, 872, 876

Microprojectile bombardment	56, 73, 87, 89, 92, 131, 135, 155, 157, 212, 258, 306, 347, 384, 398, 435, 448, 472, 609, 643, 710, 711, 712, 786, 787, 802, 862, 904, 955, 970, 1031
Model	28, 35, 47, 108, 110, 116, 124, 128, 138, 235, 238, 239, 240, 252, 284, 304, 305, 307, 312, 320, 346, 362, 369, 374, 386, 417, 418, 419, 424, 479, 486, 514, 532, 620, 635, 648, 663, 688, 695, 732, 746, 749, 769, 771, 783, 818, 833, 846, 855, 857, 865, 866, 979, 989, 999, 1015, 1016, 1023, 1029
Natural hybrid	24, 25, 36, 37, 144, 194, 206, 238, 344, 383, 390, 409, 496, 497, 519, 730, 792, 804, 805, 843, 893, 912, 920, 957, 964, 996, 1011, 1022
Nectar production	43, 51, 52, 53, 54, 55, 95, 119, 129, 169, 170, 181, 183, 219, 242, 262, 295, 298, 311, 333, 342, 352, 354, 360, 389, 426, 560, 652, 685, 692, 693, 699, 706, 707, 708, 724, 726, 728, 748, 778, 842, 873, 874, 986, 1041
Nematode tolerance	301, 328, 356, 461, 667
Outcrossing	1, 4, 47, 79, 98, 99, 108, 109, 110, 111, 113, 114, 122, 126, 128, 143, 163, 164, 176, 197, 198, 201, 208, 214, 218, 222, 224, 225, 226, 234, 235, 251, 254, 256, 275, 291, 303, 305, 312, 314, 336, 337, 339, 343, 346, 350, 362, 363, 369, 385, 387, 396, 407, 421, 457, 465, 467, 469, 480, 487, 500, 501, 503, 518, 532, 553, 554, 566, 567, 590, 594, 604, 614, 617, 619, 620, 621, 633, 648, 670, 676, 684, 685, 688, 695, 721, 758, 769, 770, 771, 772, 773, 775, 785, 817, 818, 819, 823, 833, 835, 836, 845, 856, 857, 864, 865, 866, 869, 913, 926, 929, 934, 935, 953, 979, 983, 989, 999, 1000, 1009, 1013, 1015, 1016, 1017, 1023, 1028, 1029, 1033
Ovule development	180, 471, 476, 622, 766, 895, 897
Oxidative stress tolerance	16, 32, 104, 326, 392, 694, 755, 987
Pathogen tolerance	25, 185, 210, 257, 270, 376, 399, 477, 507, 577, 610, 632, 938, 1020, 1027, 1040
Pest tolerance	46, 107, 185, 478, 526, 948, 998
Phenology	39, 44, 84, 112, 133, 134, 229, 232, 244, 249, 276, 355, 357, 358, 359, 413, 425, 456, 555, 563, 564, 568, 589, 606, 616, 630, 638, 639, 650, 654, 677, 696, 702, 719, 725, 739, 747, 778, 779, 875, 882, 884, 905, 927, 933, 956, 981, 984, 1006
Pollen competition	183, 188, 323, 681, 775
Pollen dispersal	51, 54, 55, 59, 71, 85, 113, 125, 126, 129, 169, 170, 174, 181, 183, 197, 198, 208, 219, 222, 224, 225, 226, 227, 241, 242, 262, 275, 277, 299, 311, 333, 351, 352, 353, 354, 360, 361, 388, 396, 415, 426, 457, 534, 566, 567, 653, 655, 656, 657, 658, 669, 692, 693, 698, 699, 703, 706, 707, 708, 724, 726, 727, 728, 729, 740, 748, 754, 758, 842, 864, 873, 874, 921, 965, 966, 977, 986, 997, 1012, 1041
Pollen germination	182, 330, 476, 608, 612, 622, 662, 954, 990

Pollen production   127, 180, 183, 244, 294, 483, 484, 552, 555, 622, 651, 684, 733, 776, 803, 875, 963

Pollen viability   434, 471, 555, 578, 636, 637, 776, 956, 1037

Pollinator activity   43, 51, 59, 71, 85, 95, 118, 126, 134, 160, 169, 170, 181, 183, 219, 227, 242, 265, 277, 299, 311, 333, 351, 352, 353, 354, 360, 388, 408, 415, 425, 426, 431, 432, 520, 541, 558, 585, 586, 587, 589, 630, 653, 654, 655, 656, 657, 658, 659, 669, 677, 692, 693, 699, 706, 707, 708, 724, 726, 727, 728, 729, 748, 754, 758, 777, 782, 842, 873, 874, 882, 927, 950, 965, 966, 981, 986

Pollinator foraging   51, 53, 54, 55, 59, 71, 85, 95, 112, 118, 119, 125, 129, 160, 169, 170, 174,
behaviour   181, 183, 219, 227, 242, 262, 265, 276, 277, 295, 298, 299, 311, 333, 351, 352, 353, 354, 360, 361, 388, 389, 408, 415, 425, 426, 431, 432, 520, 541, 558, 560, 585, 586, 587, 589, 630, 652, 653, 654, 655, 656, 657, 658, 659, 669, 677, 678, 692, 693, 699, 706, 707, 708, 724, 726, 727, 728, 729, 748, 754, 758, 782, 842, 873, 874, 882, 921, 927, 950, 965, 966, 981, 986, 1041

Polymorphism   1, 7, 18, 24, 29, 49, 74, 79, 82, 98, 103, 109, 110, 111, 113, 114, 123, 128, 137, 138, 141, 153, 163, 194, 195, 199, 204, 208, 218, 221, 222, 225, 236, 237, 254, 255, 256, 275, 278, 280, 291, 293, 296, 303, 304, 314, 335, 338, 343, 346, 350, 378, 385, 387, 421, 457, 465, 467, 469, 470, 487, 489, 500, 503, 532, 545, 550, 579, 596, 600, 604, 619, 624, 633, 641, 676, 684, 695, 715, 718, 737, 784, 785, 792, 818, 823, 833, 834, 845, 851, 856, 859, 861, 864, 870, 906, 913, 922, 924, 926, 929, 932, 934, 953, 972, 1001, 1009, 1010, 1014, 1022, 1023, 1024, 1028, 1029, 1032, 1035

Recombination   6, 88, 238, 394, 430, 444, 488, 512, 575, 687, 760, 767, 887, 930, 1025, 1039

Release   185, 419, 438, 634, 749, 925, 1002

Reproductive allocation 2, 31, 42, 53, 59, 71, 95, 105, 116, 119, 127, 129, 158, 160, 169, 170, 183, 191, 192, 193, 216, 227, 229, 262, 265, 295, 352, 354, 400, 426, 450, 468, 541, 560, 563, 568, 585, 586, 587, 592, 606, 638, 652, 655, 656, 663, 669, 684, 693, 697, 698, 699, 701, 706, 707, 726, 728, 733, 738, 754, 779, 803, 842, 847, 873, 874, 1041, see also vol. 1

Risk assessment   21, 22, 24, 25, 28, 62, 65, 68, 80, 117, 139, 140, 165, 177, 185, 209, 290, 300, 341, 345, 356, 372, 383, 403, 417, 418, 419, 420, 438, 446, 447, 463, 464, 486, 490, 491, 495, 511, 519, 570, 603, 634, 644, 645, 646, 647, 683, 730, 744, 749, 768, 804, 805, 844, 848, 863, 888, 896, 912, 914, 920, 925, 937, 944, 945, 947, 949, 959, 964, 974, 1002, 1003, 1007, 1022, 1030

Salt tolerance   918, 1036

Seed abortion   450, 468, 592, 612, 628, 766, 793, 897

Seed germination   106, 321, 637, 681, 811, 918, see also vol. 1

Seed production   30, 188, 200, 264, 358, 364, 400, 434, 543, 559, 622, 628, 684 739, 990, 997, see also vol. 1

Self-pollination	1, 10, 12, 105, 113, 125, 128, 143, 174, 197, 198, 201, 208, 214, 222, 224, 225, 226, 235, 251, 275, 314, 324, 336, 350, 361, 362, 363, 387, 396, 407, 411, 457, 467, 480, 500, 514, 532, 553, 554, 563, 566, 567, 568, 593, 594, 614, 633, 658, 662, 670, 675, 684, 688, 695, 697, 698, 701, 721, 777, 785, 826, 827, 833, 857, 864, 865, 866, 884, 902, 921, 935, 979, 983, 999, 1013, 1016, 1021, 1023
Sex distribution	39, 42, 43, 83, 191, 413, 593, 639, 651, 701, 774, 779, 956, 984, 997
Short-term monitoring	19, 25
Stress tolerance	102, 160, 178, 203, 317, 325, 397, 449, 506, 513, 524, 541, 562, 583, 584, 585, 586, 587, 678, 679, 798, 892, 898, 1026
Structural chromosome changes	3, 152, 159, 238, 308, 309, 380, 381, 402, 427, 430, 445, 459, 488, 493, 510, 595, 607, 613, 660, 689, 690, 722, 745, 753, 800, 887, 894, 916, 924, 930, 946
Transformation (plastid)	281, 448, 460, 515, 642, 877, 1031
Vertical gene-transfer	38, 45, 64, 79, 98, 103, 114, 162, 233, 255, 314, 329, 556, 808, 815, 823, 853, 858, 935, 964, 1010
Viral tolerance	69, 168, 217, 259, 356, 506, 508, 509, 542, 682, 799, 860, 958
Visual attractant	112, 118, 183, 242, 276, 342, 360, 408, 425, 431, 432, 520, 558, 589, 630, 654, 657, 658, 659, 677, 724, 748, 782, 882, 927, 950, 981
Water-pollination	242, 360, 724
Wind-pollination	183, 241, 242, 360, 534, 724, 740

6.2. List of categories with corresponding reviews

The list contents reviews and books which may be valuable for an introduction to the main categories listed in Chapter 3. The category "Risk assessment" has been added to the list.

Breeding system   58, 83, 142, 456

Gene-flow   5, 17, 49, 81, 82, 187, 223, 313, 314, 315, 337, 340, 401, 499, 503, 557, 756, 759, 763, 764, 795, 852, 854, 858

Genetic engineering technique   48, 86, 89, 92, 136, 157, 258, 365, 404, 448, 626, 643, 710, 711, 712, 794, 862, 872, 968, 969, 970, 994

Genome structure   184, 331, 401, 437, 572, 573, 674, 735

Hybridization   8, 37, 63, 66, 187, 246, 401, 537, 759, 761, 765, 863, 867, 940

Inserted trait   16, 46, 68, 69, 86, 107, 156, 171, 210, 217, 247, 248, 257, 259, 263, 302, 322, 356, 365, 371, 477, 494, 533, 577, 591, 623, 640, 709, 788, 794, 892, 1008

Pollen development and production 241, 534, 758, 1005

Pollination   81, 183, 242, 342, 360, 481, 724, 740, 748, 962, 980

Pollination syndrome   242, 724

Population   47, 60, 231, 310, 320, 337, 338, 345, 346, 414, 501, 536, 620, 807, 809, 854, 858, 867, 869, 923, 951, 985, 1018, 1019

Risk assessment   22, 25, 68, 139, 356, 403, 447, 511, 634, 644, 925, 1002

Seed development and production   454, 811, 897

Test procedure   See Risk assessment and Chapter 10

7. List of methods and their description

By: Francois Felber (FF), Yolande Jacot (YJ), Rikke Bagger Jørgensen (RBJ), Gösta Kjellsson (GK), Kathrine Hauge Madsen (KHM), Jens Mogens Olesen (JMO), Marianne Philipp (MP), Pia Rufener Al Mazyad (PR), Mikkel Heide Schierup (MHS) and Vibeke Simonsen (VS).

The description of the methods is organized as shown below in Section 7.1. A list of the described methods is shown in Section 7.2 and a detailed description of each method is provided in Section 7.3.

7.1. Organization of the description of the methods

Term	Remarks
Category	Categories relevant for the method.
Subcategory	Subcategories relevant for the method.
Description	A brief description of the method, including purpose and general procedures.
Assumptions and restrictions	Assumptions necessary for the method and for restrictions of the method.
Test system	Indicated, if relevant, where the method may be performed or if it is a data-analytical method.
Advantages	The specific advantages of the method.
Application	Examples of subjects for which the method is highly suitable.
Evaluation	Evaluation of the method with selective levels indicated.

	Sensitivity	1: low	5: high
	Requirements	1: limited equipment needed 5: much equipment needed	
	Time	1: short time needed	5: long time needed
	Cost	1: inexpensive	5: expensive

References	Numbers indicate the relevant literature referring to the reference list in Chapter 11.
Author	Initials of the responsible author(s), see list above.

7.2.   List of methods

1.   Amplified restriction fragment polymorphism (AFLP)

2.   Bagging of flowers

3.   Bioassay

4.   Biogeographical assay and monitoring

5.   Chromosome painting

6.   Classification of fruits, seeds and ovules

7.   Composite character index

8.   Computer programs for analysis of genetic data

9.   Diallel cross

10.   DNA sequencing

11.   Effective population-size N_e

12.   Emasculation of flowers

13.   Experimental pollination

14.   Fitness measurement

15.   Flow cytometry

16.   Fluctuating asymmetry of vegetative and floral traits (FA)

17.   Fluorescent dyes and marking of pollen

18.   Forced fertilization

19.   F-statistics

20.   Gas chromatography (GC)

21.   Gene-flow estimation with private alleles

22.   Genetic distance

23.   Genetic map construction

24.   Genetic neighbourhood-size

25.   Genet identification

26. Haplotype statistics

27. Herbarium sheet survey

28. Heterosis analysis: Over mid-parent (HMP) and better parent (HBP)

29. High performance liquid chromatography (HPLC)

30. Immunological methods

31. *In situ* hybridization to chromosomes

32. Inbreeding depression estimation

33. Karyotype analysis: Chromosome number, size and form

34. Linkage disequilibrium

35. Local adaptation analysis

36. Marking plants and flowers: Flowering phenology analysis

37. Meiotic analysis: Chromosome pairing and recombination

38. Microsatellite markers

39. Morphological character analysis

40. Nectar production: Chemical composition

41. Nectar production: Volume and rate

42. Nuclear and organelle markers combined

43. Organelle DNA analysis

44. Ovule counts

45. Paternity analysis

46. Photosynthetic efficiency

47. Plant material preservation: Cryopreservation and freeze-drying

48. Plant material preservation: Rapid drying

49. Pollen carry-over between flowers

50. Pollen counts

51. Pollen deposition on stigmas: Pollen numbers and fertility

52.   Pollen germination tests

53.   Pollen traps

54.   Pollen viability tests

55.   Pollinator attraction: Chemical cues

56.   Pollinator attraction: Visual cues

57.   Pollinator foraging behaviour

58.   Pollinator preference experiments

59.   Polymerase chain reaction (PCR)

60.   Protein electrophoresis: Isozyme analysis

61.   Random amplified polymorphic DNA (RAPD)

62.   Reproduction in alien plants: Community and invader analysis

63.   Reproductive allocation measures

64.   Restriction fragment length polymorphism (RFLP)

65.   Ribosomal DNA (rDNA) analysis

66.   Seed development analysis

67.   Seed germination tests

68.   Seed viability tests

69.   Selfing and outcrossing rate

70.   Sexual morph distribution

71.   Southern and Northern hybridization (blotting)

72.   Spatial autocorrelation analysis and Moran's I

73.   Stigma receptivity test

74.   Thin layer chromatography (TLC)

75.   Thinning of flowers

76.   Transplantation experiment

77.   Two-dimensional paper chromatography

7.3. Description of methods

M1. Amplified restriction fragment polymorphism (AFLP)

Category	Gene-flow, hybridization, genome structure, population.
Subcategory	Introgression, marker, vertical gene-transfer, hybrid, genotype, polymorphism.
Description	**AFLP** is a DNA or RNA fingerprinting technique combining restriction fragment analysis with **PCR** (M59). It is based on selective amplification of a subset of restriction fragments from a digest of genomic DNA or RNA. It involves three steps: 1) the restriction enzyme digestion of DNA or RNA and ligation of oligonucleotide adaptors, 2) the selective amplification of a subset of all the fragments in the total digest and 3) the gel-based electrophoretic analysis of the amplified fragments. Description of the procedure is supplied together with the patented **AFLP** kit from Keygene N.V.
Assumptions and restrictions	Most **AFLPs** are inherited as dominant Mendelian markers (715), and thus heterozygotes cannot be detected. However, Keygene has recently found that some markers are codominantly inherited and this may enhance the use of the method.
Test system	Laboratory experiment.
Advantages	The **AFLP** technique is **PCR** based and therefore requires only minimal amounts of starting nucleic acid. In comparison to the **RAPD** (M61) technique the **AFLP** is highly reproducible. The method requires no prior knowledge on the sequence of the target DNA or RNA as is necessary with the **microsatellite markers** (M38). The method generates many more polymorphisms than **RAPD** (261, 550).
Application	The method can be used for determination of genetic relations between populations and species (550). It may also be used for generation of highly saturated linkage maps (211) and hence the possibility of detection of introgressive hybridization. Further, the method may be useful for visualisation of differential gene expression by using RNA (cDNA) as the target for the **AFLP** analysis (49).
Evaluation	Sensitivity: 4 Requirements: 3 Time: 3 Cost: 3-4
References	49, 211, 261, 550, 715.
Author	RBJ

M2.   Bagging of flowers

Category

Pollen development and production, seed development and production, pollination, breeding system, gene-flow, hybridization.

Subcategory

Pollen competition, pollen germination, pollen production, agamospermy, seed germination, seed production, nectar production, pollen dispersal, cross-pollination, incompatibility, phenology, reproductive allocation, self-pollination, outcrossing, hybrid.

Description

Bagging of flowers is used to obtain knowledge on the mating system of a plant by excluding pollinators from flowers. The method is often employed when doing **experimental pollination** (M13). A number of different equipment designs and materials are available: Nylon nets, dialysis tubing, agricultural row cover material, stocking material, bridal-veil material, Slip-Ezy pollination bags, windowed paper pollination bags, etc. Rings and poles are used to keep the material separated from the flower and to carry the bag. A special type of isolation bag, a moisture bottle placed over the flowers, can be used to suppress the opening of anthers. The bottle is lined with moist filterpaper and fixed to a pole.

Assumptions and restrictions

Bagging of flowers reduces wind and movement of flowers. Regular shaking of flowers can be performed to compensate for the lack of wind. The temperature in the bag is higher than in the surroundings. Both wind and temperature depend on the mesh width of the bag and influence the performances of pollen, stigmas and ovules within the bag.

Test system

Field experiment.

Advantages

This method is efficient in obtaining information on the incompatibility system, ability of selfing, inbreeding, outcrossing depression and agamospermy.

Application

The method is widely used when breeding systems of plants are studied, especially in combination with experimental pollination (251, 324, 406, 675, 697, 739, 785). **Bagging** is also used to study the effects and possibilities for selfing and self-compatibility (10, 698, 778). Furthermore, it is used to investigate the nature and extent of agamospermy (684, 812, 837), fruit set without pollinators (638), duration of male and female phase (44) and outcrossing distance (982). Isolation bags can also control pollen escape of wind-spread pollen and allow the study of wind pollination by excluding insect pollination (555, 568).

Evaluation

Sensitivity: 4   Requirements: 2
Time: 3   Cost: 2

References

10, 12, 44, 251, 324, 406, 555, 568, 638, 675, 684, 697, 698, 739, 778, 785, 812, 837, 982.

Author

MP

M3.   Bioassay

Category

Genome structure, population, inserted trait.

Subcategory

Genotype, polymorphism, stress tolerance.

Description

Bioassay, the response of an organism to a substance, was already described in vol. 1 when mainly the aspect of using plants as test organisms for xenobiotic compounds were commented (316, 325, 524, 798). However, **bioassay** may be used at other trophic levels for testing phytotoxicity. Many plant species have the ability to produce secondary metabolites, which may give protection against viral, bacterial, fungal or herbivore attack.

Performance of a **bioassay** may be simple. A pest, e.g., a herbivorous insect, is fed with a plant, which may have the ability to reduce the survival or the reproduction of the pest. Bioassay may be followed by a determination of the toxic compounds by, e.g., **HPLC** (M29).

Assumptions and restrictions

The time of exposure and the life stage of the pest as well as the plant may be important for the result. The individual variation in production of a secondary metabolite, i.e., genetic environmental variation in the plant population, may also have an effect on the experiment.

Test system

Field experiment, greenhouse experiment.

Advantages

The method is simple to use, as it does not need any expensive laboratory equipment.

Application

The state of the art for herbicide **bioassays** has been described (892). The method has also been used for detecting natural resistance to various pests (260, 283). Successful transformation of plants has often been revealed by genetic markers, e.g., kanamycin resistance (see Chapter 8), which has been inserted together with trait wanted. **Bioassay** has also been used for detecting expression of an inserted trait, e.g., resistance to vira (506, 1026) or to insects (271, 288, 367, 449, 506).

Evaluation

Sensitivity: 3   Requirements: 2
Time: 4   Cost: 3

References

260, 271, 283, 288, 316, 325, 367, 449, 506, 524, 798, 892, 1026.

Author

VS, KHM

M4. Biogeographical assay and monitoring

Category

Gene-flow, hybridization.

Subcategory

Introgression, vertical gene-transfer, natural hybrid.

Description

Biogeographical assay and monitoring is a simple method to acquire knowledge about differences in gene-flow in different regions which is essential information for risk estimation of field releases. Before a field release is performed in a region, a biogeographical assay is performed to know whether wild relatives are present in the area or not. This can be checked by a comparison of distribution maps for plant species (distribution atlas for given regions or countries). For overlays of different species maps, a Geographical Information System (GIS) is useful to detect regions where species areas are overlapping and hybridization may be possible. Information of sources can also be entered into the GIS. The local and regional floristic literature have to be checked for information on distribution, wild relatives, hybridization and habitat characteristics. Furthermore, herbarium sheets from the region have to be screened for natural hybrids between the crop and wild relatives which are possible according to literature and to distribution maps. This is done by **herbarium sheet survey** (M27) and **morphological character analysis** (M39).

In regions with insufficient data, the relevant information may be derived from local agricultural expertise and by monitoring (field excursions). On excursions, populations of crop species, wild relatives and of hybrid zones are mapped for later comparison with information from other sources (e.g., comparison between old distribution maps and the present distribution to detect invasion or decay). Furthermore, field edges can be screened for investigated species from a car while driving about 40 km/h and observations noted on a detailed topographical map. Interesting populations of putative natural hybrids are collected for further examination and hybrid determination by, e.g., **morphological character analysis** (M39), **protein electrophoresis: Isozymes** (M60) or **RFLP** (M64).

Assumptions and restrictions

Detailed biogeographical information must be available and extensive field monitoring may be necessary.

Test system

Field experiment.

Advantages

Information on the historical aspects of gene-flow and hybridization is acquired and cases of successful invasion are recorded by this method. Furthermore, local regions of future risk for hybridization with GMPs may be detected.

Application

The biogeography and natural hybrids of selected weedy species have been studied in Switzerland (24). Two chromosome types of *Medicago falcata* were detected, and differences in ploidy level resulted in a regional differentiation of the risk of hybridization with cultivated alfalfa (*Medicago sativa*) (805). **Biogeographical assay**, with additional information on pollen dispersal, seed dispersal and hybridization, has been used to assess the risks that given cultivated plants could cause gene dispersal to the wild flora in the Netherlands and in Switzerland (24, 964).

Evaluation

Sensitivity: 3   Requirements: 2
Time: 5   Cost: 2

References

24, 805, 964.

Author

PR

M5.   Chromosome painting

Category   Gene-flow, hybridization, genome structure.

Subcategory   Introgression, vertical gene-transfer, hybrid, structural chromosome changes.

Description   Detection of genomic changes can be facilitated by the use of various techniques for staining or **painting** the chromosomes. Identification of specific chromosomes can be carried out by the use of methods that selectively stain specific regions or sequences of the chromosomes or all the chromosomes of one genome in plants with different genomes, e.g., interspecific hybrids. A common staining technique, for identification of individual chromosomes structural changes, is the C-banding technique for staining the heterochromatic regions of the genome (510, 660). Characterisation of the chromosome complement is also possible by silver staining of nucleolus organizing regions NORs (152, 894) or by *in situ* **hybridization** (M31) using labelled probes that mark homologous sequences of the chromosomes. An *in situ* **hybridization** with a total genomic probe will label the chromosomes of one genome in plants with different genomes, e.g., amphidiploids. A **polymerase chain reaction, PCR** (M59) using a specific primer and fluorochrome labelled nucleotides and carried out directly on a chromosome preparation (PRimed IN Situ hybridization, PRINS) will allow detection of individual chromosomes, specific genes and sequences (3, 308, 309).

Assumptions and   The most informative **chromosome painting** techniques are rather
restrictions   laborious and in some cases, e.g., PRINS, the reproducibility is low and the demand for advanced equipment high. Successful **chromosome painting** is only possible with metaphase cells of optimal quality, which is perhaps the most restrictive factor of the method.

Test system   Laboratory experiment.

Advantages   In cases where genome structural changes are difficult to reveal by traditional **karyotype analysis** (M33) or **genetic map construction** (M23) the use of **chromosome painting** is a valuable tool. Especially in combination with a genetic map, a karyotyping using chromosome paints can provide valuable information about genome structure. Genome and chromosome organisation can be revealed in a physical way by the **chromosome painting** techniques.

Application   Genomic changes like polyploidisation, structural chromosome aberrations, and introgression of genetic material can be identified by the use of **chromosome painting**. Countless studies exist that take advantage of **chromosome painting**.

Evaluation   Sensitivity: 3   Requirements: 2
                     Time: 3   Cost: 2

References   3, 152, 308, 309, 510, 660, 894.

Author   RBJ

M6. Classification of fruits, seeds and ovules

Category

Seed development and production, breeding system.

Subcategory

Fertilization, fruit abortion, seed abortion, incompatibility, phenology, reproductive allocation.

Description

Classification of fruits, seeds and ovules is utilized for studying the function of breeding systems and for evaluating different kinds of pollinations. Fruits are harvested at intervals and the status of the developing ovules is determined by visual inspection or by sectioning (see **seed development analysis**, M66). Ovules can be aborted early due to missing fertilization, and will then be very small and shrivelled. If abortion occurs later due to developmental malfunctioning or lack of resources, ovules will be larger and perhaps nearly as large as in mature seeds, but still shrunken. Fruits can be aborted due to a low number of initiated seeds, due to predation or shortage of resources. The classification should reflect these aspects. Classification of seeds can also be done with reference to the position in the fruit. This is especially useful when investigating whether certain pollen types are pollinating ovules with certain positions or whether certain positions receive more resources than others.

Assumptions and restrictions

It may be difficult to distinguish between different stages in the development of the seed.

Test system

Field experiment, greenhouse experiment, laboratory experiment.

Advantages

The partitioning of resources among fruits and seeds within individuals may be studied. The functioning of the breeding system is reflected in the abortion pattern which makes the method important.

Application

This method is often used in investigations of breeding systems. It has been used in studies of mating systems (468) and in recording the effect of experimental pollinations (592, 697, 698). Classification is utilized when investigating pollen competition (733) and factors limiting seed set (450).

Evaluation

Sensitivity: 3   Requirements: 1
Time: 4   Cost: 2

References

450, 468, 592, 697, 698, 733.

Author

MP

M7.   Composite character index

Category   Gene-flow, hybridization.

Subcategory   Introgression, vertical gene-transfer, natural hybrid.

Description   The **composite character index** has been proposed as a method to detect hybrids and introgression between species (792). It may be used when diagnostic markers (isozymes) allow the distinction of two species. For each diagnostic marker one positive point is assigned to one species and one negative point to the other. Then a total score is calculated for each individual by adding the points. Pure individuals are thus characterised by the maximal absolute value of points, either positive or negative. F_1 hybrids have a score equal to 0 while introgressive individuals have intermediate values.

Assumptions and   Individuals with intermediary values may be the consequence of the
restrictions   presence of private alleles and are not necessarily the product of hybridization or introgression. Case by case verification is thus necessary.

Test system   Data analysis.

Advantage   This index facilitates the detection of F_1 hybrids and allows an easy representation of introgression rate between two species.

Application   The **composite character index** has been used for the analysis of hybrid zones of plants (194, 1022).

Evaluation   Sensitivity: 3   Requirements: 3
                             Time: 3   Cost: 3

References   194, 792, 1022.

Author   FF

M8. Computer programs for analysis of genetic data

Category   Gene-flow, hybridization, population.

Subcategory   Introgression, outcrossing, vertical gene-transfer, hybrid, genetic diversity, genetic drift, metapopulation.

Description   **Computer programs for analysis of genetic data** of various forms have been published and the software is freely available. In the following, a selection of them is reviewed. For each program is given: the names, version number, author(s), computer-types (PC refers to IBM-PC compatibles), a short description of selected features and, if available, the WWW-address or FTP-address in double-quotes, from where the program can be downloaded.

PHYLIP vers. **3.572** (252). For PC, Macintosh and virtually any other computer system. Widely used program package for estimation of phylogenetic trees of species, populations or individuals from various kinds of genetic data, including DNA-sequences. WWW: "http://evolution.genetics.washington.edu/phylip.html".

F-STAT vers. **1.2** (312). For PC only. Estimates **F-statistics** from diploid data, using the method of Weir and Cockerham (989), with test of significance by either bootstrap over loci or permutation tests. WWW: "ftp://oracle.bangor.ac.uk/pub/fstat".

GENEPOP vers. **2.** (746). For PC only. Performs different exact tests (783) for Hardy-Weinberg equilibrium, population differentiation and genotypic linkage disequilibrium. Also included are compiled versions of the programs **DIST** (855), which performs a Mantel-test for isolation by distance, and **LINKDOS** (284), which performs pairwise linkage disequilibrium analysis in subdivided populations and calculates D-statistic (648). WWW: "ftp://ftp.cefe.cnrs-mop.fr/pub/pc/msdos/genepop".

G-STAT vers. **3.1** (846). For PC only. Estimates **F-statistics** (M19) from diploid data by the method of Weir and Cockerham (989). The program can also test for **linkage disequilibrium** (M34), Hardy-Weinberg equilibrium and compare matrices pairwise (e.g., a matrix of **genetic distance** (M22) with a matrix of geographical distances) by a Mantel-test. E-mail-address of the author: "hanss@bot.ku.dk".

MLTR, MLDT and **CSPM** (771). For PC only. **MLTR** estimates outcrossing rates from a progeny array of diploid genotypes, with bootstrap estimation of standard errors. **MLDT** performs the same task for dominant markers, e.g., **RAPDs** (M61). **CSPM** estimates the probability that two seeds of a given plant share the same father, based on the sibling-pair model (769). Additionally, programs are available at the web-site for estimations from genetic data from autotetraploids. WWW: "http://www.botany.utoronto.ca/faculty/ritland/ubc/programs.html".

Description, contd.   **MICROSAT** vers. **1.5** (304). For PC or Macintosh. Estimates R_{st} and related quantities from microsatellite data. R_{st} has been developed to resemble F_{ST} from isozyme data, but it takes account of the high mutation rate of microsatellites and their mechanism of mutation. WWW: "http://lotka.stanford.edu/microsat.html".

    RELATEDNESS vers. **4.2** (307, 732). For Macintosh only. Calculates relatedness coefficients between individuals or groups within populations from diploid data, with jack-knife estimation of standard errors. F-statistics are also estimated. WWW: "http://www-bioc.rice.edu/~kfg/GSoft.html".

    AMOVA 1.55 and **EMHAPFRE** (239, 240). For PC-Windows. **AMOVA** (Analysis of Molecular Variance) calculates the F_{ST} counterpart (Phi_ST) for haplotype data at three hierarchical levels, compares pairwise molecular variances, and tests significance of values by non-parametric permutational procedures. **EMHAPFRE** yields maximum likelihood estimates of frequencies of multi-locus haplotypes in diploid populations. WWW: "ftp://129.194.113.13/ftp/comp/win/".

    RAPD 1.04 (35) For PC only. Program for editing and analysis of **RAPD** fragment data. Several distance measures can be obtained. WWW: "http://life.anu.edu.au/molecular/software/rapd.html".

Assumptions and restrictions   Different programs use estimation based on different assumptions, which should be considered before using a program. Computer programs, though published, do sometimes contain bugs.

Test system   Data analysis.

Advantages   The programs are free and easily available via the Internet. Most of the programs are regularly updated and new features added. Some estimation procedures are very complex, and estimates would be very time-consuming to obtain without the program. Furthermore, the results can be directly compared to other published results based on the same algorithms.

Application   A computer program is usually used whenever genetic data are analyzed, since estimation procedures are often complex. Furthermore, when assumptions of standard parametric statistics are not met (e.g., small sample size or deviation from normal distribution), confidence limits of estimates may be obtained from computer programs using sampling procedures such as jacknife or bootstrap.

Evaluation   Sensitivity: 4   Requirements: 2
              Time: 2   Cost: 1

References   35, 239, 240, 252, 284, 304, 307, 312, 648, 732, 746, 769, 771, 783, 846, 855, 989.

Author   MHS

M9. Diallel cross

Category	Breeding system, hybridization, genome structure.
Subcategory	Incompatibility, heterosis, hybrid, hybrid depression, hybrid vigour, epistasis, genotype, pleiotropy.
Description	**Diallel crosses** may be used for studying the inheritance of various traits, morphological as well as molecular, see **morphological character analysis** (M39), **protein electrophoresis: isozyme analysis** (M60), **microsatellite markers** (M38), but also for investigating the genetic background of incompatibility systems. The same individuals are used as pollen donors (paternal parent) and pollen recipient (maternal parent). A scheme for doing a **diallel cross** among three plants is shown below.

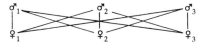

If the offspring segregates into groups in accordance with the Mendelian laws, the inheritance of the trait is obvious. The level of incompatibility is determined by the offspring production (i.e., seeds).

Assumptions and restrictions	The trait investigated has to be genetically determined. Deviation from the Mendelian laws may be due to selection or unequal distribution of the gamets. When studying incompatibility systems, which are expressed in the flowering individual, a number of selective events during germination and plant growth can influence results.
Test system	Field experiment, greenhouse experiment.
Advantages	**Diallel crosses** are simple to perform under controlled conditions. The method is of great value for pollination experiments and studies of seed production, where a general knowledge of the incompatibility system is required.
Application	The method has been used for determination of the inheritance of many traits, e.g., morphological traits (150, 213). The importance of the maternal and paternal environment and the genotype for the reproduction has been studied (597, 811). Determination of compatibility among individuals within a population has been carried out (502, 555, 820). Resistance to pests has also been elucidated (376, 478, 998).
Evaluation	Sensitivity: 4 Requirements: 2 Time: 4 Cost: 2
References	150, 213, 376, 478, 502, 555, 597, 811, 820, 998.
Author	MP, VS

M10. DNA sequencing

Category	Gene-flow, hybridization, genome structure, population.
Subcategory	Introgression, vertical gene-transfer, hybrid, genotype, genetic diversity, polymorphism.
Description	Comparison of **DNA sequence** information from genomic or plastid DNA may give proof of gene transfer between individuals or taxa. Total sequence homology between DNA fragments from two taxa is the best proof that recent introgression and not common origin is responsible for the homology. Prior to sequencing, the target DNA is inserted into bacterial or viral cloning vectors for multiplication, or the target DNA is multiplied by using the **PCR** technique (M59). The amplified target DNA is then denatured, and in a subsequent polymerase reaction, a complementary DNA strand is produced. A short primer is first annealed to the single stranded DNA template, and the reaction mixture is then split into four aliquots, and dNTPs (the four deoxynucleoside triphosphates) plus a ddNTP (dideoxynucleoside triphosphate). The polymerase will incorporate a ddNTP opposite its complementary base in the template DNA, but no further dNTPs can then be incorporated as the ddNTP lacks a 3'group. As the polymerase will terminate the complementary strand at all positions where a ddNTP is inserted, a nested set of DNA fragments is produced. In automated sequencing one of the dNTPs in each reaction is labelled with a fluorochrome, one fluorochrome for each dNTP type. The nested DNA fragments are separated on a polycrylamide gel and the four different nucleotides are distinguished according to their fluorescence. In manual sequencing one of the dNTPs is labelled with ^{32}P. The four aliquots are electrophoresed on a polyacrylamide gel and, after size separation, the nested DNA fragments are detected by X-ray film. The DNA sequence can be read directly from the band pattern on the film.
Assumptions and restrictions	Even when DNA sequencing is totally automated, the number of target sequences analyzed per day is rather low. Efficient **DNA sequencing** equipment is costly. However, many firms have specialised in DNA sequencing for clients.
Test system	Laboratory experiment.
Advantages	Detailed information on the genetic code is obtained.
Application	To date, the application of **DNA sequence** analysis in introgressive hybridization has been very limited, due to the restrictions mentioned above. However, more efficient DNA sequencing methods would have a great potential in analysis of introgression. DNA sequencing has been used in phylogenetic studies (145, 205, 828).
Evaluation	Sensitivity: 5 Requirements: 5 Time: 4 Cost: 4
References	7, 145, 205, 221, 828.
Author	RBJ

M11. Effective population-size N_e

Category

Population.

Subcategory

Genetic drift, genetic neighbourhood-area, metapopulation.

Description

The **effective population-size** N_e of a population, is the effective number of individuals in a population with respect to genetic processes. It can be defined as the size of an idealised population, where all individuals have the same chance of reproduction (i.e., a Poisson-distributed number of offspring), that would have the same genetic properties as that observed for the actual population (47, 1015). The **effective population-size** is usually (much) smaller than the **actual population-size**, N. In dioecious plant species, N_e is smaller than N if one gender is more common than the other. If the **actual population-size** fluctuates, N_e can be calculated as the harmonic mean, i.e.:

$$N_e = \frac{n}{\sum_i (1/N_i)} ,$$

where N_i is the population-size in generation, i and n is the number of generations. Variation in progeny number or clonal growth also decreases N_e. See also **genet identification** (M25).

When a plant population is subdivided in patches or as genetic neighbourhoods, see **genetic neighbourhood-size** (M24), in a continuous population, the definition of the **effective population-size** is dependent on which genetic factor is considered. Hence, the **variance effective population-size** is the size of an idealised population that has the same variance in allele-frequencies as the actual population, whereas the **inbreeding effective population-size** is the size of an idealised population that has the same correlation of genes within individuals (154, 620). In most cases, the two measures are almost identical. The **effective population-size** in a continuous population has been investigated by computer simulations (424).

A recent discussion on how to estimate the **effective population-size** from genetic and ecological data is available (386, 635).

Assumptions and restrictions

The **effective population-size** is inherently more difficult to estimate than the **actual population-size**, because it is not a physically visible property of the population. In a given population, it depends on parameters which are difficult to measure with accuracy, i.e., the variance in offspring number (especially the variance in pollen production) and the fluctuation in actual population size over time. Furthermore, as described above, the definition of the **effective population size** of a population depends on the question in focus. Because the **effective population size** is dependent on genetic drift, which is a chance process, N_e estimated from genetic data has a large variance. Therefore, the **effective population-size** should preferably be estimated from both ecological and genetic data.

Test system	Data analysis.
Advantages	The **effective population-size** is one of the most important parameters for making evolutionary predictions about genetic diversity.
Application	The **effective population-size** is larger for outbreeding plant species than for predominantly selfing species (818). Furthermore, it has been shown, that below a specific population-size, fertility is reduced to zero in *Banksia goodii* (479). The **actual population size** may fluctuate greatly between consecutive years without a corresponding fluctuation in N_e, as seen for *Phacelia dubia* (138). The authors partly attributed this to the existence of a seed-bank.
Evaluation	Sensitivity: 2 Requirements: 2 Time: 2 Cost: 1
References	47, 124, 138, 154, 386, 424, 479, 620, 635, 818, 1015.
Author	MHS

M12. Emasculation of flowers

Category

Seed development and production, breeding system, gene-flow, hybridization.

Subcategory

Agamospermy, ovule development, seed production, cross-pollination, phenology, self-pollination, outcrossing, hybrid.

Description

Emasculation (i.e., removal of anthers) is widely used in studying the functioning of the flowers. In large but closed buds, anthers are removed by a pair of forceps. This treatment is often combined with **experimental pollination** (M13) and results are measured as seed set (see **Ovule counts**, M44).

Assumptions and restrictions

Anthers are in some cases attractants for insects, and emasculated flowers therefore could act differently in attracting insects. The **emasculation** process can damage the flower and result in partial withering. If only some flowers are emasculated and treated by pollen from certain sources the allocation pattern of resources may be influenced and likewise seed set. **Emasculation** is mostly not necessary when plants are self-incompatible. It is important to inspect stigmas after emasculation to make sure that no pollen grains have been deposited by accident.

Test system

Field experiment, greenhouse experiment, laboratory experiment.

Advantages

The method is necessary when experimental pollinations are carried out in self-compatible flowers. It also gives information on the ability to develop seeds without pollination.

Application

Emasculation of flowers is widely used when studying breeding systems. Comparisons of breeding systems (324), and particulary on self-pollination versus cross-pollination (606, 636, 1021) have been performed. The method has been used to investigate flowering phenology (662, 902) and to determine the importance of outcrossing. Furthermore, **emasculation** is used in tests for agamospermy (105, 785, 812, 967) and to prevent self-pollen to be deposited on the stigma. **Emasculation** is frequently used in combination with **experimental pollination** (M13) (650, 993).

Evaluation

Sensitivity: 5   Requirements: 2
Time: 4   Cost: 2

References

105, 324, 606, 636, 650, 662, 785, 812, 902, 933, 967, 1021.

Author

MP

M13. Experimental pollination

Category

Pollen development and production, seed development and production, pollination, breeding system, gene-flow, hybridization.

Subcategory

Pollen competition, pollen germination, pollen viability, agamospermy, fertilization, ovule development, seed germination, seed production, pollen dispersal, cross-pollination, incompatibility, phenology, reproductive allocation, self-pollination, introgression, outcrossing, vertical gene-transfer, artificial hybrid.

Description

Experimental pollination is used to study the mating system of plants, often in combination with **bagging of flowers** (M2). Fresh pollen grains may be directly applied to stigmas from anthers held with a pair of forceps. Alternatively, pollen from a number of anthers from one or several individuals are mixed on a slide and applied to the stigma by a toothpick. The flowers can be untreated, emasculated, see **emasculation of flowers** (M12), or isolated from pollen transfer by bagging until experimentally pollinated. Then the flowers can be cross-pollinated, self-pollinated or pollinated by a mixture of pollen from a number of donors. A good tool for placing a specific number of pollen grains on stigmas are eyelashes glued to a pin. In some cases the corolla can be removed to prevent insect visits after **experimental pollination**. If the species possesses pairs of flowers at the same node, it is useful to assign the two flowers to different treatments, or to let one of them act as control of the treatment.

Assumptions and restrictions

The success of **experimental pollination** is dependent on the **stigma receptivity** (M73) and **pollen viability** (M54), which should be accounted for before the manipulation. If only a few flowers are hand pollinated, seed set (see **ovule counts**, M44) in these flowers may increase due to particular allocation of resources to well pollinated flowers. The number of pollen donors and pollinations should be considered as the results are influenced hereby. The resulting seed set, seed weight and seedling fitness are measures of the success of the different types of experimental treatments. Other factors do, however, influence seed set, etc. (e.g., drought, position of flowers and seed predation) and should be taken into account when the results of **experimental pollination** are evaluated.

Test system

Field experiment, greenhouse experiment.

Advantages

Controlled pollination is an efficient way of studying the breeding system, especially concerning self-pollination versus cross-pollination.

Application	**Experimental pollination** gives information on the mating system (606, 636, 697, 779, 785, 827, 884, 933, 990), the phenology of pollination (902), and how different pollen donors and type of pollinations influence the seed set and quality (543, 622, 675). The method can also be used for observing the effect of self-pollination (675, 776), outcrossing distance (982), and the effect of pollen load size (pollen competition) on seed production (330), and sporophytic characteristics (188, 681).

Experimental pollination has been used to study pollen limitations on fruit and seed set (739, 984), selective fruit abortion (466), and to test self-incompatibility and stigma receptivity (563, 719). Furthermore, it has been utilized to compare pollen tube growth after self-pollination and cross-pollination (413), and to study the inheritance of gender (216). Mixtures of pollen of two related species have been deposited on styles of the two taxa and the progeny screened for hybrids. Distortion between the ratio of pollen and that of the female type and of hybrids demonstrate pollen competition (775).

Evaluation

Sensitivity: 4   Requirements: 2
Time: 3   Cost: 2

References

188, 216, 406, 413, 466, 543, 563, 606, 622, 636, 675, 681, 697, 719, 739, 775, 776, 779, 785, 827, 884, 902, 933, 982, 984, 990.

Author

MP, FF

M14. Fitness measurement

Category   Gene-flow, hybridization, population.

Subcategory   Introgression, vertical gene-transfer, heterosis, hybrid depression, hybrid vigour, gene pool, genetic diversity, genetic load.

Description   The relative **fitness** of an individual plant or genotype in a population can be defined as the contribution of the plant's genotype to the next generation, relative to the population average (536). An accurate measure of fitness is required when making inferences on 1) inbreeding and outbreeding depression, see **inbreeding depression estimation** (M32), 2) **local adaptation** (M35), 3) **heterosis** (M28), 4) the effect of hybridization, 5) selection on specific loci, e.g., by **isozymes** (M60 and M34 in vol. 1), and 6) the adaptive values of morphological traits, see **morphological character analysis** (M39).

Relative **fitness** is inherently difficult to measure, especially for perennial plant species and some assumptions have to be made. Life-time **fitness** is often measured as the number of offspring that an individual produces throughout its life-time, assuming that the value of the offspring is independent of the parental genotype. In hermaphroditic plants, usually only the number of offspring (i.e., seeds) produced through the female function are counted. It is then assumed that the paternal contribution to the **fitness** of the individuals is proportional to the seed set, which can be checked by a **paternity analysis** (M45).

If the plant species under study is inbreeding (e.g., selfing), the maternal parent contributes more than half of its genotype to its offspring. Therefore, the **fitness measurement** must be corrected for the level of inbreeding.

In many cases, especially for long-lived plant species, life-time reproduction cannot be practically measured. Instead, components of fitness or traits which are believed to be correlated to fitness are used. **Fitness components** include survival, pollen viability and competitive ability, and seed quality, whereas growth rate, time of flowering, flower production, and visitation rate of pollinators are often measured as fitness-correlated traits. A **fitness component** does not necessarily measure life-time fitness, since other life-stages may experience much stronger selection. However, the use of **fitness components** may determine whether selection acts on the given life-stage and it may be the best option when life-time fitness is unmeasurable. **Fitness components** and fitness correlated traits may sometimes be combined into a single measure of fitness (349).

Assumptions and restrictions	Whereas **fitness components** are usually directly related to **fitness**, fitness correlated traits are assumed to be closely correlated to fitness. This assumption requires detailed experiments for verification. Furthermore, the most important life-stage for life-time fitness is very difficult to determine. When fitness-estimates are obtained in growth chamber, greenhouse or experimental gardens, it is assumed that there are no genotype by environment interactions for fitness, i.e., a given genotype would do as well in a natural population as in an artificial environment. Any maternal effects through seed quality may bias the estimation of **fitness** (1013).
	For clonal or vegetatively reproducing species it is necessary that **genet identification** (M25) is done before fitness is estimated.
Test system	Field experiment, greenhouse experiment, growth chamber experiment.
Advantages	When a reliable **measurement** of fitness or of a **fitness** component is available, it is usually easy and quick to measure a large number of individuals. By measuring fitness, predictions of successful invasion of GMPs or introgression of transgenes into natural populations can be made.
Application	**Fitness** estimation has been used in a number of studies investigating **inbreeding depression** (M32) (633, 1004). In some studies, fitness effects of inbreeding are larger in a more stressful environment (350, 633).
	Some studies have failed to find a relation between population size and average **fitness** (e.g., 671). **Fitness** of the same genotypes has been measured in different environments (e.g., 890). In this study, the author found a significant genotype by environment interaction and interpreted it in terms of spatial heterogeneous selection.
Evaluation	Sensitivity: 3 Requirements: 2 Time: 4 Cost: 1
References	349, 350, 536, 633, 671, 890, 1004, 1013.
Author	MHS

M15. Flow cytometry

Category	Gene-flow, hybridization, genome structure.
Subcategory	Introgression, vertical gene-transfer, hybrid, structural chromosome changes.
Description	**Flow cytometry** allows the rapid analysis of the amount of genomic DNA per cell. The analysis is carried out by the use of a flow cytometer. The operation of the flow cytometer involves the rapid analysis of the degree of fluorescence associated with nucleus in a labelled population of nuclei or protoplasts. This is achieved by passage of the nuclei one by one through a flow cell. The stream of nuclei intersects a laser beam and they absorb light, which is subsequently reemitted in the form of fluorescence. The emitted fluorescence comprises a rapid series of pulses that are converted into DNA amount using, as standards, the nuclei of control plants with known DNA content. The use of the method is improved, when correlation between DNA content and chromosome number can be made. Normally 5,000-10,000 nuclei or protoplasts are measured per sample.
Assumptions and restrictions	The sensitivity of the method is in many cases not sufficient to distinguish the loss or gain of a few chromosomes. However, in separating diploids from polyploids (e.g., triploids, tetraploids etc.), the method has a great potential. The access to a flow cytometer is a demand, and the price of this equipment may restrict the use of the method.
Test system	Laboratory experiment.
Advantages	The method is a fast way of estimating the ploidy level of related plant taxa. Only small amounts of plant material (e.g., leaves) are required.
Application	The method is often used in routine analysis of the genomic size of plants, especially when detailed information on the karyotype is not required.

Evaluation

Sensitivity: 3	Requirements: 4
Time: 4	Cost: 3

References	427, 607, 720, 722, 745, 924.
Author	RBJ

M16. Fluctuating asymmetry of vegetative and floral traits (FA)

Category

Pollination, inserted trait.

Subcategory

Pollinator foraging behaviour, stress tolerance.

Description

Fluctuating asymmetry (FA) can be defined as small deviations from bilateral symmetry of plant organs, e.g., leaves and flowers. **FA** is caused by environmental stress imposed on individual plants during their development. Abiotic or biotic stressors disturb the homeostatic processes that govern an individual's ability to develop identically on either side of its bilaterally symmetrical body parts. Both sides are influenced by the same genes and therefore, **FA** must reflect disturbances in the development of the individual.

Fluctuating asymmetry may be measured by **FA** indexes based on mean and variance of the distribution of right-minus-left (R - L) differences in body parts (585, 678). The two most commonly used indexes are the absolute **FA** and the relative **FA**, which is independent of size differences:

$$FA_{abs} = \frac{\sum |R_i - L_i|}{N} \quad , \quad FA_{rel} = \frac{\sum \frac{|R_i - L_i|}{(R_i + L_i)/2}}{N} \quad ,$$

where N is the number of samples. Tests for deviation from normal distribution of signed right-minus-left character values should be done in order to check for possible directional asymmetry or bimodal symmetry.

An increasing **FA** is correlated with increasing phenotypic variation. **FA** is assumed to be related to fitness, e.g., mate choice or pollination intensity, but evidence is still limited (160). In poor and stressed habitats, body parts are expected to be smaller and show high **FA**-values.

Assumptions and restrictions

Several methodological pitfalls have been found in **FA**-research methods (678, 679). It is recommended to consult the most recent literature carefully, e.g., concerning the sensitivity to sample heterogeneity and scale effects (898), the relative importance of environment and genetics to **FA**, and the correlation between **FA** and fitness (541). The choice of traits for analysis is important. **FA**-values are influenced by the importance of the trait for normal function of the organism (160). **FA**-values are often relatively small compared to the size of the trait measured, typically approx. 1%. Therefore, measurements must be precise and replicated (898).

Test system

Field experiment, greenhouse experiment, laboratory experiment.

Advantages	Inexpensive equipment and relatively small sample sizes are needed. **FA** gives a good overall measure of phenotypic quality (i.e., ability to express high fitness).
Application	During the last decade, **FA** has been used as a risk assessment tool in conservation biology (e.g., 160) and in studies of sexual selection and parasitism (e.g., 583, 584).
Evaluation	Sensitivity: 4 Requirements: 2 Time: 2 Cost: 1
References	160, 541, 583, 584, 585, 586, 587, 678, 679, 898.
Author	JMO, GK

M17. Fluorescent dyes and marking of pollen

Category

Pollination, breeding system, gene-flow, hybridization.

Subcategory

Pollen dispersal, pollinator activity, cross-pollination, heterostyly, phenology, outcrossing, hybrid.

Description

Fluorescent dyes and marking of pollen are used for estimation of pollen dispersal patterns and distances. In some species pollen grains possess different colour forms or the morphology varies within the species, and in other species male sterile plants are found. These not so common features can be used in an experimental design. A pollen donor plant with specific characteristics is placed in the middle of a large group of plants with complementary characteristics. After pollen dispersal by natural vectors, pollen grains from the centre plants can be recorded on stigmas of surrounding plants by inspection under microscope.

In other species, **fluorescent dyes** can be used as pollen analogues to study pollen dispersal. The pattern of deposition of dye particles resembles that of pollen grains and can be seen in UV-light. In flowers, in the male phase, anthers are covered with dye. Different dye colours are available, making several experiments at the same time possible. Surrounding flowers are collected after some time of exposure, they are inspected under stereo microscope and number of dye particles are recorded. Pollen can also be marked with micronized metal dusts and recorded after dispersal by backscatter scanning electromicroscopy and x-ray microanalysis.

Assumptions and restrictions

It is assumed that marked pollen grains and the dye particles behave as normal pollen grains. If possible, correlations between the movements of normal pollen and marked pollen or dye particles should be performed. It can be very difficult to localise long-distance transport of a few marked pollen grains or dye particles. Dye particles may change the appearance of the flower and consequently the way biotic pollen vectors, e.g., insects, are attracted.

Test system

Field experiment.

Advantages

Fluorescent dyes and marking of pollen give information on pollen movements which may not otherwise be available.

Application

These methods have been used in pollen flow studies (701), pollen removal and deposition studies (411), intrafloral pollen movements (1012) and to compare dispersal of pollen and dye (977).

Evaluation

Sensitivity: 2	Requirements: 3
Time: 4	Cost: 2

References

411, 701, 977, 1012.

Author

MP

M18. Forced fertilization

Category	Seed development and production, breeding system, gene-flow, hybridization.

Subcategory   Fertilization, incompatibility, outcrossing, artificial hybrid.

Description   Different barriers to pollen germination on stigmas can be by-passed by **forced fertilization**. Hereby hybridization and cross-pollination between different mating types can be performed. This can, for instance, be used for breeding purposes.

The style is cut with a razor blade 0-2 mm above the ovary. Stigmatic exudate is placed on the cut surface and pollen are immediately applied. Another method involves pollination of a compatible stigma and, after some time, this stigma is cut off and attached to an ovary of another plant helped by a piece of drinking straw.

In some genera (e.g., *Brassica*) pollination at the bud stage, or placing plants in a CO_2-rich atmosphere shortly after pollination, will also circumvent self-incompatibility.

Assumptions and restrictions   This method is most suitable for species where the styles are similar in width and length.

Test system   Field experiment, laboratory experiment.

Advantages   This method might circumvent incompatibility or incongruity reactions.

Application   **Forced fertilization** has been used in interspecific crosses (943) and to circumvent self-incompatibility (911).

Evaluation   Sensitivity: 3   Requirements: 2
Time: 4   Cost: 2

References   911, 943.

Author   MP, RBJ

M19. F-statistics

Category   Gene-flow, population.

Subcategory   Outcrossing, genetic drift, genetic diversity, metapopulation.

Description   The method, **F-statistics** is a hierarchical partitioning of the total deviation from Hardy-Weinberg proportions ($-1 < F_{IT} < 1$) into a within-population component ($-1 < F_{IS} < 1$) and a between-population component ($0 < F_{ST} < 1$). F_{IS} is a measure of the non-random mating within the population. $F_{IS} > 0$ if there is inbreeding (e.g., selfing) and $F_{IS} < 0$ if there is avoidance of matings of relatives (e.g., self-incompatibility). F_{ST} is a measure of the difference in allele frequencies between populations ("Wahlund effect"), i.e., the amount of genetic differentiation caused by genetic drift or selection in combination with limited migration. If F_{ST} is 0, allele frequencies are equal in all populations and there is no differentiation. **F-statistics** can also be extended to more hierarchical levels of population substructure (164, 518, 618, 988, 989). A simple relation between the statistics exists: $1-F_{IT} = (1-F_{IS})(1-F_{ST})$. If the studied locus is neutral (i.e., allele frequency differences are determined by genetic drift alone), the number of migrants ($N_e m$) between subpopulations per generation can be estimated from the approximate formula: $F_{ST} = 1/(4N_e m + 1)$, where N_e is the **effective population size** (M11), and m is the migration rate. This relation has produced the rule of thumb that, if $N_e m \ll 1$, populations can evolve independently through genetic drift.

                    F-statistics was defined by Wright in 1931 (1015). For application to a data set of genotypes, different estimation procedures have been suggested. The method of Weir & Cockerham from 1984 (989) is preferred by most authors, but the closely related G-statistics (621) is also widely used.

                    Several computer programs are available for estimation of **F-statistics** from genotype data (i.e., **FSTAT** and **G-STAT**), see **Computer programs** (M8).

Assumptions and   For most applications, the studied loci are assumed to be selectively
restrictions   neutral. A number of loci, with codominant alleles determined by, e.g., **RFLP** (M64), **microsatellite markers** (M38) or **protein electrophoresis: isozyme analysis** (M60) are needed, since F_{ST} is based on genetic drift which is a chance-process with a large variance. This variance is reduced more by including additional loci in an investigation than by genotyping more individuals. The variance of F_{IS} is less sensitive to the number of loci used.

Test system   Data analysis.

Advantages	Values of **F-statistics** obtained from different species can easily be compared and the amount of successful migration through pollen and seeds can be quantified by **nuclear and organelle markers combined** (M42). Inferences about **selfing rate** (M69) and local inbreeding can be made through F_{IS}. A very deviating F_{ST} value at a specific locus may indicate selection acting on the genetic region marked by the locus.
Application	**F-statistics** has been estimated for a large number of species (114, 337, 339, 869), and categorizations of species into classes of different dispersal abilities have been made (818, 1000). In some studies, a positive correlation between F_{ST} values and physical distance between populations has been reported. Differences in the values of **F-statistics** for different age-classes have revealed selection, most often against homozygotes (929). The genetic differentiation measured by F_{ST} was small, along an altitudinal gradient, even though populations differed greatly for morphological traits (676). This discrepancy between **F-statistics** and morphological differentiation may be common (1009).
Evaluation	Sensitivity: 3 Requirements: 3 Time: 2 Cost: 3
References	114, 164, 312, 337, 339, 518, 618, 621, 676, 818, 846, 869, 929, 988, 989, 1000, 1009, 1015.
Author	MHS

M20. Gas chromatography (GC)

Category

Pollination, hybridization, genome structure, population, inserted trait.

Subcategory

Chemical attractant, pollinator foraging behaviour, marker, hybrid, genotype, polymorphism, stress tolerance.

Description

The method is used for separating and identifying various chemical compounds in a similar way to **TLC** (M74).

The principle of **gas chromatography** is separation of a sample by use of an inert gas as the carrying medium, and a column consisting of fine coarse material such as silica gel or glass beads coated with various liquid chemicals. The gas used as a solvent is added to the column under pressure. Depending on the gel matrix and on the solvent, certain groups of molecules will have different retention time in the column. The method is often followed by mass spectrometry. The result can be used as a genetic marker like isozymes, see **protein electrophoresis: Isozymes** (M60 and M34 in vol. 1). See also **high performance liquid chromatography (HPLC)** (M29).

Assumptions and restrictions

It is necessary to know the retention time for a number of chemical compounds in various gas-phases. The use of the method needs a skilled person and an expensive equipment. Only a single sample can be analyzed each time and it may be difficult to collect the volatile samples.

Test system

Laboratory experiment.

Advantages

The method is very sensitive and may detect very small amounts of a chemical. It may be used both as a qualitative and a quantitative method.

Application

The method has been used for studies of floral fragrances (e.g., 81), to detect chemicals produced in plants under stress (178), for identification of hybrids (e.g., 780) and for the synthesis of sorbitol in transgenic tobacco (909).

Evaluation

Sensitivity: 5	Requirements: 5
Time: 5	Cost: 4

References

81, 90, 97, 178, 274, 599, 691, 780, 791, 909, 928, 1032.

Author

VS

M21. Gene-flow estimation with private alleles

Category

Gene-flow, population.

Subcategory

Outcrossing, vertical gene-transfer, genetic diversity, metapopulation.

Description

The average number of migrants (Nm) exchanged between local populations may be estimated with **private alleles** (852, 853). N represent the **effective population size** (M11) and m is the migration rate. The principle is that isolated populations will accumulate private alleles, i.e., alleles present in only one of the populations, while gene-flow will prevent genetic differentiation. The logarithm of Nm is approximately linearly related to the logarithm of the average frequency of **private alleles**, $\bar{p}(1)$, according to the formula:

$$\log_{10}(\bar{p}(1)) = a \, \log_{10}(Nm)+b,$$

where a and b depend on the number of individuals sampled from each population (64).

Assumptions and restrictions

Linearity is lost for very large and small values of Nm (852, 853). **Private alleles** may be more subject to misidentification than widespread alleles (858).

Test system

Data analysis.

Advantage

The F_{ST}, see **F-statistics** (M19), and **gene flow estimation with private alleles** have very similar properties which are based on the assumptions of an island model (64).

Estimates are quite robust because Nm depends only weakly on both the mutation rate at a locus and possible selection affecting a locus, but is very sensitive to average migration rate in the population. Moreover, $\bar{p}(1)$ reaches its equilibrium value relatively rapidly (852, 853). This measure of gene-flow is also independent of the assumptions of diploid inheritance and can be thus used for polyploids.

Application

The method has been used in conjunction with the method derived from F-statistics for estimating gene-flow (114, 1010) from isozyme data.

Evaluation

Sensitivity: 3   Requirements: 3
Time: 2   Cost: 3

References

64, 114, 852, 853, 858, 1010.

Author

FF

M22. Genetic distance

Category Gene-flow, population.

Subcategory Introgression, outcrossing, vertical gene-transfer, genetic diversity, genetic drift, metapopulation.

Description The **genetic distance** is a measure of the genetic difference based on the use of one or several polymorphic genetic markers (e.g., proteins and nucleic acids). **Genetic distance** can be calculated at different hierarchical levels, e.g., between individuals, populations or species. Information on the genetic distance can indicate whether crosses between two different populations or species (e.g., a GMP and a wild relative) are viable or will suffer from outbreeding depression, see **inbreeding depression estimation**, (M32). Furthermore, **genetic distance** may give important information on the evolutionary history (e.g., phylogeny), colonisation pattern and dispersal history of the sampled populations or species.

Different measures of genetic distance have been suggested, with Nei's **D** (617) being the most commonly used for codominant genetic data. **D** is defined as:

$$D = - \log_e \left(\frac{J_{xy}}{\sqrt{J_x J_y}} \right), \qquad \text{where}$$

$$J_x = \sum_i p_i^2, \quad J_y = \sum_i q_i^2, \quad J_{xy} = \sum_i p_i q_i \,,$$

and where p_i and q_i are the frequencies of allele i in population **X** and **Y**, respectively. Estimation of **D** from experimental data requires a correction for sample error (619, 620).

This statistic has the desired property that it is proportional to the expected time of divergence between the samples. From the set of samples (e.g., allele frequencies for a set of populations), **D** is calculated for all pairwise combinations. This matrix of **D**-values can then be used to construct a genetic phylogeny (i.e., a phylogenetic tree) of the samples.

Different measures of **genetic distance** can also be computed from multilocus, dominant markers obtained from **RFLP** (M64), **RAPD analysis** (M61) and **AFLP** (M1). These **genetic distance** measures are usually based on the sharing of bands, **S** between a pair of haplotypes **X** and **Y**, defined as (47):

$$S = \frac{2N_{XY}}{N_X + N_Y} \,,$$

where N_{XY} is the number of bands shared, and N_X and N_Y are the number of bands in individual **X** and **Y**, respectively (926).

Assumptions and restrictions	For an accurate determination of **genetic distance**, several polymorphic loci are needed. When interpreting the results, it is usually assumed, that the polymorphisms under study are selectively neutral. Furthermore, when **genetic distances** are used for phylogenetic analysis, a molecular clock is usually assumed.
Test system	Laboratory experiment, data analysis.
Advantages	The **genetic distance** is easy to calculate when genetic data are available and it can be compared directly between species. Furthermore, phylogenies of populations can be constructed. The use of computer programs like **PHYLIP** (see M8) makes phylogeny reconstruction an easy task.
Application	**Genetic distances** have been used to establish the phylogenetic relationship between crop cultivars (225 with **isozymes**, 926 with **RFLP**), and phylogenetic trees of natural populations of a single species (122), or phylogenetic trees of closely related species (256). These phylogenetic trees can be compared to the geographical distribution of the populations for inferences of dispersal and colonisation patterns (e.g., 385). From a **genetic distance** matrix of populations of *Deschampsia caespitosa*, a specific population deviated and it was concluded that this population was founded by a recent invasion (291).
Evaluation	Sensitivity: 3 Requirements: 3 Time: 3 Cost: 2
References	47, 122, 225, 256, 291, 385, 617, 619, 620, 926.
Author	MHS

M23. Genetic map construction

Category

Genome structure.

Subcategory

Genotype, recombination, structural chromosome changes.

Description

A **genetic map** is a map of the location of a number of polymorphic loci on the chromosomes. The distances between these loci on the chromosomes are given in map units, which are defined as the probability of a crossing-over between the two loci during meiosis. A **genetic map** of a plant species is made by assigning different polymorphic loci to specific locations in the species genome. When a **genetic map** has been obtained, it is easy to insert additional genes into the map.

A **genetic map** is often made by crossing two different homozygous lines or ecotypes. These lines have been inbred for a number of generations, rendering them homozygous for almost all their genes, but different lines are homozygous for different alleles. The resulting F_1 plants are then either selfed or backcrossed to one of the parents to obtain a large number of F_2 plants in which all loci which carried different alleles in the original lines segregate. From the segregation of these marker loci it is then possible statistically to estimate recombinational distances between each set of markers. This analysis is most often performed using various commercial computer programs. A reliable **genetic map** is obtained when the segregating loci fall into a number of linkage groups that corresponds to the haploid number of chromosomes.

If isogenic lines do not exist, the same approach can be used, but then only markers which are homozygous for different alleles in the two parents can be used.

A variety of polymorphic markers can be used, including morphological markers obtained from **morphological character analysis** (M39) and markers from **protein electrophoresis: Isozyme analysis** (M60), **RAPDs** (M61), **microsatellites** (M38), **AFLPs** (M1) and **RFLPs** (M64). Published maps often contain a combination of these markers.

Assumptions and restrictions

A large number of markers (>100, depending on the genome size) are necessary for construction of a reliable map. Furthermore, a large number of F_2-progeny have to be analyzed in order to obtain good estimates of gene locations. This makes the method expensive and time consuming. The use of bulked DNA (bulked segregant analysis) allow more individuals to be screened at the same time (1039). **Genetic maps** of different species are often based on different polymorphic marker loci. This complicates the use of maps in studies of hybridization between two species.

Test system

Greenhouse experiment, laboratory experiment, data analysis.

Advantages

Once a genetic map has been constructed it is relatively easy to map a locus of interest, i.e., the location of the incorporation of a foreign gene.

Application

Genetic maps have primarily been constructed for agriculturally important species, e.g.: barley (*Hordeum* spp.), maize (*Zea mays*), rice (*Oryza sativa*), wheat (*Triticum* spp.), tomato (*Lycopersicon* spp.) and potato (*Solanum tuberosum*) (394); sorghum (*Sorghum bicolor*) (1025); pearl millet (*Pennisetum glaucum*) (512); alfalfa (*Medicago sativa*) (444) and common bean (*Phaseolus vulgaris*) (6). Markers used for a genetic map of maize have been used to create a map of *Sorghum bicolor* (88). The genetic map of common bean has been utilized to localise resistance genes against fungal attack (6). An **RAPD** based map has been applied for studying the pattern of hybridization between wild sunflowers (760).

Evaluation

Sensitivity: 4   Requirements: 4
Time: 5   Cost: 5

References

6, 88, 394, 444, 512, 760, 1025, 1039.

Author

MHS

M24. Genetic neighbourhood-size

Category

Pollination, gene-flow, population.

Subcategory

Pollen dispersal, outcrossing, genetic drift, genetic neighbourhood-area, metapopulation.

Description

The concept of **genetic neighbourhood-size (N_e)** was defined by Wright (1016, 1017) as the size of the random mating unit in a continuous population with restricted gene-flow. The genetic neighbourhood-size is thus an important determinant of independent evolution in different parts of a population. Wright defined the **genetic neighbourhood area** as:

$$N_a = 4\pi\sigma^2,$$

where σ^2 is the expected axial variance of the dispersal distance given a two-dimensional normal distribution of dispersal distances. The axial variance can be computed from the observed dispersal distances along an axis. The mean dispersal distance is assumed to be zero. Both parents of an individual are found within the **genetic neighbourhood-area N_a** with a probability of 86.5%. N_e is then the number of individuals within N_a, i.e., $N_e = N_a d$, where **d** is the density of plants.

The above formula can be extended to separate gene dispersal from pollen and seeds:

$$N_a = 4\pi(\sigma_s^2 + 0.5t\sigma_p^2) \text{ and } N_e = N_a d,$$

where σ_s^2 is the variance of seed dispersal distance, σ_p^2 is the variance of pollen dispersal distance, and t is the outcrossing rate (176, 503). Further extensions have been made for cases when dispersal distances are not normally distributed (99, 685, 1032).

Seed dispersal can be directly measured from **seed traps** (M67 in vol. 1) placed at different distances from a focal plant (590), e.g., by seeds marked with **fluorescent dyes** (M17) (99, 218). Indirectly, dispersal can be measured through genetic markers, e.g., by **nuclear and organelle markers combined** (M42) and **organelle DNA analysis** (M43).

Pollen dispersal can be estimated directly by either **fluorescent dyes and marking of pollen** (M17) or by observation of **pollinator foraging behaviour** (M57) (254, 685, 1032), or indirectly from genotypic data at marker loci (98, 218). The direct and indirect measures of seed and pollen dispersal are expected to differ if there is considerable **pollen carry-over between flowers** (M49) by pollinators or if apparent pollination is different from effective pollination.

Description, contd.   A related measure of neighbourhood-size is Levin's concept of the paternity pool size (501), defined as the number of individuals in an area from which 99% of all pollen grains arriving to the focal plant derives.

Assumptions and restrictions   For accurate estimation of the **genetic neighbourhood-size**, genotypes with several loci have to be included for a large number of individuals. Using observation data it is assumed that all pollen and seeds contribute equally to the next generation. Pollen and seed dispersal may differ greatly between years, making evolutionary inferences uncertain if the results are based on a single year.

Test system   Field experiment, data analysis.

Advantages   The method makes it possible to identify the random mating unit (i.e., the panmictic unit) in a population. Furthermore, the relative importance of seed and pollen dispersal can be quantified.

Application   Genetic neighbourhood-areas and sizes have been measured for a number of species (e.g., 98, 218, 254, 465). Neighbourhood-area has been shown to be inversely correlated to plant density (254, 1032). By comparing estimates of N_e from direct and indirect methods the amount of pollen carry-over can be quantified (254, 590). See also **pollen carry-over between flowers** (M49).

Evaluation   Sensitivity: 3   Requirements: 3
Time: 4   Cost: 2

References   98, 99, 176, 218, 254, 465, 501, 503, 590, 685, 1016, 1017, 1032.

Author   MHS

M25. Genet identification

Category	Population.
Subcategory	Gene pool, genetic diversity, metapopulation.
Description	In clonal, perennial plants it is often difficult to identify individual genets morphologically, since each genet may have become fragmented through vegetative growth and cloning. The **identification of genets** is necessary when knowledge of population dynamics, **effective population size** (M11), genetic diversity and the importance of sexual versus vegetative reproduction is sought.
	From a population, fresh material from a number of ramets (e.g., shoots) are sampled for genetic analysis. Any technique of genotype determination can be used, since dominance of markers is unimportant. Different genetic markers and techniques, however, vary in their power of identification, with **AFLP** (M1), **RAPD analysis** (M61) and **microsatellite markers** (M38) usually being more informative and thus more powerful than **isozyme analysis** (M60).
Assumptions and restrictions	It is assumed that if two sampled ramets display different genotypes, by any of these methods, they belong to different genets. It is, however, impossible to prove that two ramets belong to the same genet, but if high-resolution markers are used, it can be made very likely.
Test system	Laboratory experiment.
Advantages	With these methods, it is possible to study genets which grow intermingled. The alternatives, excavations of ramets or morphological measurements, are less precise and often impossible to do.
Application	**Isozyme analysis** (M60) have been used to infer population structure in *Fragaria chiloensis* (18), cultivar type in sugarcane (280) and geographical origins of invaded clones in Australia of the apomictic *Chondrilla juncea* (141). In these studies, the resolving power exceeds 90%. **RAPD analysis** (M61) has been used to show that some *Rubus saxatilis* populations consist of single genets (236). DNA-fingerprinting was used to show that vegetative reproduction is limited in *Rubus idaeus* (29) and that some *Phragmites* stands are depauperate in genetic variation (624). Clones of *Solidago altissima* in fields of varying ages have been mapped (527).
Evaluation	Sensitivity: 4 Requirements: 3 Time: 4 Cost: 3
References	18, 29, 141, 236, 280, 527, 624.
Author	MHS

M26. Haplotype statistics

Category

Gene-flow, population.

Subcategory

Introgression, outcrossing, vertical gene-transfer, genetic diversity, genetic drift, metapopulation.

Description

Haplotype statistics are used for data analysis in genetic studies of haploid DNA. These methods are used when genetic information obtained from **RFLP analysis** (M64) or **DNA sequencing** (M10) is available for non-recombining and haploid DNA, e.g., chloroplast-DNA or mitochondrial-DNA. A haplotype can then be regarded as an allele at a haploid locus, and haplotypes therefore differ from diploid, codominant genotypes obtained through **isozyme analysis** (M60) or **microsatellite markers** (M38), because heterozygotes cannot be revealed. A few commonly used statistics are listed below, but extensive treatments of the topic are also available (47, 620, 809, 988).

The **nucleon diversity** is defined as the probability that two randomly drawn haplotypes are different and thus resembles the expected heterozygosity for diploid data (620).

The **nucleotide diversity** π for DNA-sequence data is defined as the average number of nucleotide differences between two sequences, i.e.:

$$\pi = \sum_{i,j} x_i x_j \pi_{ij} \quad ,$$

where x_i and x_j are the population frequency of the ith and jth type of DNA-sequence, and π_{ij} is the proportion of different nucleotides between the ith and jth types of DNA-sequences (620). π can also be estimated from **RFLP**-data (620).

Phi_ST, which resembles **F-statistics** (M19), can be calculated by assuming that each haplotype is an allele and that the subpopulations are in Hardy-Weinberg equilibrium (240). The computer program **AMOVA** (M8) performs this calculation.

Another statistic, g_{st}, also resembles F_{ST}, but can be used to infer the hierarchical level of substructure rather than testing a known hierarchy. Furthermore, it can be used for constructing a phylogenetic tree of the samples (375).

Genetic distance (M22) can also be computed for haplotypes and be used for constructing phylogenetic trees of the sample.

For DNA-sequences, a large number of methods are available for constructing phylogenetic trees (47, 620). In the computer package **PHYLIP** (M8), a number of these methods have been implemented.

Assumptions and restrictions	Most of the **haplotype statistics** are designed for organelle DNA which is non-recombining. The statistics are, however, also used on sequences from nuclear genes with the assumption that recombination can be considered negligible.
Test system	Data analysis.
Advantages	**Haplotype statistics** provide measures for organelle DNA which are equivalent to, e.g., **F-statistics** (M19) and Nei's **genetic distance** (M22) for nuclear loci. Because there is no recombination, the phylogenetic trees constructed from organelle markers may be more accurate than those from nuclear markers. Haplotype data has a smaller **effective population-size** (M11) than nuclear loci and, therefore, may be better suited for detection of genetic bottlenecks or colonisation events.
Application	**Haplotype statistics** were used to investigate the distribution of nucleotide diversity in cpDNA in *Eucalyptus nitens* and revealed regional differentiation in agreement with nuclear isozyme data (123). Other cp-DNA surveys have also been used to infer genetic differentiation (544, 580), and in the latter study no correlation was found between the cp-DNA variation and the morphological variation. Another study revealed high levels of mtDNA polymorphism within and between populations of *Thymus vulgaris* (75).
Evaluation	Sensitivity: 3 Requirements: 2 Time: 2 Cost: 1
References	47, 75, 123, 240, 375, 544, 580, 619, 620, 809, 988.
Author	MHS

M27. Herbarium sheet survey

Category   Gene-flow, hybridization, population.

Subcategory   Introgression, vertical gene-transfer, hybrid, genetic diversity, polymorphism.

Description   Herbarium specimens can be used for DNA analysis, although this is not yet routine. Methods for DNA extraction have been described (519, 804, 912), mostly for phylogenetic studies. The DNA content of a herbarium specimen depends on its age (often decades) and how it has been dried and stored. The DNA content will often be too small for analysis. In old material (up to 151 years), **PCR** (M59) of **cpDNA** (M43) is often inhibited and special treatment of the amplification mixture may be necessary (804).

Insect protection treatments, e.g., poisoning or fumigating the herbarium, accelerates DNA degregation seriously (206). In modern herbaria, these methods are usually replaced by freezing the incoming material, which seems to have no effect on DNA quality.

Classical methods for studies on herbarium specimens include **morphological character analysis** (M39), e.g., on rare hybrids and biogeographical studies (383, 964). See also **biogeographical assay and monitoring** (M4).

Assumptions and   **Rapid drying** (M48) of the plant material is essential for preserva-
restrictions   tion of DNA content. Old material differs greatly in DNA content. Extraction techniques are not yet fully developed, so routine analysis on old vouchers is not yet possible.

Test system   Laboratory experiment.

Advantages   DNA extraction techniques may allow hybrid identification of old and recent material and molecular analysis of rare natural hybrids.

Application   Risk assessment studies regionalising the potential of gene-flow may use herbarium collections for screening of regions (25, 964). Herbarium vouchers are also useful for the localisation of natural populations, for identification of hybrids and for morphological analyses (383, 453).

Evaluation   Sensitivity: 2   Requirements: 3
                              Time: 3   Cost: 2

References   206, 344, 383, 453, 519, 804, 912, 964.

Author   PR

M28. Heterosis analysis: Over mid-parent (HMP) and better parent (HBP)

Category	Hybridization.
Subcategory	Heterosis, hybrid depression, hybrid-vigour.

Description

Heterosis is an increased vigour of a hybrid compared to the two parental plants. It can be calculated in two ways: Over the mean of the two parents (OMP) or over the best parent (OBP). The OBP analysis is the method most often used in agronomic studies. Some morphological, phenological, pathology resistance and yield characters have to be recorded for each plant, the parents and the offspring (the hybrids). The means of the characters for the F_1 hybrids and for the parents are then compared. **Heterosis over mid-parent (HMP)** is calculated as:

$$(M_{F1} - M_P)/M_P \times 100,$$

where M_{F1} is the mean for the F_1 hybrids and M_P the mean for the two parents. **Heterosis over better parent (HBP)** is calculated as:

$$(M_{F1} - M_{BP})/M_{BP} \times 100,$$

where M_{F1} is the mean for the F_1 hybrids and M_{BP} the mean for the better parent. When **heterosis over better parent** is calculated on the yield data, it is also called "**heterosis over better yielding parent (HBYP)**". Heterosis may be calculated character by character, plant by plant or population by population, when comparing hybrids against parents.

Assumptions and restrictions

When heterosis is calculated at the population level, it is lower than that observed at the single plant level. The real potential of the hybrids can only be evaluated from field plots and not from single plant data because heterosis is closely dependent on the environmental conditions.

Test system

Field experiment.

Advantages

The method is simple, depending on the character studied.

Application

This method has been widely applied for improvement of agricultural crops and, for the last decade, often in combination with **RFLP** (M64) or with **PCR** (M59) methods. When heterosis values are negative, they are measures of hybrid depression.

Evaluation

Sensitivity: 4	Requirements: 1
Time: 5	Cost: 1

References

15, 478, 492, 588, 673, 716, 743, 801, 1034.

Author

YJ

M29. High performance liquid chromatography (HPLC)

Category

Pollen development and production, seed development and production, breeding system, gene-flow, hybridization, genome structure, population.

Subcategory

Pollen viability, seed production, incompatibility, marker, hybrid, genotype, polymorphism.

Description

The method is used for separating and identifying various chemical compounds such as **TLC** (M74).

The principle of **HPLC** is separation of a sample by a solvent (eluent) added to a column containing silica gel or other gel matrices. The solvent is added to the column under pressure. Depending on the gel matrix and on the solvent, certain groups of chemical compounds will have different retention times in the column. The detection of the molecules in the separated sample may be registered by a spectro-photometer and the peak of a compound is an expression of the concentration of this particular molecule. The result may be used as a genetic marker like **isozymes** (M60 and M34 in vol. 1). See also **gas chromatography** (M20).

Assumptions and restrictions

It is necessary to know the retention time for a number of compounds in various solvents. The use of the method needs a skilled person and an expensive equipment. Only a single sample can be analyzed each time. The method may be performed on freeze dried material, see **plant material preservation: Cryopreservation and freeze-drying** (M47).

Test system

Laboratory experiment.

Advantages

The method is very sensitive and may detect very small amounts of a chemical compound. It may be used both as a qualitative and as a quantitative method.

Application

HPLC has been used for detection of change in concentration of chemical compounds, which are important for pollen germination and for seed development. It has also been used to detect variation in floral nectar-sugar composition, identification of compounds important for self-incompatibility, stress response to insects (e.g., 317) and environmental factors (e.g., 960). The method has been applied to study hybrids (957) and for differentiation among plants it has been combined with **RAPD** (M61). Analysis of genomic DNA in transgenic plants has been performed by using **HPLC** (582).

Evaluation

Sensitivity: 5	Requirements: 5
Time: 4	Cost: 3

References

189, 200, 203, 296, 297, 317, 561, 582, 600, 957, 960, 1014.

Author

VS

M30. Immunological methods

Category	Gene-flow, hybridization, genome structure, population.
Subcategory	Introgression, marker, vertical gene-transfer, hybrid, gene expression, gene stability, genotype, polymorphism.
Description	**Immunological methods** depend on the reaction between an antigen and a specific antibody for this particular antigen. A brief description of **immunoassay** and **immunoblot (Western blot)** is presented in vol. I (M30, M31). The use of gold particles coated with an antibody in combination with electron microscopy has shown to be useful for cytological detection of expression of a gene in transgenic plants. More details about immunogold labelling are available (78, 370).
Assumptions and restrictions	These methods rely on the specificity of the antigen-antibody reaction. The best result will be obtained by using monoclonal antibodies. It is necessary to have the facilities to produce the proper antibody, which is the time consuming part of the procedure.
Test system	Laboratory experiment.
Advantages	The methods are very sensitive and quick when an adequate antibody is available. Many samples can be analyzed per day.
Application	The methods have mainly been used for studying gene expression in transgenic plants (especially Western blot). Immunogold labelling has widely been applied for localization of tissue, expressing a certain transgene. Besides expression of nuclear genes, plastid genes may be followed by these methods (281). A combination of these methods (which determine a gene product) and DNA methods (e.g., **Southern blotting** (M71)), which determine the introgression of a transgene, is valuable for detection of failing expression (e.g., silencing) of the transgenes (573).
Evaluation	Sensitivity: 5	Requirements: 4
	Time: 4	Cost: 3
References	32, 78, 121, 161, 190, 281, 334, 370, 373, 447, 482, 538, 544, 548, 572, 573, 581, 822, 829, 838, 908, 915, 1038.
Author	VS

M31. *In situ* hybridization to chromosomes

Category	Gene-flow, hybridization, genome structure.
Subcategory	Introgression, vertical gene-transfer, hybrid, genotype, structural chromosome changes.

Description

***In situ* hybridization** to chromosomes is a molecular and cytological technique that allows visualisation of specific parts of the genome. A nucleic acid probe is labelled, denatured and applied to a likewise denatured chromosome preparation. Subsequently, the preparation is incubated at conditions allowing renaturation between probe and target DNA if homology exists. The probe can be a total genomic DNA-probe or smaller probes from cloned sequences, synthetic oligonucleotides or PCR-generated sequences. The site of **hybridization** is detected by the position of the labelled probe. The probe is normally labelled by biotin, fluorochromes or radioactivity. The probe is applied to dividing cells in metaphase or prophase, but with some probes results may also be obtained in interphase cells. Cell wall material is removed to improve the access of the probe.

Assumptions and restrictions

The procedure is laborious and technically demanding. Cell preparations of high quality are needed. Only relatively large segments of the genome can be revealed as a target size of 5-10 kb is needed for reliable identification.

Test system

Laboratory experiment.

Advantages

Genome and chromosome organisation can be revealed in a physical way. Especially in combination with **genetic map construction** (M23), the physical map obtained by the *in situ* technique gives valuable information about genome structure.

Application

***In situ* hybridization** has been used to integrate physical and genetic chromosome maps (689). Used on hybrids or amphidiploids the method may provide identification of parental species (601). Identification of introgression and chromosomal rearrangements is also possible. Identification of the position of transgenes in the genome is another application of the technique. However, technical improvements are needed to fully serve this purpose.

Evaluation

Sensitivity: 3	Requirements: 2
Time: 4	Cost: 3

References

401, 493, 961.

Author

RBJ

M32. Inbreeding depression estimation

Category

Breeding system, gene-flow, population.

Subcategory

Self-pollination, outcrossing, genetic load.

Description

Inbreeding depression is the decrease in **fitness** (M14) caused by inbreeding. Inbreeding, either caused by selfing or by mating between related individuals, is a common phenomenon in many plant species. The amount of **inbreeding depression** is important for the evolution of the mating system, e.g., the outcrossing rate (480).

Inbreeding depression, δ, is often estimated from experimental crosses and comparing the fitness, see **fitness measurement** (M14), of selfed and outcrossed seeds, that is:

$$\delta = 1 - (W_{self}/W_{outcross}),$$

where W_{self} denotes fitness of a selfed plant and $W_{outcross}$ the fitness of an outcrossed plant (480). δ can be calculated by comparing outcrossed and selfed seeds from either the same seed family (1) or from different seed families, which may give δ different statistical properties (407).

The level of inbreeding is measured by the inbreeding coefficient **F**, which is the probability that the two alleles at a locus are identical by descent. Thus for selfing, the inbreeding coefficient, $F=0.5$. For self-incompatible species, **inbreeding depression** can be estimated after biparental inbreeding, e.g., sib-crosses which correspond to an inbreeding coefficient $F=0.25$ (363).

Inbreeding depression can also be estimated directly from genetic data without the need for experimental crosses. Hence, by estimating the inbreeding coefficient **F** and selfing rate of adults in one generation and comparing this with the estimated inbreeding coefficient of adults in the next generation (201). If dispersal is very restricted, inferences on inbreeding depression can also be made from crosses of neighbouring plants (983).

The effect of **inbreeding depression** is often estimated on several **fitness**-related traits throughout the life-span of the F_1 plants, e.g., pollen tube growth rate, seed size, germination rate, see **fitness measurement** (M14). In this way, the life-stages of the plant which are most sensitive to inbreeding depression can be identified. The fitness-related traits can be measured either in a natural population or in greenhouse trials. By making an appropriate design of artificial crosses, it is possible to get progeny with different inbreeding coefficients. From the fitness of this progeny, it is then possible to predict the relationship between the level of **inbreeding depression** and the level of inbreeding. From the level of **inbreeding depression**, the amount of genetic load can also be estimated (e.g., 467).

Description, contd.   **Outbreeding depression** (sensu 721), is the decrease in fitness caused by crossing plants separated by a large geographic distance (different populations or sub-species). It is estimated from experimental crosses similarly to **inbreeding depression**. The amount of outbreeding depression is important in determining the amount of successful gene-flow between agricultural cultivars and their wild relatives.

Assumptions and restrictions   It is assumed that a reliable **fitness measurement** (M14) can be obtained. Caution should be used when interpreting greenhouse derived estimates of **inbreeding depression**, which may differ from **inbreeding depression** in natural populations because of genotype by environment interactions. Furthermore, it is assumed that plants used for outcrosses are unrelated.

Test system   Field experiment, greenhouse experiment.

Advantages   When appropriate **fitness**-related characters have been chosen, it is often easy to measure the amount of **inbreeding depression**, especially if the inbreeding coefficient is known from experimental crosses.

Application   The extent of **inbreeding depression** measured by crossing experiments has been reported for a number of species (143), with fitness measured either under field conditions (594, 670) or in the greenhouse (1, 214, 336, 387, 467, 554, 935). The effect of environmental factors (e.g., stress) on **inbreeding depression** has been investigated (e.g., 349, 633, 1013). The relationship between the inbreeding coefficient and δ has been studied (554, 614) and has been found to be close to linear. **Inbreeding depression** has also been related to population size (e.g., 670) and breeding system (143). If a population is subdivided, **inbreeding depression** may decrease with distance between parents used for crossings (500, 553).

Evaluation   Sensitivity: 3   Requirements: 2
                  Time: 4   Cost: 2

References   1, 143, 201, 214, 336, 349, 363, 387, 407, 467, 480, 500, 553, 554, 594, 614, 633, 670, 721, 935, 983, 1013.

Author   MHS

M33. Karyotype analysis: Chromosome number, size and form

Category	Gene flow, hybridization, genome structure, population.
Subcategory	Introgression, vertical gene-transfer, hybrid, genotype, structural chromosome changes, genetic diversity.
Description	Number, size and form of the chromosome complement of a cell constitute the **karyotype**. All somatic cells of an individual have the same karyotype and different taxa can normally be identified by their specific karyotype. The **karyotype** is revealed by microscopy of dividing cells at mitosis or meiosis. Observations can be improved by different chromosome stains (e.g., C- and N- banding of chromosomes) or staining with fluorochromes (159, 932), see **chromosome painting** (M5). Important diagnostic characters of the chromosome form are the position of the centromeres and the presence of satellites. Generally, hybrids will reveal a chromosome morphology and number intermediate between the parental taxa. Backcross plants from wide-species crosses are often aneuploids, some with chromosomes carrying structural changes.
Assumptions and restrictions	The **karyotype analysis** is a rather simple and crude technique which provides information on exchange of genetic material between karyotypically well separated taxa (e.g., in wide-species crosses). The **karyotype** can only be revealed in plant organs during vigorous growth. The method is laborious and will only allow analysis of a small number of plants compared to, e.g., molecular marker analysis. The power of **karyotype analysis** is improved substantially by combination with chromosome banding (M5), *in situ* **hybridization** techniques (M31) and linkage of molecular markers to specific chromosomes.
Test system	Laboratory experiment.
Advantages	Except for the requirement of a microscope, the technical demands are small and the costs low.
Application	The method reveals introgressive hybridization between individuals with different diagnostic **karyotypes**. The method has been used to indicate hybridization, subsequent steps of the introgression process and chromosome doubling involved in the origin of amphiploids (402, 595, 607, 916). Larger chromosomal rearrangements involved in these processes can be detected, e.g., translocations, deletions, insertions, duplications, etc (e.g., 380, 459).
Evaluation	Sensitivity: 3 Requirements: 3 Time: 4 Cost: 2
References	159, 380, 402, 459, 485, 595, 607, 916, 932.
Author	RBJ

M34. Linkage disequilibrium

Category

Gene-flow, population.

Subcategory

Outcrossing, genetic drift, metapopulation.

Description

A pair of polymorphic genetic loci are said to be in linkage equilibrium if alleles at the two loci are randomly associated. Any deviation from this is termed **linkage disequilibrium**, **D**, and for two alleles at each locus it is measured by:

$$D=P_{11}P_{22}-P_{12}P_{21},$$

where P_{ij} denotes the frequency of gametes carrying allele **i** at locus 1 and allele **j** at locus 2 (346). When there is linkage equilibrium, **D**=0, because P_{ij} is then the product, $P_{ij}=P_{1i}P_{2j}$, of the frequency P_{1i} of allele **i** at locus 1 and the frequency P_{2j} of allele **j** at locus 2.

Linkage disequilibrium can be caused by some forms of selection, by sampling effects (i.e., genetic drift) and by non-random mating. From estimation of **linkage disequilibrium**, inferences of these processes may therefore be made. Most population genetic statistics such as **F-statistics** (M19) assume linkage equilibrium. For each generation of random mating, the value of **linkage disequilibrium** (M34) decreases by the fraction 1-**r**, where **r** is the recombination frequency between the two loci. Therefore, the probability of observing **linkage disequilibrium** increases when loci are close together, e.g., nearby nucleotides in a DNA-sequence. The existence of close inbreeding (e.g., selfing) lowers this rate of convergence to linkage equilibrium.

Population subdivision with restricted migration can be powerful in creating **linkage disequilibrium** especially when allele frequencies are very unequal between populations (250). A very large **linkage disequilibrium** is observed if the populations are subsequently completely mixed.

The P_{ij} values cannot be directly determined from diploid genotypes, but have to be estimated. Most of the methods proposed are based on Hill's method of gene counting (108, 369). When genetic data from more than one subpopulation are available, the **linkage disequilibrium** can be divided into a within and between subpopulation component, analogous to **F-statistics** (M19) (648).

Computer programs are available for estimation of **linkage disequilibrium**, e.g., **G-STAT** and **GENEPOP** (M8).

Assumptions and restrictions

A large sample size is usually required for sufficient power to detect **linkage disequilibrium**, especially when recombination between loci is large as for most isozymes, see **protein electrophoresis: Isozyme analysis** (M60 and M34 in vol. 1).

Test system	Data analysis.
Advantages	Analysis of **linkage disequilibrium** is relatively easy to perform and may reveal evidence for epistatic selection or restricted dispersal. **Linkage disequilibrium** analysis adds information on the multilocus structure of genetic data.
Application	**Linkage disequilibrium** analysis was used to infer migration patterns in *Triticum dicoccoides* (305) and gametic disequilibrium was found to correlate with latitude in *Pinus contorta*, where gametes can be assessed directly in the female meta-gametophyte (1029). Evidence for founder events and selection in barley has been found (110).

Evaluation   Sensitivity: 3   Requirements: 3
                     Time: 2   Cost: 1

References   108, 110, 250, 305, 346, 369, 348, 1029.

Author   MHS

M35. Local adaptation analysis

Category   Population.

Subcategory   Cline.

Description   **Local adaptation** is caused by genetic traits that are advantageous to a specific plant population in a specific environment. Specific environmental conditions may have selected for genes coding for advantageous traits during evolutionary time. The effect of this selection is most obvious when a cline exists for a changing environmental factor, e.g., shifting altitude, moisture or nutrient availability.

The existence of **local adaptation** can be investigated by making reciprocal **transplantation experiments** (M76). These tests are made by moving plants between populations and record the fitness of the introduced plants relative to the native plants with a proper **fitness measurement** (M14). If introduced plants are less successful than the native plants in all transplant tests, there is evidence for **local adaptation**.

Assumptions and   **Local adaptation** is defined in relation to a specific set of environ-
restrictions   mental variables. Many of these may be unknown or not present at the time of an investigation.

Test system   Field experiment.

Advantages   **Transplant experiments** are often quite easy and inexpensive to perform. **Local adaptation** is an important parameter when the probability of invasion of GMPs in a given environment is evaluated, especially if the inserted gene itself has an advantageous phenotype in the specific environment (e.g., herbivore-, pathogen- and herbicide-tolerance).

Application   **Local adaptation** is often found when reciprocal transplants are done between widely separated populations (e.g., along an altitudinal gradient (278)), but not always (757). Evidence for **local adaptation** within 12 meters in a population of *Impatiens capensis* was found (814). Indirect evidence for **local adaptation** has been found by relating genetic differences to environmental differences (e.g., 33, 718, 1023).

Evaluation   Sensitivity: 3   Requirements: 2
                        Time: 4   Cost: 2

References   33, 230, 278, 470, 718, 757, 814, 978, 1023.

Author   MHS

M36. Marking plants and flowers: Flowering phenology analysis

Category

Pollen development and production, seed development and production, pollination, breeding system, gene-flow.

Subcategory

Pollen production, pollen viability, agamospermy, fruit abortion, ovule development, seed abortion, seed germination, seed production, nectar production, pollen dispersal, pollinator activity, cross-pollination, incompatibility, phenology, reproductive allocation, self-pollination, sex distribution, outcrossing.

Description

Marking of plants and flowers is used in order to follow the fate of a number of plants or flowers during a certain period or to obtain values for reproductive features. The marked individuals can be positioned on a transect, they can be chosen randomly or selected for special purposes. **Marked plants and flowers** are generally randomly assigned to various observations. The plant parts are marked with coloured thread, tape, etc.

Investigating flower phenology involves that each flower is marked when opening and followed daily and processes are described in detail until withering: Opening of corolla, order of dehiscence of anthers, movements of anthers, period of stigma receptivity, etc. Not less that 10-20 flower buds are marked on different plants at a number of times during the flowering period. By studying phenology of inflorescences, marked plants are visited every day and newly opened flowers are individually marked. The study of seed ripening and removal includes marking of flowers and observation of the development and disappearance of fruits and seeds. Special care concerning the design of the study should be taken if the plant is clonal. See **genet identification**, (M25). Phenology of populations of the same species and closely related taxa may be compared on a given day. The phenological stage of each individual is scored and used for comparisons.

Assumptions and restrictions

Marking and handling of plants always influence the fate of the plant or the part of plant marked. It is very important to mark buds instead of flowers as otherwise a random selection of flowers are difficult. Best results come from field observations under natural conditions, but, in some cases, greenhouse experiments are just as appropriate.

Test system

Field experiment, greenhouse experiment.

Advantages

Marking of plants or plant parts such as flowers is appropriate when a part of the population is to be studied.

Application   This method is widely used when studying flower and flowering phenology both at the individual level and at the population level (84, 324, 355, 357, 358, 359, 413, 555, 638, 639, 696, 697, 884, 905). The method is also used to study seed and fruit set (188, 232, 702), and to see if seed set is pollen limited (997). Furthermore, **marked plants** have been used to compare breeding systems (324), to record the effect of different kinds of experimental pollinations (208) and to observe the distribution of sexual morphs on plants (229).

Evaluation   Sensitivity: 5   Requirements: 1
                      Time: 4   Cost: 1

References   84, 188, 208, 229, 232, 249, 324, 355, 357, 358, 359, 413, 555, 564, 638, 639, 696, 697, 702, 884, 905, 997.

Author   MP, FF

M37. Meiotic analysis: Chromosome pairing and recombination

Category	Gene-flow, hybridization, genome structure.
Subcategory	Introgression, hybrid, genotype, recombination, structural chromosome changes.
Description	The **chromosome pairing** will provide information on hybridization between taxa with homologous or homoeologous (i.e., partly homologous) chromosomes. In diploid or polyploid plants, homologous and homoeologous chromosomes will pair at meiosis and the chromosome pairing is studied in the first pro- and metaphase. The homologous parts of the chromosomes will align in order to exchange alleles through chromosomal **recombination**. The meiotic configurations as chromosome pairs (bivalents), unpaired chromosomes (univalents) or several chromosomes aligned (multivalents) indicate the extent of homology between chromosomes. When genetic maps of parental individuals exist, the recombination frequency can be calculated from segregating offspring in the next generation (number of recombinants divided by the total number of progeny, see **genetic map construction**, M23). Co-segregation of linked donor markers in offspring from a potential recipient is an indication of introgressive hybridization (765). The absence of homologous chromosome pairs can result in odd chromosomal configurations. Chromosomal rearrangements may also result in aberrant meiotic chromosome configurations; the morphology of configuration indicates the type of rearrangement (e.g., translocation, inversions, insertions of larger DNA fragments (445, 887, 946).
Assumptions and restrictions	**Pairing of chromosomes** may not always reflect homology, as the meiotic configurations can be controlled by other genetic factors.
Test system	Laboratory experiment.
Advantages	Except for the requirement of a microscope the technical demands are small and the costs low.
Application	Chromosome pairing has been used to reveal phylogenetic links between taxa (238, 488, 930) which in turn provide information on risks of inter-taxa introgression of transgenes (430).

Evaluation

Sensitivity: 3	Requirements: 3
Time: 2	Cost: 2

References	238, 430, 445, 488, 765, 887, 930, 946.
Author	RBJ

M38. Microsatellite markers

Category

Gene-flow, hybridization, genome-structure, population.

Subcategory

Introgression, marker, vertical gene-transfer, hybrid, genotype, polymorphism.

Description

Microsatellites are used as codominant polymorphic **markers** in the same way as **isozymes**. **Microsatellites** are short stretches of DNA which contain a number of short (2-6 base pairs) repeated sequence elements (e.g., ATATATAT is 4 repeats of the two base pair element AT). **Microsatellites** are abundant in all investigated genomes (596). Most of them have a high level of polymorphism, sometimes exceeding 20 alleles, where the alleles differ in the number of repeated sequence elements. **Microsatellites** are also denoted simple sequence repeats (SSR).

A single **microsatellite** can be visualised following **PCR** analysis (M59) with specific primers that can attach to sequences flanking the microsatellite. The product of the PCR-reaction is run through a separating gel (e.g., a sequencing gel), which separates fragments by size. Thus the number of repeats of the alleles of the **microsatellite** can be inferred.

The function of **microsatellites** in the genome is unknown, but their length (number of repeats) is believed to be selectively neutral. The high level of polymorphism is due to a very high mutation rate. Most mutations only cause a change in length of one repeat unit (343, 953). Special statistical methods, which consider this mutation mechanism, are currently being developed for analysis of **microsatellite** data. A computer program, **MICROSAT** (M8), performs data analysis with these methods (304, 845, 856).

Microsatellites are also used as probes in DNA fingerprinting and are powerful for **genet identification** (M25) and **paternity analysis** (M45).

Assumptions and restrictions

The main restriction in use of **microsatellites** is the need of specific primer sets for each **microsatellite**. Though the same primer set often can be used for a group of closely related species, most primer sets have to be developed for each species separately. This involves cloning and is both a time consuming and expensive procedure. The high mutation rate violates assumptions of most population genetics theory and, therefore, measures like **F-statistics** (M19) and **genetic distance** (M22) can be biased.

Test system

Laboratory experiment.

Advantages	**Microsatellites** offer an almost unlimited number of polymorphic, codominant loci for a given species. The level of variation is generally very high (heterozygosity often exceeds 90%), with a number of alleles up to 30. Therefore, a **microsatellite** marker used in a population study often gives more information than an **isozyme** (M60) marker, especially when **paternity** (M45) patterns are investigated.
Application	Primer sets have been published for a few species, e.g., *Arabidopsis thaliana* (195), *Dioscorea tokoro* (913), *Quercus rubra* (204), *Pinus radiata* (859) and *Brassica campestris*, and primers for a large number of other species are expected to appear very soon. **Microsatellites** have been used to characterise cultivars of rice (1028). Furthermore, microsatellites have been used as probes in a number of studies of DNA-fingerprinting (e.g., 335, 378, 737, 834).
Evaluation	Sensitivity: 5 Requirements: 4 Time: 4 Cost: 4
References	195, 204, 304, 335, 343, 378, 596, 737, 834, 845, 856, 859, 913, 953, 1028.
Author	MHS

M39. Morphological character analysis

Category

Gene-flow, hybridization, population.

Subcategory

Introgression, outcrossing, vertical gene-transfer, heterosis, hybrid, hybrid depression, hybrid vigour, metapopulation.

Description

Analysis of morphological characters of the phenotype, is a classical method used by taxonomists and plant breeders to identify hybrids and to describe strains. **Morphological character analysis** may also be used in combination with molecular, cytological and chemical methods for the study of gene-flow in populations.

Both fresh plants and dried herbarium specimens can be used. A character set is determined from the floristic characters which distinguish the studied species. The set may include up to 30-40 different metric, semi-quantitative and nominal characters. Samples to be measured must include at least 50 individuals in order to use multivariate statistics. A good method description of measurement is available (453). Large characters as total height of the plant are measured to the nearest cm. Shape of leaf is measured using dial callipers to the nearest 0.1 mm. Using a 30X dissecting binocular with a calibrated ocular micrometer, measurement of morphological details is performed to the nearest 0.01 mm.

The biometrical data may be processed in a database and standardized for use in a statistical program. Some multivariate statistics are well adapted to hybrid identification. A critical comparison of classical statistical methods inferring hybridization from morphologically intermediate characters exists (1007), recommending the use of pictorialized scatter diagrams (27) and a character count procedure. Both procedures allow differentiation between hybridization and divergence.

For the detection of hybridization, Well's hybrid distance diagram (991) or the Principal Coordinate Analysis (PCO) using weighted Gower metrics have been recommended (8). Common methods such as **principal component analysis (PCA)** (M54 in vol. 1) and the Canonical Variate Analysis (CVA) lead to a higher percentage of misidentification, which is detected by other methods such as molecular analyses (8). Overview of multivariate morphometrics may be found in textbooks on statistics (537, 756, 867).

Assumptions and restrictions

Morphological character analysis has to be combined with molecular techniques, as morphological characters alone may, in some cases, lead to misidentification. An appropriate choice of statistics according to data structure is essential for good results.

Test system

Field experiment, greenhouse experiment, laboratory experiment.

Advantages	Only standard laboratory equipment is required plus the relevant statistical programs. **Morphological analysis** may be used in many different types of test systems, for comparison of results. Long-term studies and the comparison with historical material is possible using **herbarium sheet survey** (M27). In a field test, it might be useful to combine this method with **fitness measurement** (M14).
Application	**Morphological character analysis** and **isozyme analysis** (M60) are widely used in combination (e.g., 390, 429, 1022). Natural hybrids of *Leucaena* has been detected by morphological, molecular and biogeographical evidence using dried material as a source of DNA, see **plant material preservation: Rapid drying** (M48) (383). The morphological variation of somatic hybrids of *Solanum* has been analyzed (717) and variation in F_1 hybrids and backcrosses of *Carduus* has been studied (976).

Evaluation	Sensitivity: 4	Requirements: 2
	Time: 3	Cost: 2

References	8, 27, 383, 390, 429, 453, 537, 717, 756, 867, 976, 991, 1007, 1022.
Author	PR

M40. Nectar production: Chemical composition

Category   Pollination, breeding system.

Subcategory   Chemical attractant, nectar production, pollinator foraging beha-
                      viour, reproductive allocation.

Description   The **chemical composition** of floral nectar shows adaptations to
                      pollinators. Thus, nectar from plants pollinated by the same pollina-
                      tor group is often - to some extent - similar in composition. Nectar
                      is a solution of sugars (especially sucrose, fructose, and glucose) (51,
                      52). Other compounds are mainly amino acids, proteins (enzymes),
                      lipids, alkaloids, vitamins, phenolics and antioxidants (219).

                      Nectar is extracted from flowers with a microcapillary tube. Sugar
                      concentration (in sucrose equivalents) of the sample is then
                      measured with a hand refractometer modified to small volumes >0.5
                      µl (95, 169). The various sugars and amino acids may be separated
                      and identified by paper chromatography (693, 986). Determination
                      of concentrations of amino acids can be done by simple use of
                      ninhydrin (54, 55) or by **HPLC**, (M29) (311). See also **Nectar produc-
                      tion: Volume and rate** (M41).

Assumptions and   Composition of nectar demonstrates a tremendous variability in
restrictions   space, time and between plant species (183, 692). Composition varies
                      more in flowers with an open morphology than in closed flowers
                      (708). Non-sugary components also add to the refractive index (389).
                      Many north temperate flowers contain much less than 0.5 µl nectar.

Test system   Field experiment, greenhouse experiment.

Advantages   In general, the sugary compounds are easy to determine. This is
                      often done in combination with an estimation of **nectar production**
                      (M41).

Application   Knowledge of nectar composition is important in many studies, e.g.,
                      of pollinator energetics (426, 874). Information on sugar concentra-
                      tion is used to determine the energy content of nectar (95, 333).
                      Nectar variation affects behaviour and movement of flower visitors
                      (726) and pollen- and gene-flow within and between patches and
                      populations of plants (129).

Evaluation   Sensitivity: 4   Requirements: 3
                      Time: 3   Cost: 2

References   51, 52, 54, 55, 95, 129, 169, 181, 183, 219, 298, 311, 333, 389, 426, 692,
                      693, 706, 708, 726, 842, 873, 874, 986.

Author   JMO

M41. Nectar production: Volume and rate

Category

Pollination, breeding system.

Subcategory

Nectar production, pollinator foraging behaviour, reproductive allocation.

Description

Nectar is, together with pollen, the most important floral reward to pollinators. It is produced by nectaries often situated around the ovary. The standing crop of nectar per flower is the volume of nectar when the flower is available to visitors. It is equal to rate of production multiplied by the time passed since last pollinator visit (295, 707). The production rate is the volume produced per time unit and measurements requires that the flowers are bagged, see **bagging of flowers** (M2).

Volume of nectar is usually measured with calibrated microcapillary tubes of various sizes (range: 1-25 ml, min. volume measured approx. 0.1 ml), by centrifugation of flowers or with small filter-paper strips (53, 560). See also **Nectar production: Chemical composition** (M40).

Assumptions and restrictions

The standing crop of nectar varies tremendously in space and time. A large sample of flowers is often needed because distribution of nectar may be spatially clumped (706, 842, 873). Nectar secretion also varies diurnally and with temperature (169, 170), with flower age (652) and with weather (183, 262). Flowers may reabsorb produced nectar if not harvested by visiting pollinators (119). Bags for insect-exclosure may alter humidity and temperature of flowers (169, 170). Many flowers are destroyed when nectar is extracted with microcapillary tubes (699) either because of their complex morphology or small size.

Test system

Field experiment, greenhouse experiment.

Advantages

In general, methods are easy to handle and offer a lot of information about the amount of energy available to pollinators.

Application

Knowledge of **nectar production** is vital in many studies, e.g., of pollinator energetics (95, 352, 354, 426, 874), behaviour and movement (726) and in studies of pollen and gene-flow within and between patches and populations of plants (129). Measures of nectar production is a standard procedure for studies of pollination biology.

Evaluation

Sensitivity: 4   Requirements: 1
Time: 3   Cost: 1

References

53, 95, 119, 129, 169, 170, 183, 262, 295, 352, 354, 426, 560, 652, 693, 699, 706, 707, 726, 728, 842, 873, 874, 1041.

Author

JMO

M42. Nuclear and organelle markers combined

Category   Gene-flow, hybridization, population.

Subcategory   Introgression, outcrossing, vertical gene-transfer, hybrid, genetic diversity, metapopulation.

Description   When genetic information is available from both diploid, nuclear loci and from haploid loci (chloroplast or mitochondrial DNA), a number of methods exist that make use of the different inheritance of the different markers. Gene-flow inferred from **nuclear markers**, e.g., through **F-statistics** (M19), **spatial autocorrelation** (M72) or rare allele, see **gene-flow estimation with private alleles** (M21), include both gene-flow through seeds and through pollen. A maternally inherited marker provides information about gene-flow through seeds only. Therefore, the two different sources of gene-flow can be separated by combining nuclear and maternally inherited markers. For the inheritance of organelles, see **organelle DNA analysis** (M43).

Larger differentiation between populations is expected for maternally inherited markers than for nuclear markers which are biparentally inherited. For paternally inherited markers, a large differentiation is expected between populations because each individual only carries one copy of the gene. The expected differentiation of the three types of markers has derived (233).

Alternatively, information can be gathered from the study of cytonuclear disequilibria, i.e., linkage disequilibria between cytoplasmically inherited markers (i.e., organelle DNA) and nuclear markers (45, 815).

The method is very suitable for studies of introgression processes, because these may be asymmetrical, i.e., one species acts primarily as pollen donor and the other as ovule donor. Therefore, maternally inherited DNA may introgress directionally (38).

Assumptions and restrictions   **Genetic markers** need to be available for both nuclear and organelle DNA. Since organelle DNA is non-recombining, it can only be regarded as a single locus and the genetic analysis is therefore more sensitive to chance processes than genetic analysis based on a number of nuclear loci. Furthermore, the inheritance of the organelles need to be established for each species, since, e.g., mtDNA may be biparentally inherited in some species (233).

Test system   Data analysis.

Advantages   Differences in gene-flow estimates, from the two different kinds of markers, can yield information on the relative importance of gene-flow through seed and pollen. Organelle DNA in dioecious species may be more sensitive to population bottlenecks and may yield an estimate of the **effective population-size** (M11).

Application   It was observed (556), that the F_{ST}-value, see **F-statistics** (M19), bet-
                           ween populations was 5-fold as large for maternally inherited
                           cpDNA than for nuclear isozymes in *Silene alba*, suggesting that
                           pollen move much more freely than seeds in this species. A com-
                           parison (38) of cpDNA-markers with nuclear markers, in an investi-
                           gation of interspecific gene-flow in *Iris* species, showed that pollen
                           flow between populations of the species was more important than
                           seed flow and some evidence for the direction of the introgression
                           process was found.

Evaluation   Sensitivity: 3   Requirements: 3
                           Time: 2   Cost: 2

References   38, 45, 233, 556, 815.

Author   MHS

M43. Organelle DNA analysis

Category

Gene-flow, hybridization, genome structure, population.

Subcategory

Vertical gene-transfer, hybrid, genotype, genetic diversity.

Description

Analysis of **organelle DNA**, i.e., plastids (including chloroplasts) and mitochondria, is useful as markers for detecting gene flow and hybridization. Whereas genomic DNA is biparentally inherited, mitochondrial DNA (mtDNA) is usually maternally inherited in plants, but recombination does occur (515). Chloroplast DNA (cp-DNA) can be either paternally (most gymnosperms) or maternally inherited (angiosperms) (172, 331). So far, recombination in cpDNA has not been observed, so historical information is well preserved in cpDNA sequences.

The organelle DNA may be analyzed by **Southern hybridization** (M71). Chloroplast or mitochondrial DNA labelled probes are used to characterise the **organelle DNA**. Differences between the restriction patterns allow parents and hybrids to be distinguished. **Nuclear markers and organelle markers combined** (M42) may be used as a tool to analyze gene-flow and introgression.

Gene-transfer of plastid DNA (ptDNA) via pollen only occurs in case of biparental or, rarely, paternal inheritance (172, 331). Regular biparental inheritance occurs when 5% or more of progeny contain biparental ptDNA (863), whereas occasional biparental inheritance is when less than 5% of progeny contain biparental ptDNA. Potential capability of biparental inheritance of ptDNA can be determined by rapidly screening pollen for the presence of pt-DNA using the DNA fluorochrome 4',6-diamidino-2-phenyl-indole (DAPI) in conjunction with epifluorescence microscopy (172). Alternatively, the parental origin of ptDNA may be determined by *in situ* **hybridization** (M31) of ptDNA in the pollen or **RFLP analysis** (M64) combined with **Southern hybridization** (M71) using cloned ptDNA as a probe (1030).

Details about chemistry, structure, growth and inheritance of plastids are available (437).

Assumptions and restrictions

Detection of even occasional biparental ptDNA inheritance requires a large test sample, especially when close relatives are included.

Test system

Laboratory experiment.

Advantages

By **Southern hybridization** (M71) fragment patterns for many individuals may rapidly be obtained in the same electrophoretic procedure. By epifluorescence microscopy (172), a large number of individuals can be quickly screened for their potential to transfer plastids via pollen.

Application	**Organelle DNA analysis** is useful to identify hybrids, the maternal parent and to reconstruct the phylogenetic relationships between the parental chloroplasts. cpDNA is the organelle sequence of choice for evolutionary studies in higher plants because of its high copy number, its evolutionary conservatism and the rarity of large insertions, deletions and rearrangements.

The method of epifluorescence microscopy has been applied to more than 235 species including about 50 crop species of agronomic importance (172). Reviews are available (345, 863).

Evaluation	Sensitivity: 4 Requirements: 3 Time: 3 Cost: 3
References	172, 196, 233, 331, 345, 437, 439, 440, 515, 863, 893, 996, 1030.
Author	YJ, PR

M44. Ovule counts

Category	Seed development and production, breeding system.
Subcategory	Agamospermy, fruit abortion, ovule development, seed abortion, seed production, cross-pollination, phenology, reproductive allocation, self-pollination, sex distribution.

Description

Ovule counts are used to study seed set (e.g., the proportion of ovules maturing into seeds), to compare female function among samples of plants and to evaluate different pollen donors.

Ovaries are collected and dissected at different times after pollination. Even in mature fruits it should be possible to count undeveloped ovules plus mature seeds to get the total number of ovules. It is often possible to make a **classification of ovules** (M6) into non-fertilized ovules, early aborted ovules and mature seeds. Image analysis equipment facilitates the counting process.

If ovaries contain a large number of ovules it will be necessary to count ovules only in parts of the ovary and make extrapolation of the total number of ovules. In orchids, for instance, seeds and unfertilized ovules have been counted in 0.5 mm sections of semimature capsules.

Assumptions and restrictions

When counting only parts of an ovary it is assumed that it is possible to extrapolate from the chosen part to the whole ovary.

Test system

Laboratory experiment.

Advantages

These methods are useful for collecting information on the female function of the plant.

Application

Ovule counts are widely used to determine potential seed set (698, 905) and to compare breeding systems in closely related species or different morphs within a species (738, 774) for instance by determination the Pollen/Ovule-ratio (180, 622). Ovules are also counted in studies of pollen limitation to seed set (739).

Evaluation

Sensitivity: 4   Requirements: 2
Time: 3   Cost: 2

References

180, 622, 698, 738, 739, 774, 905.

Author

MP

M45. Paternity analysis

Category

Pollination, breeding system, gene-flow, population.

Subcategory

Pollen dispersal, cross-pollination, self-pollination, outcrossing, genetic diversity, genetic neighbourhood-area, metapopulation.

Description

The goal of **paternity analysis** is to assign paternity to seeds of a maternal plant of known genotype from a number of possible fathers (pollen donors). This is done by determining the genotypes of possible fathers and of seeds of chosen plants for a number of codominant, polymorphic loci, e.g., isozymes, see **protein electrophoresis** (M60 and M34 in vol. 1), or **microsatellite markers** (M38).

Paternity analysis (M45) provides information of exact, successful transfer of pollen within the population, and can identify pollen flow into a population, thus giving a much more detailed pattern of gene-flow than is revealed by **F-statistics** (M19) or **spatial autocorrelation** (M72). Furthermore, by assays of several seeds from the same fruit, the degree of multiple **paternity** (i.e., more than one father to the different seeds of a fruit) can be determined (226).

If all individuals in a population can be genotyped, it is possible to perform simple **paternity** exclusion to the seeds of the maternal plants (222). With this method, the maternal contribution to a seed is deducted to obtain the possible paternal contribution. Potential fathers can then be excluded by genotype comparisons. If all but one father can be excluded, **paternity** is assigned to this father. If all possible fathers are excluded, gene-flow from outside the population has been detected.

In most cases, several possible fathers within the population cannot be excluded and the fractional **paternity** assignment method is used (198). In this method, the remaining possible fathers are assigned probabilities of **paternity** conditional on their genotype. A statistical model is also available for estimating the amount of migration when all fathers in the local population cannot be excluded (197).

Assumptions and restrictions

The genetic markers employed need to be codominant. Usually both a large number of loci and a large number of individuals need to be screened for a reliable estimation. **Microsatellites** (M38) are ideal markers since they usually have a large number of alleles and thus a large discrimination power between fathers. The method yields insight on gene-flow through pollen only and gives a snapshot picture of gene-flow.

Test system

Field experiment analysis, laboratory experiment.

Advantages   The pattern of gene-flow through pollen can be investigated in detail. Inferences on the fitness, see **fitness measurement** (M14), through the male function of hermaphrodites can be investigated, the distribution of dispersal distances can be determined and migration events through pollen can be directly observed.

Application   **Paternity analysis** has been used to investigate multiple paternity of individual fruits (e.g., 208, 222) and to make inferences on the amount of **pollen carry-over** (M49). Maps of matings within a population have been drawn (566) and paternal contributions to reproduction have been correlated to various morphological characters (567, 864). In some cases, plants with specific rare marker alleles have been introduced in experimental populations to gain more power in the paternity assignment (e.g., 275, 396). Migration rates into small populations from surrounding populations have been estimated (224, 225, 457). These rates were, in some cases, found to be higher than expected from observing **pollinator foraging behaviour** (M57) (e.g., 113, 224, 225), indicating gene-flow over large distances, which is an important observation in the context of risk assessment of GMPs.

Evaluation   Sensitivity: 4   Requirements: 3
                      Time: 4   Cost: 4

References   113, 197, 198, 208, 222, 224, 225, 226, 275, 396, 457, 566, 567, 864.

Author   MHS

M46. Photosynthetic efficiency

Category

Hybridization, inserted trait, adult plant (see vol. 1).

Subcategory

Heterosis, hybrid depression, hybrid vigour, stress tolerance, plant competition (see vol. 1).

Description

The generation of oxygen in the photosynthesis of plants has been utilized in photoacoustic spectroscopy. The principle of the method is in short, that a leaf disc is placed in an acoustic cell and exposed to light. A depiction of an acoustic cell has been published (115). Due to emission from the oxygen generated, the energy released will increase the temperature in the cell. The increase is transduced to an acoustic pressure wave, which may be registrated via a microphone and an amplifier. Procedure for separation and calculation of the oxygen-evolution and photothermal signals is available (714).

Another method for measuring **photosynthetic efficiency** may be the utilization of the fluorescence transient produced under the activity of the photosynthesis. The fluorscence may be measured by a fluorometer.

Assumptions and restrictions

The method depends on the leaf discs selected. When a comparison among plants shall be performed, it is important to get leaves of the same age, growing under the same environmental conditions.

Test system

Laboratory experiment.

Advantages

The method is simple, rapid and highly sensitive. Moreover, the destruction of the plant, or a part of it, is not required.

Application

Photosynthetic efficiency has been applied to detect hybrids, as hybrids may have a different photosynthetic rate compared to the parents (569, 849). It was found that a more intense oxygen production was associated with the activity of PSII (photosystem II). The method has also been used for studying the effect of herbicides (530). Enhanced ethylene production has been investigated by the method (102). Stress tolerance, which is connected to oxygen production, may be detected by applying **photosynthetic efficiency** (397, 694).

Evaluation

Sensitivity: 4   Requirements: 4
Time: 2   Cost: 3

References

102, 115, 397, 530, 569, 694, 714, 849.

Author

YJ, VS, KHM

M47. Plant material preservation: Cryopreservation and freeze-drying

Category

Gene-flow, hybridization, population.

Subcategory

Introgression, vertical gene-transfer, hybrid, gene pool, genetic diversity.

Description

Cryopreservation techniques are widely used to store plant material for later genetical and **chemical analysis**. Metabolism of biological material is inhibited by freezing in liquid nitrogen, viability is maintained and regeneration of plants is possible. It has been applied to many types of structures (e.g., pollen, protoplasts and meristems) for many species. For successful **cryopreservation**, it is essential to avoid lethal intracellular freezing. In addition, cells and meristems have to be sufficiently dehydrated before being immersed into liquid nitrogen at -196°C.

In general, the following steps are performed:

a) Cold hardening of the donor plant over several weeks (not always performed),
b) Tissue preparation,
c) Applying cryoprotective solutions which dehydrate the cells before and during the freezing process,
d) Further dehydration by a vitrification solution, or freezing with a specific cooling rate to approx. -30°C (795, 973),
e) Plunging in liquid nitrogen (-196°C),
f) Storage under constant conditions,
g) Rapid thawing by plunging in water (approx. 40°C),
h) Removal of the cryoprotective solution,
i) Exposition in culture medium for regeneration to plantlets.

Freeze-drying is a very gentle method to preserve compounds as phenolic compounds, proteins and antibiotics for photochemical analysis. Immediately after the harvest, the plant material has to be frozen to avoid decomposition of the compounds. The freeze-drying is performed on frozen or fresh material at -60° to -80°C in vacuum, protected against light. The dried plant material can be stored dry at room temperature protected against light until use. See also **plant material preservation: Rapid drying (M48)**.

Assumptions and restrictions

Cryopreservation conditions are highly variable. Each species and each tissue type may need a specially adapted protocol (931). Adaptation of the protocol to new cases is highly time consuming and therefore expensive. The phenotype of regenerated plants should be confirmed by cytological, biochemical and morphological analysis (549). Maintenance of embryogenic cell suspensions is difficult but needed for gene transfer into cereal crops and forage grasses (973).

Test system

Field experiment, greenhouse experiment, laboratory experiment.

Advantages	Cryopreserved material remains genetically stable, which is crucial for analysis and breeding experiments. Therefore, it is the ideal method for preservation of genetic material. **Freeze-drying** is gentle to the plant material and can be used for both biological and photochemical investigations.
Application	**Cryopreservation** is an important tool for *ex situ* conservation of endangered flora (931) and long-term preservation of genetic resources for future agricultural breeding. In general, cryopreservation of plant tissues has been more successful with cell suspension cultures than with organised tissues. For conservation purposes, where genetic stability is of greatest concern, apical meristem or shoot-tip cultures are preserved. *In vitro* grown apical meristems of *Wasabia japonica* were successfully cryopreserved by vitrifiation and subsequently regenerated to plants (549).

Microspore **cryopreservation** has been used for long-term storage of germplasm for in vitro embryo production of plant species such as *Brassica napus* (151). **Cryopreservation** of singlegenotype derived embryogenic cell suspensions of forage grass species (e.g., *Festuca* and *Lolium*) has been described (973).

Effects of tissue storage conditions (freezing) and storage times have been compared for allozymes analysis of tropical tree species (433).

A comparison of the content of rosmarinic acid and caffeic acid in eelgrass (*Zostera marina* L.) using **freeze-drying** and drying at +80°C, has been made. In freeze-dried tissues rosmarinic and caffeic acids were well preserved. Heating generated substantial losses and higher variability among replicates (742). Vacuumdrying prevents the degradation of phenolic glycosides and condensed tannins in leafs of willow (*Salix*) and avoids the need for cooling in liquid nitrogen (668).

Evaluation	Sensitivity: 5	Requirements: 5
	Time: 4	Cost: 4
References	15, 215, 433, 549, 649, 668, 742, 752, 795, 931, 973.	
Author	PR	

M48. Plant material preservation: Rapid drying

Category Gene-flow, hybridization, population.

Subcategory Introgression, vertical gene-transfer, hybrid, gene pool, genetic diversity.

Description **Rapid drying** procedures are essential to preserve plant material collected in the field and to preserve the DNA for molecular analyses, e.g., by **PCR** (M59) later. Samples from a field trip of several months can be analyzed within one year or can be frozen at -70°C for longer-term storage (206). An assessment of different drying methods is available (206, 344, 730). The following techniques have been used:

1. Silica-drying where approx. 1 g of young leaves is put into a snap-top plastic bag with 25 g of self-indicating silica gel (144).

2. Contact paper-drying where fresh plant material is placed between sheets of blotting paper in a plant-press and papers are changed daily until the specimens are dry.

3. In corrugate-drying, the plant material is sandwiched between two layers of blotting paper and separated from adjacent samples with aluminium corrugates. Stacked samples are clamped and dried overnight over a low heating kerosene burner (classical method for the tropics). For contemporary effective methods for preservation of genetic material, see **cryopreservation and freeze-drying** (M47).

It is recommended (344) to sample one voucher per individual, rapidly pressed and dried for **morphological character analysis** (M39), and one leaf sample dried with silica gel ideally preservating the DNA for **PCR** (M59) and **RAPD** (M61) (920). The increased effort associated with drying a few grams of leaves for a molecular study is negligible.

Pressed plants can also be used for **morphological character analyses** (M39) and be conserved in a herbarium collection. For the treatment of herbarium specimens to preserve DNA for future analyses see **herbarium sheet survey** (M27).

Assumptions and restrictions If the drying period is too long, the cell content is injured, which rapidly endures phenolic compounds and free radicals production, resembling the responses to senescence, making DNA extraction and amplification difficult (804).

Test system Field experiment.

Advantages **Rapid drying** makes it possible to collect samples of plant material for later analysis far from freezing facilities. Compared with other preservation methods, they are very cheap and easily done.

Application	These techniques can be used to collect material of natural populations for morphological and molecular studies, e.g., for hybrid identification (large-scale release, long-term monitoring). Silica-drying has been used for confirming hybrids of *Leucaena* in Mexico (383). DNA from the dried leaf material was used for identification of sterile hybrids.

Evaluation

Sensitivity: 4   Requirements: 1
Time: 1   Cost: 1

References   144, 206, 344, 383, 730, 804, 920.

Author   PR

M49. Pollen carry-over between flowers

Category	Pollination, pollination syndrome, breeding system, gene-flow, population.
Subcategory	Pollen dispersal, pollinator foraging behaviour, bird-pollination, insect-pollination, mammal-pollination, cross-pollination, self-pollination, outcrossing, genetic neighbourhood-area.
Description	In most plant species, the pollen collected by a pollinator is not all deposited on the first flower visited, but some pollen is carried further. This movement of pollen from a donor flower to flowers beyond the first visited is termed **pollen carry-over**. If **pollen carry-over** is taking place, pollen- and gene-flow calculated from pollinator flight patterns alone will create underestimates. The importance of pollen carry-over is reduced if the flight becomes random (658). Models of **pollen carry-over** are available (e.g., 125): **Carry-over** = Y_x/Y_1, where Y_x and Y_1 are the numbers of pollen grains deposited on the first and the x'th flower visit. **Pollen carry-over** can be measured directly by using **fluorescent dyes and marking of pollen** (M17) in combination with observation of **pollinator foraging behaviour** (M57). Indirectly, it can be estimated by comparing the observed pollinator flight patterns with the amount of gene-flow as determined from genetic markers.
Assumptions and restrictions	Some plant species need a certain amount of pollen on their stigmas before fruit set is initiated. A few pollen grains transported a long distance through **pollen carry-over** will therefore be of minor or no importance (361). Carry-over loses much of its importance in plant species with multi-flowered inflorescences where many flowers are visited per plant. **Pollen carry-over** may be impossible to estimate in some plant-pollinator systems, e.g., systems involving long-flying butterflies and sphingids (174).
Test system	Field experiment, greenhouse experiment.
Advantages	Pollinators with a linear flight are needed in order to get more exact estimates of pollen flow. Pollen carry-over data are also crucial for estimating the relative importance of geitonogamy and xenogamy (361).
Application	The very few studies of **pollen carry-over** all show a tremendous variation. A study of **pollen carry-over** in relation to pollinator behaviour has been made (921). **Pollen carry-over** has also been studied in relation to multiple paternity (222).
Evaluation	Sensitivity: 2 Requirements: 2 Time: 4 Cost: 2
References	125, 174, 222, 361, 658, 921.
Author	JMO

M50. Pollen counts

Category

Pollen development and production, breeding system, gene-flow, hybridization.

Subcategory

Pollen germination, pollen production, pollen viability, cross-pollination, incompatibility, phenology, reproductive allocation, self-pollination, outcrossing, hybrid.

Description

Methods for estimating the number of pollen grains are necessary in analyzing the breeding system. Several methods for **pollen counts** are available:

1. Mature but undehisced anthers are collected in 50% ethanol and anthers are emptied into a Petri dish or on a slide in a drop of glycerol. The pollen grains can be stained by acetocarmine, cotton blue in lactophenol, or in Alexander's stain, which improve the contrast in the preparation and reveal colour differences between morphologically normal and abnormal pollen grains. Pollen grains are counted over a grid in a specific area of the preparation.

2. Pollen can be counted in a haemacytometer after one of the following treatments:
    A) Anthers are put into a microcentrifuge tube containing 0.5 ml H_2O with Triton 1% w/v and agar 0.1% w/v. After stirring with a vortex mixer for 60s, subsamples are immediately deposited in a haemocytometer.
    B) Anthers are dissected out and placed into vials containing 0.3 ml 0.7 M mannitol solution coloured with a little acetocarmine to stain the cytoplasma. Pollen are counted in a haemocytometer after shaking thoroughly.
    C) Anthers are crushed in a 3:1 mixture of lactic acid:glycerine, containing methylene blue. After thorough shaking for several minutes, samples of pollen are transferred to a haemocytometer.
    D) Closed anthers are placed in vials and are left to dry in order to get the anthers to dehisce. Then they are stained with aniline blue (0.10 g aniline blue, 0.17 g K_3PO_4, 100 ml H_2O), 45% acetic acid, and 80% ethanol. Vials are placed in a sonicator for 2 min to separate clumps of pollen grains and samples are transferred to a haemocytometer.

3. Number of pollen grains and their size can be measured with a Elzone 180 XY electronic particle counter (Particle Data Inc., Elmhurts, Illinois, USA), or a Coulter TAII particle counter. Before analysis, anthers need one of the following treatments:
    A) Pollen samples in 70% ethanol are suspended in 0.63% NaCl solution and sonicated for 5 min.
    B) Pollen from anthers are oven-dried for 3 days and later rehydrated for 1 day in 1% NaCl solution.

Description, contd.

C) Undehisced anthers are placed in a 1.5 ml polypropylene centrifuge tube and are allowed to dehisce by air drying. Then 70% ethanol can be added to preserve the grains. The pollen grains are then suspended in 100 ml of 0.1% NaCl.

D) Anthers are stored in an open plastic microcentrifuge tube to dry for approximately 1 week. After drying, anthers are transferred to a glass vial containing a known quantity of 20% NaCl solution, sonicated for 1 min to facilitate the release of pollen from the anthers.

4. Indirect ways of estimation of pollen numbers are also available. In orchids the number of massulae in pollinia and the number of tetrads per massula can be counted. The product estimates the number of pollen per pollinia. Also the length and weight of some male flower part can be correlated with the pollen production. Weight of anthers at different stages during anthesis can be compared to weight of unopened mature anthers. Pollen are then counted by one of the above procedures.

Assumptions and restrictions

The various methods available reflect the fact that methods have to be adapted to each species investigated. Variations between flowers with different positions on the plant and variations between anthers in different whorls should be accounted for. If counting is performed on slides with samples of suspended pollen grains, the distribution of the pollen grains is important for the precision of the results. By an analysis with Electronic particle counter or Culter counter the separation of anther wall material from the grains is important for exact counts of pollen.

Test system

Field experiment, laboratory experiment.

Advantages

These methods are efficient in determining the number of pollen grains in an anther or in a flower.

Application

These methods are used to estimate the production of pollen (251, 294, 552, 622, 651, 733, 875, 963, 984), in anthers within the same flower (182), at various positions in the inflorescence (134), and in different flower morphs (774). **Pollen counts** has been used to compare pollen production in different species or ecological races (738) and the relationship between pollen production and environmental factors such as nitrogen and phosphorous (483, 484). The method has been utilized to record release of pollen (244, 616), pollen removal rate (84) and as a measure of Pollen/Ovule-ratios (44, 180).

Evaluation

Sensitivity: 4	Requirements: 4
Time: 5	Cost: 4

References

44, 84, 134, 180, 182, 244, 251, 294, 483, 484, 552, 616, 622, 651, 733, 738, 774, 776, 803, 875, 963, 984.

Author

MP

M51. Pollen deposition on stigmas: Pollen numbers and fertility

Category

Pollen development and production, pollination, pollination syndrome, breeding system, hybridization.

Subcategory

Pollen germination, pollen production, pollen dispersal, pollinator activity, bird-pollination, insect-pollination, mammal-pollination, water-pollination, cross-pollination, incompatibility, phenology, self-pollination, hybrid.

Description

Counts of pollen grains are used to study **pollen deposition on stigmas** at different times during anthesis, look for relationships between flower morphology and number of pollen deposited and to determine whether pollen competition occur. Counts of pollen on stigmas are useful to determine the ratio between number of germinated and total number of pollen grains on stigmas after experimental pollination and to compare it with ratios after natural pollination. This comparison may infer on what processes are occuring in natural populations (e.g., self-fertilization and hybridization).

Exposed stigmas are mounted on slides in glycerin-gelatine where fuchsine can be added. The number of pollen adhering to the stigma is then counted under a microscope. Stigmas are cut off and stained in Alexander's stain which allows the identification of the morphologically normal pollen (red) from aborted pollen (green). In some species with large pollen grains and slender stigmas, the number of grains can be counted directly on the stigmas by a magnifier. This makes direct counts of pollen deposition in the field possible.

Assumptions and restrictions

The number of pollen adhering to the stigmas are most likely the pollen grains which have germinated. This method is most applicable to species with long slender stigmas.

Test system

Field experiment, laboratory experiment.

Advantages

Knowledge on **pollen deposition on stigmas** under various conditions is important for the understanding of the male function in the breeding system. Furthermore, the method is useful to determine the relationship between total number of pollen in the population and the number actually arriving at stigmas.

Application

Pollen deposition on stigmas has been used to study the phenology of pollination. The number of pollen needed to fertilize ovules (330), the pollen-pistil interactions (568) and the effect of floral display size on pollen deposition (777) have been investigated by counting pollen grains on stigmas.

Evaluation

Sensitivity: 4   Requirements: 3
Time: 4   Cost: 3

References

43, 134, 330, 568, 777.

Author

MP, FF

M52. Pollen germination tests

Category   Pollen development and production, seed development and production, breeding system, hybridization.

Subcategory   Pollen competition, pollen germination, pollen viability, fertilization, cross-pollination, incompatibility, phenology, self-pollination, hybrid.

Description   Germination of pollen grains determines how fertilization will proceed and influences pollen competition, etc. The germination process can be studied either on a germination medium or by application of pollen on styles. If stigmas have more than one lobe, two different types of pollen grains can be applied and their germination percentage and growth rate can be compared.

Several types of **pollen germination tests** are available:

1. Tests on pollen in a germination medium:

    A) A sample of pollen, suspended in 1 ml 30% sucrose solution, is spread on a glass slide covered with a thin film of 1% agar containing 30% sucrose, 100 ppm H_3BO_4, 100 ppm $Ca(NO_3)_2 \cdot 4H_2O$, 100 ppm $MgSO_4$, and 100 ppm KNO_3. The slide is incubated at 25°C for 24 hrs. in a closed Petri dish. The germination is examined under a microscope.

    B) Pollen can be put directly onto a germination medium (1% w/v agar, 20% w/v sucrose and 20 ppm w/v boric acid in Petri dishes and then incubated at 20°C for 30-40 min.

    C) Pollen can be dispersed on a medium containing (w/v) 0.6% Bacto-Agar, 0.1% H_3BO_3, 0.03% $Ca(NO_3)_2 \cdot 4H_2O$, 0.02% $MgSO_4 \cdot 7H_2O$, 0.01% KNO_3, 0.01% KH_2PO_4, 0.003% vitamin B_1, 0.005% vitamin B_6, and 15% sucrose (monocotyledons) or 40% sucrose (dicotyledons) in Petri dishes and incubated at 25°C under white light for 14 hrs. The plates are stored after conservation with a solution comprising water, glycerine, formaldehyde and glacial acetic acid (72:20:5:3 v/v) at 2°C.

    D) Pollen are placed on a germination medium containing 15 g of sucrose and 2 g of agar in 100 ml of distilled water. The medium is poured in 5 cm Petri dishes and pollen deposited on the medium. The dishes are placed in an oven at 30°C with water for 5 hrs. The pollen are stained with a drop of acetocarmine or malachite green, covered with a slide and observed under a microscope at low magnification (x100).

    E) Pollen are germinated on a medium of 2 g Difco Bacto agar and 15 g refined cane sugar per 100 ml distilled water. The dishes are incubated for 26 hrs. at 22°C in a growth chamber.

Description, contd.   2. Tests on pollen applied to styles:

A) Exposed styles are treated directly with 2M NaOH at 55°C for 8 hrs. and rinsed in water and mounted in 0.1 M aniline blue in tribasic potassium phosphate (pH 11.6), or 8M NaOH and stained with aniline blue, or fixed in 70% ethanol, or 96% ethanol:acetic acid 3:1 before the treatment in NaOH. The pollen tubes can be analyzed under UV-light.

B) Exposed pistils can be sampled in ethanol (100%):lactic acid (2:1) and later prepared for fluorescence microscopy by pressure cooking (15 min, 101.3 kPa) in sodium sulphite (10%) for softening, clearing and stained and then mounted in decolorized aniline blue. Pollen tubes are observed under a fluorescence microscope.

C) Exposed styles can be fixed in 2.5% glutaraldehyde in 0.1 M phosphate buffer, pH 7.2. Pollen tubes are studied after embedding in acrylic resin, slicing (2-3 µm) and staining in toluidine blue O (0.05% in 0.5 M acetate buffer, pH 4.5) or photine HV (0.01% in aqueous solution).

D) Pollen tubes in styles can be studied by staining with naphthol blue-black stain (10% propionic acid, 4% lactic acid, 0.2% naphthol blue-black). A small amount of pollen is placed in an aqueous germination medium containing 20% sucrose, 15% Knox gelatin, and 5% nutrient solution (containing boric acid, 100 ppm; calcium, 300 ppm; potassium, 100 ppm) spread on a microscope slide. The slide is placed in a Petri dish containing moistened tissue paper, and kept in the dark at room temperature for 48 hrs. The slides are stained with Alexander's stain and mounted by melting the gelatine mixture before adding a cover slip.

E) Pollen tubes in styles are studied by fixing the exposed styles in 70% ethanol, slitting styles longitudinally and flattening them. Then they are stained with 1:1 mixture of acidified 0.1% aniline blue and acetocarmine (40% acetic acid saturated with carmine) for 10 min.

F) Pollinated flowers are fixed in Carnoy's solution and afterwards the pistils are treated with 0.8 N NaOH at 60°C, washed in deionized water, squashed on a microscope slide in a drop of 0.1% aniline blue in 0.1 N K_3PO_4 and the pollen tubes are observed under a microscope using UV-light.

G) Exposed stigmas can also, after fixation in Carnoy's solution, be stained in a mixture of 2 ml 1% aqueous acid fuchsine, 2 ml 1% aqueous light green, 40 ml lactic acid and 46 ml distilled water and pollen tubes observed.

Assumptions and restrictions	The condition under which a **pollen germination test** is conducted is assumed to be compared to natural conditions. Hence care should be taken that the results are not badly influenced. Pollen germination on intact styles can be difficult to score on very broad styles.
Test system	Field experiment, laboratory experiment.
Advantages	Counts of pollen tubes in stigmas come close to scoring the real number of germinated pollen grains. Also pollen tube growth rate can be studied with these methods.
Application	Observation on pollen tubes in styles can be used to compare what happens after different types of experimental pollination (725) such as self- and cross-pollination (826, 827, 954), pollination with different pollen donors (182), and hybridization (132). Pollen germination tests have been used to evaluate the percentage of viable pollen grains at different times during anthesis (413, 990). Furthermore, to study the effect of pollen size on germination ability (182) and as to whether highly heterozygous pollen have a higher percentage of germination (323). Also, the phenology of pollen tube growth (348) and the relationship between environmental factors such as salinity, drought and UV-B-radiation and pollen germinability have been studied by these methods (434, 476, 608).
Evaluation	Sensitivity: 4 Requirements: 4 Time: 5 Cost: 4
References	132, 182, 323, 348, 413, 434, 4776, 608, 725, 826, 827, 954, 990.
Author	MP

M53. Pollen traps

Category

Pollen development and production, pollination, pollination syndrome, breeding system.

Subcategory

Pollen production, pollen dispersal, wind-pollination, cross-pollination, phenology.

Description

Airborne pollen can be caught by a number of different traps to measure pollen flow and production. In gravimetric methods, the falling pollen are caught, whereas volumetric methods sample pollen in a volume of air.

Horizontal traps to collect pollen falling to the ground has been used especially for pollen analysis. Glass slides covered with silicon oil can be placed around and among target plants at different heights and angles. The Trauber sampler consists of a cylindrical container with a layer of glycerol as pollen sampling liquid in the bottom. The container is covered by an aerodynamically shaped collar with a central orifice, through which particles enter the trap. Calculation of pollen concentration in air volume is not possible with gravimetric methods.

Traps using the volumetric method collect pollen in a specific amount of air on a receiving medium (e.g., a sticky tape). Burchard **pollen traps** collect pollen in a specific amount of air on sticky tape, which is changed weekly for analysis of hourly and daily pollen concentrations. Swing-shield rotoslide air samplers are another possibility. The slides collecting pollen can rotate at adjustable speeds (Rotorod and Rotoslide sampler). The use of **seed traps** is described in M67 in vol. 1.

Assumptions and restrictions

The physical position of the trap is important and wind movements (i.e., speed and turbulence) should be considered. The methods are restricted to wind-pollinated plants.

All the pollen trap methods depend on particle size and density. The gravimetric methods are cheap and easy to handle but give only a rough estimation of the amount of pollen. The Rotorod sampler is efficient in collecting pollen but the removal from the rods is not easy. The rotoslide sampler is efficient and easy to handle but has limited capacity, is sensitive to wind and requires electricity. The efficiency of the Trauber sampler is low. It is mainly used for pollen sampling, over an entire pollen-release period at sites where no electricity is available. The Burkard trap is reasonably efficient, easy to handle, but the efficiency depends on the wind speed and electricity is needed for the vacuum pump. Volumetric methods are reliable in still air, but errors from wind and different settling velocities of pollen are difficult to overcome. It may be very difficult to distinguish pollen from different species, e.g., grass species cannot be identified.

Test system	Field experiment.
Advantages	**Pollen traps** are efficient in obtaining information on the three-dimensional dispersal pattern of pollen grains. Additionally, it is possible to separate events in spatial and temporal components.
Application	**Pollen traps** have been applied to measurements of pollen production (244, 1006) and wind dispersal of pollen (698). Pollen traps are also important to help produce prognosis for pollen allergic humans and in studies of palynology (pollen analysis), see p. 58.
Evaluation	Sensitivity: 3 Requirements: 4 Time: 5 Cost: 4
References	183, 241, 244, 534, 698, 703, 740, 1006.
Author	MP, YJ

M54. Pollen viability tests

Category   Pollen development and production, seed development and production, pollination, breeding system, hybridization.

Subcategory   Pollen germination, pollen viability, agamospermy, pollen dispersal, phenology, hybrid.

Description   **Pollen viability** influences the success of pollination. A direct way to investigate pollen viability is by **experimental pollination** (M13) of stigmas of known age with pollen from anthers dehisced at different dates or with pollen of known age. Development of seeds or fruits is used as a measure of viability of pollen grains and consequently of successful pollination.

Pollen viability can also be studied in vitro following different procedures:

A) All pollen grains from anthers are spread out and stained on a microscope slide and the stained and unstained pollen grains are counted to a predetermined total number. The coloured grains are the morphological normal pollen grains. The viability is determined as the percentage coloured to non-coloured grains. Stains: Lactophenol-aniline blue, lactophenol blue (cotton blue), Alexander's stain which gives red colour to morphological normal pollen grains and green to aborted pollen grains.

B) Anthers are dissected out and placed into vials containing 0.3 ml 0.7 M mannitol solution coloured with a little acetocarmine to stain the cytoplasma. Dead pollen grains appear transparent in contrast to the densely stained viable grains.

C) Anthers are squashed in a drop of fluorescein diacetate in 10% sucrose and observed with a UV-fluorescence microscope.

D) Pollen can be treated with a drop of freshly-prepared medium containing tetrasolium salt 3-(4,5-dimethylithyazolyl)-2,5-diphenyl monotetrazolium bromide at a concentration of 0.9% w/v in sucrose (54% w/v aq) for 10-15 min at room temperature. Viable pollen grains are red.

Assumptions and   **Pollen viability** is highly influenced by environmental events. Low
restrictions   and high temperatures in critical stages during meiosis can change viability. **Pollen viability** can also depend on the position of flowers on individual plants. *In vitro* methods are generally not so well suited as *in vivo* methods.

Test system   Field experiment, greenhouse experiment, laboratory experiment.

Advantages   These methods are useful when evaluating pollination and breeding system.

Application	**Pollen viability tests** have been used in studies of floral biology (650, 662, 956) and breeding systems (44, 84, 134, 636, 785). Studies on **pollen viability** record the production of morphological normal pollen (428, 471, 776) and the possibility of long-distance pollen dispersal (134). The methods have revealed how environmental factors such as salinity (434), spaceflight (471), temperature and photoperiod (1037) and storing (555) influence the viability of pollen grains.
Evaluation	Sensitivity: 4 Requirements: 4 Time: 5 Cost: 4
References	44, 84, 134, 428, 434, 471, 555, 636, 637, 650, 662, 776, 785, 956, 1037.
Author	MP

M55. Pollinator attraction: Chemical cues

Category

Pollination, pollination syndrome, breeding system.

Subcategory

Chemical attractant, nectar production, pollinator activity, pollinator foraging behaviour, insect-pollination, mammal-pollination, phenology, reproductive allocation.

Description

Many plants have flowers which emit scents to attract pollinators. This is one of the main **chemical cues** which secures pollination together with nectar production, see **nectar production: Chemical composition** (M40). However, flower colour and shape are also essential for securing pollinator visits, see **pollinator attraction: Visual cues** (M56). The way a visitor approaches a flower may tell whether visual or chemical cues are the most important, see **pollinator foraging behaviour** (M57). Visual stimuli tend to produce a more straight flight towards the flower than when the stimulus is mainly olfactory and the approach may follow a zigzag route.

Experimental designs for study of pollinator attraction using flower scents (183) include:

A) The flower or the plant is hidden in a non-transparent cover from which scent is emitted through suitable holes. The insect reaction is observed and compared to the reaction when the flowers is placed in a clear cover which prevents scent transmission.

B) A simple glass-tube test reveals whether a given insect is attracted by sight or by odour (242). The blossom is enclosed in an open glass tube from the ends of which the odour emerges. Insects generally fly towards the flower in the tube indicating that they react on visual attraction.

C) Application of one or several scent chemical on flower models or pieces of paper or cloth are followed by observation of attraction and behaviour of the visitors. Preference experiments can also be performed in cages under controlled experimental design.

D) Glass olfactometers (two connected V-shaped transparent tubes, analyzing the preference of the insects between two odour sources) are used to study the response of small insects to flower scents in different composition and quantity.

The chemical composition of flower scents can be analyzed using **gas chromatography** (M20). Odour glands (osmophores) can be localized by staining with neutral red. Experimental methods that may be used together with **pollinator attraction: Chemical cues** (M55) include: **Bagging of flowers** (M2), **experimental pollination** (M13), **stigma receptivity tests** (M73).

Assumptions and restrictions	Flower scents may be a mixture of over one hundred substances. Detecting the substances which are the key attractor to pollinators are difficult and time-consuming. Tests with hidden flowers or perfumed models give only a general idea about the scent's relative importance to the pollinator reaction. The scented source has to be replenished frequently. Pollinator behaviour in cage experiments may be different than under natural conditions.
Test system	Field experiment, greenhouse experiment, laboratory experiment.
Advantages	Animal behaviour may be tested under both natural and semi-natural conditions. Perfumed flower models or chemical scent sources may attract unknown visitors under natural conditions. Many different chemicals may be tested simultaneously. Using preference experiments in cages, the animals can be observed individually under controlled conditions.
Application	Capillary gas chromatographic analysis of the composition of volatile oils produced by *Euphorbia fragifera* in two different habitats has been used to show chemical polymorphism induced by the environment with two variants of the species (517). Chemosensory attraction of fig wasps to substances produced by receptive fig-trees has been studied (377). In the mutualism between figs (*Ficus carica*) and their species-specific fig wasp pollinators, evidence for long-distance olfactory attraction by volatile substances (pentane) was found. The perfume syndrome and pollination by fragrance-collecting euglossine bees in the neotropic solanaceous genus *Cyphomandra* has been confirmed by field observations and scent analysis (806).

Evaluation

Sensitivity: 4	Requirement: 4
Time: 3	Cost: 3

References   183, 242, 342, 360, 377, 481, 517, 806, 962.

Author   PR

M56. Pollinator attraction: Visual cues

Category	Pollination, pollination syndrome, breeding system.
Subcategory	Pollinator activity, pollinator foraging behaviour, visual attractant, bird-pollination, insect-pollination, mammal-pollination, phenology, reproductive allocation.
Description	Flower colour, pattern and shape are the main visual cues which attract pollinators to plants. Quantification of colour spectrum of flowers is needed in order to evaluate effects of selection on plant lines and to sort out any discriminatory floral behaviour of pollinators, see **pollinator foraging behaviour** (M57). In order to make colour (incl. ultraviolet) photographs of flowers special techniques for spectral separation, using different types of filters, are available (558). A thorough understanding of visual attraction to a flower requires an analysis of the entire flower colour reflectance spectrum. A spectrophotometer with an integrating sphere can be used to generate a spectral reflectance curve (657). A photographic analysis using a set of monochromatic filters is a very common procedure in many pollination studies (118, 431, 432).
Assumptions and restrictions	Reflectance data need to be quantified, e.g., by comparing reflectance with a calibrated grey-scale. Background coloration and ambient lighting may also influence the colour perception of flower-visitors, but very little is known about this.
Test system	Field experiment, greenhouse experiment.
Advantages	Quantification of **visual cues**, including photographs and analysis of the ultraviolet part of the reflectance spectrum of a flower, gives valuable information to any pollination study.
Application	Quantitative analysis of colour spectrum of flowers is routine in studies of floral biology and pollination (e.g., 408). Surveys of colour composition of entire floras have also been made (e.g., 782, 950).
Evaluation	Sensitivity: 4 Requirements: 4 Time: 3 Cost: 2
References	118, 408, 431, 432, 520, 558, 657, 782, 950.
Author	JMO

M57. Pollinator foraging behaviour

Category

Pollination, pollination syndrome, breeding system.

Subcategory

Nectar production, pollen dispersal, pollinator activity, pollinator foraging behaviour, bird-pollination, insect-pollination, mammal-pollination, cross-pollination, self-pollination.

Description

Pollinator foraging behaviour determines the pattern of pollen flow. Analysis of pollinator foraging may include, e.g., visitation rate (i.e., number of visitors or visits per time unit) and movement pattern. **Foraging behaviour** may be observed directly in the field or in experimental gardens or flight cages. Video-recording is also a possibility (183). The movement pattern can be analyzed from the interfloral flight distance and the interfloral change in flight direction. The angular change in direction may be given in degrees, i.e., straight out is 0°, flying back to the previously visited flower is a change of 180° (966). This procedure may be done directly in the field or from a map of flower position with insect routes indicated (659). Observations of pollinator visits often require that insects are marked. Methods to mark flower-visiting insects make use of: Trace elements, paint, numbered tags (e.o. Opalith-Plättchen) and fluorescent dusts (426). Tags may be bought in most apicultural stores.

Assumptions and restrictions

Many flower-visitors perform several types of tasks on a single flight, e.g., food harvesting, mate-searching and territory defence, thus masking the true pollen flow pattern. Studies of **pollinator foraging behaviour** alone is not exact enough for estimating gene-flow but, used together with studies of **pollen carry-over** (M49), it can produce good estimates of pollen flow.

Test system

Field experiment, greenhouse experiment.

Advantages

Only studies which include data and analysis of pollinator movement pattern are able to give true estimates of pollen flow in plant populations. Studies of **foraging behaviour** may be used to estimate the relative importance of geitonogamy and xenogamy.

Application

A vast amount of literature on **pollinator foraging behaviour** has been published, mostly concerning bees and hummingbirds. Many studies have analyzed bee movements in a patch of flowering plants (299, 415, 726, 727, 729, 965, 966). A few have compared foraging of nectar and pollen harvesting bees (e.g., 277) and the movements within vertical inflorescences (e.g., 85). Some classic studies have been made on foraging of social bees (351, 353). Studies of foraging by non-Hymenopteran insects, e.g., flies, are very rare (657, 658).

Evaluation

Sensitivity: 3   Requirements: 2
Time: 5   Cost: 2

References

85, 183, 277, 299, 351, 353, 388, 415, 426, 653, 657, 726, 727, 729, 966.

Author

JMO

M58. Pollinator preference experiments

Category

Pollination, pollination syndrome, breeding system.

Subcategory

Pollinator foraging behaviour, bird-pollination, insect-pollination, mammal-pollination, heterostyly, phenology.

Description

Many plant species have dimorphic or polymorphic flowers, or show variation in their breeding system (unisexuality, heterostyly), flower morphology, colour and scent (276). Some of these traits can be analyzed by **sexual morph distribution** (M70).

Studies of **pollinator preferences** may be performed in the field or on artificial flower arrays in cages. A thorough description of how to build various artificial flowers and design choice experiments has been given (426). Through various types of manipulation, it is possible to discover which types of cues are important in visitor discrimination, e.g., colour, scent and morphology. Inflorescences may be covered with a plastic cylinder (630). If insects sought the opening of the cylinder, their cue was scent and if they went straight towards the flowers, ignoring the plastic wall, their cue was colour or maybe shape. Pollen-collecting visitors may also discriminate between male and female flowers and, in heterostylous species, between short- and long-styled flowers.

Assumptions and restrictions

In nature, it is often difficult to sort out the various signals affecting flower visitors. Thus a combined study of natural observations and controlled experiments is recommended. Experiments are very time-consuming and captivity-reared, inexperienced pollinators are often needed.

Test system

Field experiment, greenhouse experiment.

Advantages

Preference experiments performed during risk assessment may give valuable insight into the expected outcome of a later release of the transgenic plant.

Application

Preference experiments and studies of flower polymorphism have a long tradition. Flower colour polymorphism has been reviewed (425). How variation in colour affects gene-flow has been studied (112) and choice experiments with bombylid flies have been done (927). Many studies of pollinator colour preferences have been performed (e.g., 630, 677, 882, 981).

Evaluation

Sensitivity: 4   Requirements: 3
Time: 4   Cost: 3

References

112, 276, 425, 426, 589, 630, 654, 677, 882, 927, 981.

Author

JMO

M59. Polymerase chain reaction (PCR)

Category	Gene-flow, hybridization, genome structure, population.
Subcategory	Marker, hybrid, genotype, polymorphism.

Description

The **PCR** method is based on *in vitro* amplification of previously chosen DNA sequences by simultaneous primer extension of the complementary strands of DNA. Two specifically designed primers are needed, each matching the 5' end of the complementary strands of the target sequence. The primer is a short oligonucleotide (often 20-22-mer). In the *in vitro* assay the primer is added to the plant DNA and the primer acts as a template for the amplification of the target sequence. Amplification is carried out by a thermostable DNA polymerase in the presence of the four DNA nucleotides. After termination of the reaction, the amplified fragments can be visualized by DNA staining following a gel-electrophoresis (237, 565).

Information about the transfer of a target sequence from a donor to a recipient can be improved by the inverse **PCR** technique that allows amplification and, thereby, subsequent identification of unknown DNA sequences flanking a core region of known sequence (179, 237).

Assumptions and restrictions

Generally heterozygous individuals cannot be detected as **PCR** markers are codominantly inherited. Information about the sequence of the target gene is necessary to be able to design the two specific primers and to know the size (in bp) of the amplified fragment to be identified after electrophoresis.

Test system

Laboratory experiment.

Advantages

A reproducible and relatively fast method in comparison to **DNA sequencing** (M10). The method reveals coding as well as non-coding regions. Only small amounts of DNA or RNA are needed for analysis.

Application

The method may be used for detection of known DNA insertions, e.g., transgenes and introgressed genes (167). **PCR** directly on chromosome preparations is called **PRimed IN Situ hybridization (PRINS)** and is a technically demanding method which can reveal the physical integration site of an insert (936). A **PCR** is part of many other methods providing genetic information, e.g., of **microsatellite markers** (M38) and **AFLP** (M1). Expression of genes can be analyzed by **PCR** with reverse transcriptase and RNA as the template, i.e., RT-PCR (871).

Evaluation

Sensitivity: 5   Requirements: 4
Time: 4   Cost: 4

References

167, 179, 237, 565, 797, 871, 936.

Author

RBJ

M60. Protein electrophoresis: Isozyme analysis

Category

Gene-flow, hybridization, genome structure, population.

Subcategory

Introgression, marker, vertical gene-transfer, hybrid, gene expression, gene stability, genotype, polymorphism.

Description

Isozyme analysis (see also M34 in vol. I) utilizes the electric charge and the catalytic ability of a specific group of enzymes, i.e., **isozymes** having identical catalytic ability but different electric charge. **Isozyme analysis** consists of a non-denaturant extraction of proteins, followed by an electrophoretic separation according to the molecular weight and electric charge on, in general, starch or polyacrylamide gels. Visualization is carried out by the immersion of the gel in the substrate of an enzyme, followed by the staining procedure for the product of the reaction (535). Phenotypes are revealed as bands of a zymogram. If the bands are products of the same locus, they are not only isozymes but "allozymes". Crossing experiments as, e.g., **diallel cross** (M9) may confirm the inheritance of the allozymes. The genetic status of bands as isozymes or as allozymes may also be inferred from knowledge of closely related species or from literature (e.g., 869). Not only isozymes, but also proteins in general, may be visualized by applying protein-specific stains.

Assumptions and restrictions

Nuclear inheritance is assumed. Null alleles, corresponding to alleles either with activity, not revealed under the electrophoretic conditions used, or without activity may complicate genetic interpretation. Only part of the genetic variation may be detected with this technique. Expression of some enzyme systems may vary according to developmental stage or environmental conditions.

Test system

Laboratory experiment.

Advantage

Isozymes analysis is a relatively cheap and quick method for analyzing natural populations. A number of enzyme systems may be tested, allowing a conservative estimate of genetic diversity. The method is valuable when comparing several genetical units. Codominant expression facilitates genetic interpretation of genotypes.

Application

Isozyme analysis has been used in a wide area of research. Numerous descriptions and comparisons of the genetic diversity of species have been produced (e.g., 310, 338, 851). Outcrossing rates (1, 303), gene-flow (79, 823) and neighbourhood size (98) have been estimated. Moreover, hybridization and introgression have been evaluated in natural hybrid zones (36). In breeding experiments, the recognition of introgression is facilitated with isozymes (602).

Evaluation

Sensitivity: 4   Requirements: 4
Time: 3   Cost: 3

References

1, 36, 79, 98, 303, 310, 338, 340, 535, 602, 686, 823, 851, 868, 869, 993.

Author

FF

M61. Random amplified polymorphic DNA (RAPD)

Category

Gene-flow, hybridization, genome structure, population.

Subcategory

Introgression, marker, vertical gene-transfer, hybrid, genotype, polymorphism.

Description

The **RAPD** technique can be used for studies of hybridization and introgression. Furthermore, it is suited for studying the genetic structure of plant populations, e.g., for **genet identification** (M25). The method is based on an *in vitro* amplification of randomly selected DNA sequences (735, 1001). Amplification of the sequences takes place by simultaneous primer extension of complementary strands of DNA. The primer is a short oligonucleotide (often a 10-mer) with a random sequence. In the *in vitro* assay, the primer is added to the plant DNA and the primer acts as a template for the amplification of DNA fragments with which it has homology. Amplification is carried out by a thermostable DNA polymerase in the presence of the four DNA-nucleotides. After termination of the reaction, the amplified fragments are separated by size in a gel-electrophoresis and detected by staining. Specific computer programs for analysis of **RAPD** data exist, e.g., **RAPD 103** (M8).

Assumptions and restrictions

Heterozygous individuals cannot be detected, as **RAPD** markers are dominantly inherited. It is assumed that fragments amplified by the same primer are homologous if they are of identical size. Especially when comparing taxa from different species this may not always be correct (734). Homology of fragments can be tested by **Southern hybridization** (M71), digestion by restriction enzymes or by DNA sequencing. Reproducibility demands high laboratory standard.

Test system

Laboratory experiment.

Advantages

Provided that isolation of good quality DNA from the plant is easy, the method is fast and cheap. The DNA fragments amplified represent coding as well as non-coding regions. The method usually reveals a high extent of DNA polymorphism. The plant DNA used as template can be a bulk from several individuals which gives a more efficient way of searching for specific markers (153, 906, 1035).

Application

The **RAPD** technique has been used in numerous studies on different biological aspects. Evolutionary relationships (922), genetic variability within and between populations and species (137), hybridization and introgression (175, 579) and gene mapping (149, 784) (see also **genetic map construction** M23) are some of the topics that have been elucidated by **RAPD** analysis.

Evaluation

Sensitivity: 4	Requirements: 3
Time: 4	Cost: 3

References

137, 175, 734, 735, 922, 1001.

Author

RBJ

M62. Reproduction in alien plants: Community and invader analysis

Category	Pollination, breeding system.
Subcategory	Chemical attractant, pollinator activity, pollinator foraging behaviour, visual attractant, cross-pollination, phenology, reproductive allocation.
Description	Reproductive success of an invading plant species may be affected by the resident plant and pollinator communities and by existing plant-pollinator interactions. The properties of the plant-pollinator interactions may be evaluated from the interaction matrix, e.g., the number and variance of interactions per resident plant or pollinator species, the strength of the interactions and the degree of asymmetry in dependence values between plant and pollinator.

A plot should be selected in an expected receiver habitat. A list of the two interacting communities should be made, i.e., plants and their flower visiting insects or, if preferable, flower foraging insects and their host plants. Each community is the resource utilization axis of the other. A number of characteristics of the receiver habitats and their plant and insect communities should be measured and treated independently, including plant and insect diversity and abundance interactions for invader versus resident plants and flowering phenology (655, 656). At the end of the season, seed set of the invader species in the tested habitats is scored and later related to the characteristics of the various habitats. |
Assumptions and restrictions	In Olesen et al. (655, 656), it is assumed that visitation reflects importance and thus fitness, which is only true to some extent. A community approach to invasibility is very time-consuming.
Test system	Field experiment.
Advantages	Studies, combining analysis of breeding system of invaders with a community analysis of the receiver habitats, are expected to give good estimates of invasibility. If such studies are combined with experiments testing recruitment and herbivory, a forecast of the outcome of release of exotic species into natural environments may be possible.
Application	A community level approach to invasibility has rarely been done (655, 656). In these studies, the variation in reproduction is explained by the variation in habitat parameters and invasibility was greatest in the community with the highest number of plant species.
Evaluation	Sensitivity: 2 Requirements: 2 Time: 4 Cost: 4
References	59, 71, 227, 265, 655, 656, 669, 754.
Author	JMO

M63. Reproductive allocation measures

Category

Pollen development and production, seed development and production, pollination, breeding system.

Subcategory

Pollen production, fruit abortion, seed abortion, seed production, chemical attractant, nectar production, visual attractant, reproductive allocation.

Description

Allocation measures of resources to different flower parts under different conditions (biotic as well as environmental) are used when the reproductive functioning of flowers is studied. Flowers in one or several different phenological stages are harvested and dissected into parts: pedicel, calyx, corolla, pistil, stamens. These parts are then dried and weighed according to standard methods (e.g., at 80°C for 24 hrs.). See also **dry weight allocation** (M13, vol. I).

Assumptions and restrictions

Decisions on delimitation of which flower parts belong to female and male function may be difficult. Allocation of resources to different flower parts can be related to flower function but, in addition, knowledge on flower morphology and measurements of the function of the sexual organs are necessary.

Test system

Field experiment, greenhouse experiment, laboratory experiment.

Advantages

Reproductive allocation measures of flower parts are very useful for understanding the function of the sexual morphs.

Application

Allocation of resources has been studied when comparing breeding systems in different species or ecological races (738) or when comparing biomass of flowers in selfing and outcrossing species (684). Furthermore, the method has been used for investigating variation in gender within a population (127, 216), comparing masses of reproductive structures between sexual morphs (2, 31, 229) and studying seasonal variation in biomass investment in floral structures (42, 193).

Evaluation

Sensitivity: 4   Requirements: 4
Time: 5   Cost: 4

References

2, 31, 42, 127, 193, 216, 229, 684, 738.

Author

MP

M64. Restriction fragment length polymorphism (RFLP)

Category	Gene-flow, hybridization, genome structure, population.
Subcategory	Introgression, marker, vertical gene-transfer, hybrid, genotype, polymorphism.

Description

The **RFLP** is a valuable technique for studying population structure and hybridization. Extracted plant DNA is digested by restriction enzymes which recognise and cut the DNA at specific sequences. Differences between individuals in the length of restriction fragments result from differences in their DNA sequence. The restriction fragments are separated by electrophoresis according to their size. After electrophoresis the separated fragments are transferred to a filter and a specific DNA- or RNA-probe with a label is added to the filter. The labelled probe will hybridize with those restriction fragments that hold sequence homology to the probe. DNA polymorphism between individuals is revealed as a different position on the filter of the hybridization signal. The probe can be a cDNA-clone, a genomic clone or a synthetic oligonucleotide. The probe is usually labelled with radioactive tracers. As various restriction enzymes cut at different sequences, different combinations of enzyme and probe will result in different polymorphisms (see also M62 in vol. 1). **Restriction fragment length polymorphism** is a characteristic of the **AFLP** technique (M1).

Assumptions and restrictions

A relatively large amount of DNA is needed per individual (approx. 15 µg). Probes which have repetitive sequences are not suitable, as they will hybridize to a high number of fragments and produce a smear on the filter. The method is relatively laborious and will only allow analysis of a limited number of individuals per day, but efficiency can be improved by bulking DNA from several individuals (199, 641, 972). Introgression is best revealed by co-segregation of markers known to be linked in the potential donor.

Test system

Laboratory experiment.

Advantages

The DNA fragments represent coding as well as non-coding regions. The method usually reveals a high extent of DNA polymorphism.

Application

The **RFLP** technique has been used for studies of variation within and between populations (74, 436), phylogenetics (381, 613, 922) and gene mapping (293, 489). It can also be used for detecting inserts in GMP by **Southern hybridization** (M71).

Evaluation

Sensitivity: 4	Requirements: 4
Time: 3	Cost: 3

References

74, 199, 293, 381, 436, 455, 489, 613, 641, 753, 870, 907, 922, 972, 975.

Author

RBJ

M65. Ribosomal DNA (rDNA) analysis

Category   Gene-flow, hybridization, genome structure, population.

Process   Introgression, vertical gene-transfer, hybrid, genotype, genetic diversity, metapopulation.

Description   **Ribosomal DNA (rDNA)** is used as a molecular marker, mainly in species where there is no or a limited prior knowledge of the genome. This is because rDNA is very well conserved between taxonomically widely separated species and, therefore, information from one species can be applied to a new species of interest. The rDNA genes are present in multiple copies in both the nuclear and organelle genomes of the plant species. The genes are arranged in tightly linked clusters, separated by short non-coding sequences called internal transcribed spacers (ITS). Two types of variation usually exist between individuals within the gene cluster: 1) Sequence variation within the ITS (especially between species) and 2) Variation in the number of repeated rDNA elements (both within and between species).

The sequence variation of ITS can be determined from **PCR** (M59) followed by cycle-sequencing or traditional **DNA sequencing** (M10). The primers for **PCR** are placed within the highly-conserved rDNA genes on either side of the ITS region, which is usually 300-400 base pairs wide. Universal primer sets exist which can be used in any plant species.

The number of repeat rDNA genes is usually determined by **southern hybridization** (M71) with a radiolabelled probe. From the restriction patterns, the number of repeated elements can be inferred.

Assumptions and   While the coding region of the repeat is extremely conservative, the
restrictions   intergeneric spacer varies among the closely related individuals and is useful for evolutionary study at lower taxonomic level. Care should be taken that the plant material is not infected by fungi, since the **PCR** markers will also amplify the fungi DNA, when using the first method. The second method is expensive and only few individuals can be handled per day. The methods should be preferentially used if none of the alternative molecular markers (e.g., **microsatellite markers** (M38), **isozymes** (M60), **RFLP** (M64), **RAPD** (M61), **AFLP** (M1)) can be used.

Test system   Laboratory experiment.

Advantage	The main advantage of these methods is that they can be used as molecular markers without any prior genome knowledge. Furthermore, it may be determined from which of the parental species, the rDNA genes of a putative hybrid have descended. An advantage of this method over **isozyme** analysis is that, working directly on DNA, it is not sensitive to differences in plant tissue, plant age, physiological state, etc. as **isozyme** analysis is, working with gene product.
Application	Sequence variation in the ITS-region have been used primarily for taxonomical studies. The number of rDNA repeats have been used for traditional gene-flow studies. As rDNA gene clusters are usually found on several chromosome pairs, genomic *in situ* **hybridization** (M31) of rDNA probes may be used for karyotyping (689, 690). It is also used for the detection of interspecific hybrids and amphiploids (601) and also to study gene introgression.
Evaluation	Sensitivity: 4 Requirements: 3 Time: 3 Cost: 3
References	103, 162, 255, 329, 781, 808, 903, 996.
Author	YJ, MHS, RBJ

M66. Seed development analysis

Category
Seed development and production, breeding system

Subcategory
Agamospermy, fertilization, fruit abortion, ovule development, seed abortion, seed germination, seed production, cross-pollination, incompatibility, phenology, reproductive allocation, self-pollination.

Description
Fruit and **seed development** are influenced by the position of the flower on the plant and by the time at which the anthesis occurs. The development is followed by marking buds in an early stage and harvesting fruits and seeds at intervals during the period of development. Flowers are collected at intervals of a few days from the date of meiosis. The development of ovules is observed to reveal the type of embryo sac, the nature and extent of apomixis, to determine the time of abortion during the maturation of ovules, or to study competition between developing ovules. The status of the developing seed can be recorded after sectioning of the seed or after direct visual inspection. See also **ovule counts** (M44) and **classification of fruits, seeds and ovules** (M6).

Several methods for fixation and staining of ovules are available:

A) Ovules are fixed in FAA (18:1:1 (or 10:2:2), 700 g L^{-1} ethanol:glacial acetic acid: 370 g L^{-1} formaldehyde). The material is then dehydrated in a tertiary butyl alcohol series, embedded in paraffin and sectioned at 10-15µm and stained with safranine 0-fast green or safranine/aniline blue.

B) Ovules can be fixed in 3% glutaraldehyde in 0.025 M phosphate or cacodylate buffer (pH 6.8) for 2 hrs. at 22°C followed by 24 hrs. at 4°C in the same fixative. Postfixation in 2% osmium tetroxide in 0.025 M phosphate or cacodylate buffer (pH 6.8) for 4 hrs. Dehydration followed by gradual change to propylene oxide before infiltration with a mixture of propylene oxide and Spurr's low viscosity resin followed by daily changes of 100% fresh Spurr resin for 7 days. The material is then polymerized at 70°C under partial vacuum for 16 hrs. The sections are stained in 2% ethanolic crystal violet in 0.05 M ammonium oxalate buffer (pH 6.7) at 85°C for 0.5-1.0 min.

C) The dissected embryo sacs may be stained with DNA-specific fluorochrome DAPI solution for 10 min. and embedded in glycerine-gelatine.

D) Cleared pistil technique: Embryo sacs are fixed in FAA (70% ethanol/formalin/acetic acid 18:1:1), and cleared in an ethanol/methyl salicylate series and viewed in contrast microscopy.

Description, contd. E) The mature seeds are fixed in 3% glutaraldehyde buffered with 0.1 M sodium phosphate buffer (pH 6.8), or fixed in 2.5% glutaraldehyde + 2% paraformaldehyde in 0.025M phosphate buffer at pH7 (16-24 h), dehydrated and embedded in glycol methacrylate. After sectioning (2 μm), the material is stained with periodic acid-Schiff's reagent and with toluidine blue buffered with benzoate buffer (pH 4.4).

F) Flowers are fixed in FPA50 (formalin, propionic acid, 50% ethanol, 5:5:90 by volume) and transferred to 70% ethanol for storage at 4°C. Ovules are then cleared in Herr's solution and observed with Nomarski optics or with confocal scanning laser microscope.

Direct inspection can, in special cases, be carried out on almost fully developed fruits. In the fruit wall of one locule a slice can be cut with a sharp razor blade and the development of the ovules may be observed directly. This operation should not affect the development of the ovules compared to controls.

Assumptions and restrictions Abortion of developing ovules occurs at different stages, which makes it important to score the embryo viability and quality as late as possible. Inspect at intervals to see when abortions take place. The origin of embryo sacs can be difficult to determine. Sometimes the origin has to be deduced from the position within the ovule.

Test system Field experiment, laboratory experiment.

Advantages Knowledge of the development and fate of seeds and developmental failure is important in order to evaluate the functioning of the breeding system and effects of different treatments such as watering and **experimental pollination** (M13).

Confocal scanning laser microscopy on unstained cleared ovules, in some cases, gives a more accurate picture compared to paraffin serial sectioning technique.

Application These methods have been used in studying the phenology of fruit and seed development (612), to determine the embryo sac development and causes of abortion (267, 330, 615, 766, 837) and to investigate the ratio between sexual and aposporous embryo sacs (120, 428, 704, 731, 837, 895).

Evaluation Sensitivity: 3   Requirements: 4
Time: 3   Cost: 4

References 120, 267, 330, 428, 612, 615, 628, 704, 731, 766, 837, 895.

Author MP

M67. Seed germination tests

Category

Pollen development and production, seed development and production, breeding system, hybridization.

Subcategory

Pollen competition, fruit abortion, fertilization, ovule development, seed abortion, seed germination, seed production, cross-pollination, phenology, reproductive allocation, self-pollination, hybrid.

Description

Viability of seeds can most conclusively be evaluated by **germination tests**. Germination on wet filter paper on Petri dishes (or saturated with a solution of a fungicide) or in pots with soil effectively distinguish viable from non-viable seeds, if the right conditions for germination are provided. Different treatments may be necessary to induce seed germination: Stratification with different periods of cold treatment, scarification with sand paper, puncturing of seed coat with a needle or dipping seeds in sulphuric acid. The seeds can also be soaked in 10^{-3} M gibberellic acid for 3 days at 20°C or, after hydrating seeds overnight, seed coats can be removed and the embryo be placed on moist filterpaper. **Seed germination rate** has been increased for some species by storing seeds in damped coarse silica sand for one month at 5°C. Also treatment against attack by fungi can be necessary, e.g., 9% Clorox. Varying temperatures and length of daylight, in many cases, increases the germination success. General requirements for germination of many NW European plant species are available (321). See also M19 in vol. 1.

Assumptions and restrictions

Seed germination is a complicated process, with a large number of influential factors. It is difficult to obtain optimal conditions for germination, so background information is important. Germination trials in the field are often most appropriate but will, in many cases, be inconvenient. Seed dormancy and burial may cause seeds to germinate very late, one to several years after start of trials. A low germination rate can also be the result of inbreeding.

Test system

Field experiment, greenhouse experiment, laboratory experiment.

Advantages

Seed germination tests are important to determine the fitness of seeds or of the success of a certain pollination event.

Application

Seed germination has been used to evaluate crossings between populations and species (637, 827). The method is used when outcrossing rates are determined (214), to test seed viability after selfing (105, 593, 684) and to assess germination probabilities for seeds resulting from different pollen load size on stigmas (681).

Evaluation

Sensitivity: 3   Requirements: 2
Time: 2   Cost: 3

References

105, 214, 593, 637, 681, 684, 827.

Author

MP

M68. Seed viability tests

Category

Seed development and production, breeding system, hybridization.

Subcategory

Fruit abortion, ovule development, seed abortion, seed germination, seed production, cross-pollination, phenology, reproductive allocation, self-pollination, hybrid.

Description

Seed viability tests are important when evaluating procedures for **experimental pollination** (M13) and seed handling. Plants nearly always initiate more ovules than those resulting in mature seeds. Due to lack of pollen or to resource limitation a number of ovules are normally aborted. Aborted seeds can be detected by visual inspection. They are normally tiny and shrivelled compared to normal seeds. Individual seed weights correlate with germination ability and are useful in determining the seed quality. The visual inspection can be strengthened by a **seed germination test** (M67). A **seed viability test** includes treatment with 1% tetrazolium chloride solution and incubation at 30°C for 8-12 hrs. Viable seed tissue is identified by a red colour. Protocols for evaluation of tetrazolium tests are available for most crop species through ISTA (International Seed Testing Association). The condition of the seed can further be evaluated by an embryological investigation of sectioned ovaries, see **seed development analysis** (M66) and **ovule counts** (M44).

Assumptions and restrictions

The quality of seeds can be influenced by the position of the seed in the fruit and on the individual plant. Predation of developing seeds and fruits sometimes strongly influence the number and quality of seed production. Using tetrazolium staining on small seeds, the results can be very difficult to evaluate.

Test system

Laboratory experiment.

Advantages

In studying breeding systems, much information can be gained from examination of the viability of ovules at different stages and the viability of mature seeds. The tetrazolium method is technically quite simple to use, but it is not always easy to judge the results.

Application

Counts of aborted ovules and **seed viability tests** have been important in understanding breeding system (884), pollen competition and allocation patterns. The weight of individual seeds may identify aborted seeds and the position at which abortion most probably will take place (522). Measurements of **seed viability** have revealed how the temporal and spatial position of the flower influence the abortion rate of seeds (612, 793) and have been used to estimate the amount of seeds viable at different times after harvesting (133).

Evaluation

Sensitivity: 4   Requirements: 4
Time: 3   Cost: 3

References

133, 348, 522, 612, 793, 884.

Author

MP

M69. Selfing and outcrossing rate

Category   Gene-flow, population.

Subcategory   Outcrossing, genetic diversity, genetic drift.

Description   Estimation of the **outcrossing rate** (and thus the **selfing rate**) is a
                              key element in understanding the mating system. The **selfing and
                              outcrossing rates** are important determinants of the level and dist-
                              ance of gene-flow and the amount of inbreeding depression.

                              Estimation of **selfing rate** and **outcrossing rate** for a single plant or
                              a population, is most accurately done from a progeny array of
                              genotypes at a number of genetic loci. The seeds of a given maternal
                              plant constitute such a progeny array and is a mixture of half sibs
                              (different fathers) and full sibs (same father).

                              Statistical methods have been developed to simultaneously estimate
                              the **outcrossing rate**, t_m, the panmictic rate (or single-locus outcros-
                              sing rate), t_s, and the genotype of the mother plants (109, 772, 836).
                              For each individual, t_m uses the genotypes of all loci simultaneously
                              to estimate the proportion of seeds that have not been self-fertilized
                              and thus, the **selfing rate**, s is equal to $1-t_m$. t_s estimates the deficit
                              of heterozygotes caused by non-random mating within the popula-
                              tion and the quantity t_m-t_s therefore estimates the amount of inbreed-
                              ing of other kinds than selfing, e.g., sib-mating. This quantity is also
                              termed biparental inbreeding (487).

Assumptions and   The method applies only to hermaphroditic species. Depending on
restrictions   the expected **outcrossing rate**, from 1 to 10 codominant polymorphic
                              markers are needed for accurate estimation (835). Any codominant
                              marker such as **RFLP** (M64), **microsatellites** (M38) or **isozymes**
                              (M60) can be used. One important assumption is that the pollen
                              taking part in outcrossing come from independent sources. Values
                              of t_m and t_s can be biased, if pollen arrives in discrete packets (e.g.,
                              pollinia) (819), or if the population is substructured (234). This bias
                              can be estimated from the genotype data (769, 770, 773) and if large,
                              an alternative model should be used (817).

Test system   Laboratory experiment, data analysis.

Advantages   The method yields an estimate of the proportion of seeds set after
                              outcrossing in a given season. Estimates from different seasons can
                              be compared to direct observations of pollen dispersal. The effective-
                              ness of structures promoting outcrossing can be evaluated and the
                              existence of a cryptic self-incompatibility system (i.e., discrimination
                              between self pollen) can be inferred. Furthermore, the method can
                              yield the genotypes of the maternal plants, when these cannot be
                              directly obtained (e.g., isozyme expression and successful extraction
                              can be difficult in old plants). When comparing **outcrossing rates**
                              between fruits of a plant or between populations, the effect of

Advantages, contd.	different factors (e.g., pollinators, fruit position on plant, flowering time and population size) can be evaluated. **Computer programs (M8)**, which can be applied to data, are available (771).
Application	**Outcrossing rates** have been determined in a large number of species (e.g., 109, 111, 303, 487, 604). **Outcrossing rates** have been related to population size and density (421, 469, 934), and to ecological gradients (163, 676). Different **outcrossing rates** have been obtained for different morphological forms of *Senecio vulgaris* (4).

Evaluation

Sensitivity: 4   Requirements: 3

Time: 4   Cost: 4

References

4, 109, 111, 163, 234, 303, 421, 469, 487, 604, 676, 769, 770, 771, 772, 773, 817, 819, 835, 836, 934.

Author   MHS

M70. Sexual morph distribution

Category   Breeding system.

Subcategory   Phenology, reproductive allocation, sex distribution.

Description   **Distribution of sexual morphs** on individual plants are studied in
                              dioecious, monoecious, andromonoecious and gynomonoecious
                              species. Branches are cut off or marked and the sex expression of
                              each flower noted. The branch can be subdivided into sections and
                              the sex ratio within the sections calculated. This is repeated on a
                              number of branches and a number of individuals.

Assumptions and   This method requires that branches can be compared. Different local
restrictions   light regimes and other environmental factors can influence the
                              development of the flowers.

Test system   Field experiment, laboratory experiment.

Advantages   The position of the sexual morph can be important for its function.
                              Knowledge of the **distribution of sexual morphs** is therefore impor-
                              tant for understanding the breeding system.

Application   This method has provided information on the connection between
                              position on the plant and the sexual morph of the flower which is
                              needed to understand the breeding system (229, 638).

Evaluation   Sensitivity: 4   Requirements: 1
                              Time: 4   Cost: 1

References   229, 638.

Author   MP

M71. Southern and Northern hybridization (blotting)

Category

Gene-flow, hybridization, genome structure, population.

Subcategory

Introgression, marker, vertical gene-transfer, hybrid, gene expression, gene stability, genotype, polymorphism.

Description

Southern hybridization (blotting) is a much used molecular technique that combines electrophoretic separation of DNA fragments with nucleic acid hybridization. The **Southern hybridization** technique can demonstrate homology between DNA restriction fragments and a DNA or RNA probe.

The DNA is cut by restriction enzymes and an electrophoresis is carried out. After separation the fragments are denatured in the gel and transferred (e.g., by capillary action) to a sheet made from nitrocellulose or another matrix. The single stranded DNA binds to the sheet and a probe labelled by ^{32}P or a biotinylated probe is added under renaturing conditions. The probe will bind to the DNA fragments with a complementary DNA sequence. The hybridized fragments are detected by autoradiography (probe labelled with ^{32}P) or by a colour reaction (biotinylated probe). **Northern hybridization (blotting)** is a method analogous to Southern blotting, but RNA not DNA is the target of hybridization. **Northern hybridization** can thus reveal the expression of specific genes. **Southern and Northern hybridization** have been described in detail (797).

Assumptions and restrictions

The methods are quite laborious, and the amount of target DNA/RNA needed is rather large (approx 2-10 μg). Estimates of sequence homology is provided by **Southern hybridization**, but the ultimate test of homology is a sequence analysis.

Test system

Laboratory experiments.

Advantages

Southern hybridization is a sensitive method which allows the detection of heterozygotes.

Applications

Southern hybridization has been intensively used, e.g., for detection of transgenic inserts and their copy number in the genome (435, 475) and demonstration of homology between endogenous DNA fragments (734. The use of **Southern hybridization** is broadened through the **RFLP** technique (M64) which is based on this technique. **Northern blotting** is applied in studies of expression of specific genes (379, 423, 885).

Evaluation

Sensitivity: 3   Requirements: 3
Time: 4   Cost: 3

References

379, 423, 435, 475, 734, 797, 885.

Author

RBJ

M72. Spatial autocorrelation analysis and Moran's I

Category

Pollination, breeding system, gene-flow, population.

Subcategory

Pollen dispersal, cross-pollination, self-pollination, outcrossing, genetic diversity, genetic drift, genetic neighbourhood-area, metapopulation.

Description

Spatial autocorrelation analysis is used for detailed investigation of restricted gene-flow in continuous populations. The method can thus be used to investigate isolation by distance (1016).

Spatial autocorrelation measures the correlation between values of variables at different distance classes. These variables can typically be allelic frequencies or morphological measurements and the distance class can be an interval of physical distance between plants or populations. Commonly, each sampled plant in a population is considered a unit and the full matrix of pair-wise distances between the plants is divided into 5-10 distance classes.

The most common definition of this correlation is termed **Moran's I** and is defined for each distance class k as:

$$I_k = \frac{n \sum_{i \neq j} w_{ij}^{(k)}(x_i - \bar{x})(x_j - \bar{x})}{\sum_{i \neq j} w_{ij}^{(k)} \sum_i (x_i - \bar{x})^2},$$

where n is the number of individuals or populations sampled, x_i is the value of the character for individual (population) i, $w_{ij}^{(k)} = 1$ if i and j are both in class k, and 0 otherwise. For genetic data of an allele at a single locus, $x_i = 1$ if plant i is homozygous for this allele, 0.5 if the plant is heterozygous for the allele and 0 if the plant does not carry the alleles.

The set of values of I_k can be plotted as a function of k, and the graph is called a corellogram. From these corellograms, inferences can be made of the structure of the investigated parameter in the population, e.g., restricted migration through seeds and pollen or selection on specific genetic loci. Specific values of I_k can be tested for significance through various methods (865, 866), with testing against a randomized data set being preferred by most authors (857).

Recent theoretical studies of spatial autocorrelation claim that join-counts of genotype-pairs (865, 866) are statistically superior to Moran's I-statistics (235), but this method has only been applied to data in a few cases (833).

Assumptions and restrictions

A number of codominant marker loci are needed for an accurate estimation of **Moran's I**. Sampling of individuals should be from several different areas, because I is sensitive to stochastic events at the single site (362). Caution should be exerted in testing for statistic significance of I-values (857).

Test system	Field experiment, laboratory experiment, data analysis.
Advantages	**Spatial autocorrelation analysis** may detect very fine-scaled genetic structures in a continuous population. A deviating value of I for a specific locus may indicate some form of selection acting on the locus. Information on the **genetic neighbourhood-area** (M24) may be obtained and the results can be compared to a **paternity analysis** (M45).
Application	**Spatial autocorrelation analysis** have been performed for a large number of plant species in natural populations, commonly based on **isozyme** data (M60) (362, 532, 695, 1023). A general pattern is that isolation by distance increases with the level of self-fertilization (362). Isolation by distance in the self-incompatible *Ipomopsis aggregata* (128) was found and the values of I were used for estimating neighbourhood-size. Ant-dispersed populations of *Sclerolaena diacantha* had more fine-scaled spatial genetic structure than populations without ant-dispersal (688) and spatial genetic structure was larger in low-density populations of *Cryptotaenia canadensis* (999). Moran's I for nectar volume in *Delphinium nelsonii* was analyzed (979), but no evidence for spatial patterns was found.
Evaluation	Sensitivity: 3 Requirements: 2 Time: 2 Cost: 1
References	128, 235, 362, 532, 688, 695, 833, 857, 865, 866, 979, 999, 1016, 1023.
Author	MHS

M73. Stigma receptivity tests

Category

Pollen development and production, seed development and production, breeding system.

Subcategory

Pollen germination, fertilization, seed production, incompatibility, phenology.

Description

The period of **stigma receptivity** is very important to determine if **experimental pollination** (M13) shall succeed. The stigmas are pollinated at intervals from opening to withering of flowers. Each flower is marked and pollen germination or fruit set is a measure of pollination success. Available methods are described under **pollen germination tests** (M52). **Stigma receptivity** can also be tested by using a 30% H_2O_2 solution. If stigmas are receptive, bubbles can be seen where H_2O_2 has been applied. Receptivity can, in some cases, be reliably indicated by the stigma surface being visibly sticky with a mucilaginous substance.

Assumptions and restrictions

The most exact method is to germinate pollen grains on stigmas. The chemical methods can sometimes be difficult to interpret. Also the temperature influences the results.

Test system

Field experiment, greenhouse experiment.

Advantages

The method is necessary for understanding the success of natural as well as experimental pollinations. The length of the receptive period determines how many different pollen donors will arrive and germinate on the stigmas.

Application

Investigation of **stigma receptivity** is part of a study of the general phenology of the individual plants. The method has been used to study spatial and temporal variation in flower function (84, 413, 555, 568, 639, 719, 884, 902), to compare different morphs within a species (39) and to see how salinity affected **stigma receptivity** (434).

Evaluation

Sensitivity: 5   Requirements: 2
Time: 2   Cost: 2

References

39, 84, 413, 434, 555, 568, 639, 719, 884, 902.

Author

MP

M74. Thin layer chromatography (TLC)

Category   Pollination, gene-flow, hybridization, genome structure, population.

Subcategory   Chemical attractant, marker, hybrid, genotype, polymorphism.

Description   **Thin layer chromatography (TLC)** is used for separating and identifying various chemical compounds such as amino acids and hexoses.

A glass plate coated with a silica gel, alumina (aluminum oxide) or cellulose is the carrying medium. The samples are applied in one end of the gel and separated by a solvent migrating into the gel matrix. The samples will migrate into the gel depending on the solubility of the molecules in solvent and the affinity to the gel matrix. By using specific stains an array of groups of chemical compounds may be revealed and the result may be used as a genetic marker like **isozymes** (M60).

Assumptions and restrictions   The performance of the method depends strongly on the selected solvent and the solubility of the molecules in question. Before applying the method, it is necessary to know which chemical compounds are to be detected in order to choose the relevant staining procedure.

Test system   Laboratory experiment.

Advantages   The method is simple and not expensive in equipment. It is possible to analyze many samples per day for a given group of compounds.

Application   The method has been applied to analysis of isomers, e.g., malonyl-tryptophan (513) and hormones, e.g., phytoestrogenes. An array of phenolic compounds has been analyzed for identification of hybrids in *Equisetum* (957). Expression of chloramphenicol acetyl transferase in transgenic *Actinidia deliciosa* has been confirmed by **TLC** (661).

Evaluation   Sensitivity: 3   Requirements: 2
                     Time: 2   Cost: 2

References   513, 531, 661, 957.

Author   VS

M75. Thinning of flowers

Category

Pollen development and production, seed development and production, breeding system.

Subcategory

Pollen germination, pollen production, fruit abortion, seed abortion, seed germination, seed production, phenology, reproductive allocation.

Description

Thinning of flowers is used for evaluating the influence of resource allocation to fruit and seed set. The inflorescence of the plant can be divided into two equal parts, one part is thinned and the other part is control. Alternatively, the number of flowers in an inflorescence is reduced to a certain level, by removing flowers in the bud stage at random. In another procedure, every other immature fruit is removed along the reproductive branches and the subsequent size, etc. is measured on remaining fruits as well as on controls. See also **experimental pollination** (M13).

Assumptions and restrictions

A wide range of factors will influence the seed set: Position of flower, general nutritional and water status, quality of deposited pollen, etc. These factors also influence the results from the **thinning** procedure.

Test system

Field experiment, laboratory experiment.

Advantages

The method does not require much equipment.

Application

This method gives information on allocation to fruit and seed production (158, 191, 192). It may reveal when seed set is limited by allocation (400, 847) and when selective seed abortion enhances seed quality (886). The method has also been used to study the relation between sexual and asexual reproduction (990).

Evaluation

Sensitivity: 3   Requirements: 1
Time: 3   Cost: 1

References

158, 191, 192, 400, 847, 886, 990.

Author

MP

M76. Transplantation experiment

Category	Seed development and production, breeding system, hybridization, population.
Subcategory	Seed production, cross-pollination, phenology, reproductive allocation, self-pollination, hybrid, cline.
Description	Individual plants can be **transplanted** from one area to another in order to study how different environmental regimes affect plant performance. **Transplantations** may also be used for plants in an experimental field, where a number of collections can be grown under standard conditions in order to detect differences between genotypes, (see also M78 in vol. 1). Reciprocal transplantations between sites is a classical method for studies of adaption to local environments, see **local adaptation analysis** (M35), and development of clines. See also **reproduction in alien plants: Community and invader analysis** (M62).
Assumptions and restrictions	It is assumed that the standard conditions correspond to similar environments for all samples. The process of transplanting might change the condition of the individual for some time. Transplants need acclimatization for a period.
Test system	Field experiment.
Advantages	Reciprocal transplant experiments will allow tests of differential selection and adaptation under semi-natural conditions. Transplants in experimental gardens can be nursed and herbivory can be prevented.
Application	**Transplantation experiments** has been used for comparisons of female reproductive success (997). Flowering phenology of several populations of ericaceous shrubs was censused in a common garden (742). Differential survival and growth of ecological races were observed at different altitudes (364). Transplant experiments have been used in order to test small-scale environmental variability (30, 264). Reciprocal transplant experiments allowed to demonstrate local adaptation (278, 559).
Evaluation	Sensitivity: 3 Requirements: 3 Time: 2 Cost: 4
References	30, 264, 278, 364, 559, 747, 980, 997.
Author	MP, FF

M77. Two-dimensional paper chromatography

Category

Gene-flow, hybridization, genome structure, population.

Subcategory

Marker, hybrid, genotype, polymorphism.

Description

Paper chromatography is a useful technique in the investigation of hybrid populations and polyploid taxa, in elucidating evolutionary pathways and in determining the parentage of hybrids. Extract material is applied on Whatman 3MM chromatographic paper and then developed by the descending method with two sets of solvents.

Assumptions and restrictions

The chromatographic patterns of the F_1 hybrids are a complement of those of their parental species. This is also true for natural hybrids, although all plants with complementary patterns need not be true F_1 hybrids. In the presence of the appropriate genic balance, backcross and advanced generation derivates may synthesise constituents of both genomes and thus be confused with F_1 hybrids. Conversely, plants lacking all markers of one or the other species may be eliminated from consideration as F_1 hybrids. Novel compounds may occasionally appear in the hybrid profiles whereas, in other cases, parental compounds are found to be missing in hybrids. These phenomena are generally considered to be rare and are not a serious limitation to the application of chromatography to the analysis of hybrids.

There may be some quantitative variation in the expression of a chromatographic character among different hybrids and even within parental species.

Test system

Laboratory experiment.

Advantages

Chromatographic studies may detect back-crosses which would have been overlooked on the basis of gross morphology. The chromatographic and morphologic data are generally reliable. Another advantage, assuming their reliability, is that the presence of a chromatographic marker indicates the presence of a specific marker gene in the population.

Application

This method is used in determining hybrid status, introgression and phylogenetic relationships.

Evaluation

Sensitivity: 3   Requirements: 2
Time: 2   Cost: 2

References

19, 76, 409, 496, 497, 498, 761, 1011.

Author

YJ

8. Genetic engineering techniques

By: Vibeke Simonsen and Rikke Bagger Jørgensen

Introduction

This chapter gives a brief introduction to the biotechnological techniques which have been used for transforming genomes of plants and plastids. The aim of the engineering methods is to transfer a piece of a species-alien DNA molecule into the genome of a selected species providing the species with a new function or an altered function. The transfer methods all depend on the successful incorporation of the alien DNA into the genome of the species, on the expression of the gene and on the transfer of the gene to the next generation.

Transformation methods

The use of transformation of plant or plastid genomes by means of genetic engineering has increased much during the last two decades. Delivery methods can be divided into direct transfer, which is obtained by **electroporation** (T2), **chemical poration** (e.g., T3), **biolistics** (T4) or **microinjection** (T5) and indirect transfer, which uses **biological vectors** (T1), mainly *Agrobacterium spp.* The natural host for these bacteria are dicot plants and many dicots have been transformed by the biological vectors. Comparison of methods and description of results have been reviewed by several authors (e.g., 258, 643, 710, 970).

An alternative to the direct DNA transfer technique is the use of silicon carbide fibre, coated with DNA, as a transferring agent (41, 872). The method may be classified as **microinjection** (T5), as it uses micromanipulation. However, **microinjection** (T5) has mainly been applied to animal cells. Electrophoresis has also been suggested as a possible method for gene transfer (872). The use of a laser beam, for penetrating the cell wall for transport of DNA into the cell, may also be a possible method (710). A guide-book of methods and protocols for different applications has been published (712). A survey of techniques, applied for transformation of various plants, has shown that the indirect transport of DNA is still the most widely used (258). Two recent reviews state that monocots can be transformed by *Agrobacterium spp.* (48, 962). A comparison and evaluation of the transfer methods via *Agrobacterium* (T1), via **electroporation** (T2) or via **biolistics** (T4) is available (89). *Agrobacterium* mediated transfer and direct delivery of DNA have been compared for *Brassica nigra* among others (327). The construction of vectors and of promoters are of great importance for a successful transformation of a certain plant (969).

Effect of the transformation

The position effect in the genome of the plant has been discussed (969). Position effects may be due to the location in the genome, where the alien gene is incorporated and may cause a reduced production of the protein determined by the alien gene. Other negative effects from transformation of the genome such as gene silencing or epistasis have been elucidated (92).

Applications

The applications of the transgenic plants are numerous (86, 712) and the number of inserted traits in various plants are increasing tremendously. The most widely transferred traits are genes coding for resistance to herbicides, to pests, to pathogens, etc. These and many other traits are discussed in Chapter 9. The transfer methods are highly valuable tools for obtaining more information about gene expression in various tissues and at different life stages.

Detection of transformation and expression

The presence of an alien DNA molecule in the genome may be checked at two levels: Incorporation in the genome of the transgenic plant and expression of the alien gene. The incorporation can be detected by **Southern blotting** (M71) or *in situ* **hybridization** (M31). The expression of the incorporated gene can be quantified either as the production of m-RNA corresponding to the alien gene or as the production of the protein. The presence of the m-RNA may be proved by **Northern blotting** (M71) and the production of the protein by histochemical methods, e.g., staining of selected tissue, **bioassay** (M3) or **immunological methods** (M30) as Western blotting and immunogold labelling.

Genetic markers

Genetic markers have been used for rapid detection of successfully transformed plants. A genetic marker can be attached to the gene of interest. The two most extensively used markers is kanamycin resistance, which is caused by the production of the enzyme neomycin phosphotransferase II (NPTII), and β-glucuronidase (GUS). The presence of NPTII may be revealed by using a **bioassay** (M3) and GUS by a histochemical procedure. An excellent protocol is available, which also mentions the possible pitfalls of using genetic markers (273). The use of firefly luciferase as marker is another quick detection method for transgenic plants, as the luciferase catalyses light-emitting reactions.

Knocking out marker genes

There is an increasing concern that some marker genes could have detrimental effects on health and environment. Therefore, methods for knocking out marker genes after the transformation event have become of interest. The process of knocking out the marker genes can be accomplished through a second transformation of the plant carrying the gene. The marker gene is eliminated or mutated by homologous recombination between the transformation vector, the targeting vector and the target site. The original engineered insert harbouring the marker gene is positioned in the target site. The targeting vector is carrying sequences homologous to the target site and is therefore able to be integrated by homologous recombination. With correctly designed targeting vectors, the marker gene will be excised or mutated during this process. Gene targeting (e.g. knocking out marker genes) in higher plants is at its very beginning and with the present state of the art the use of gene targeting is generally very low.

Reviews

41, 48, 86, 89, 92, 258, 273, 327, 575, 643, 674, 687, 710, 712, 767, 862, 872, 969, 970.

8.1. Methods of DNA transfer

Method description

In this section the most common methods used for genetic engineering of plants are described. The methods are described in a similar manner to the methods in Chapter 7 and indicated with a **T**.

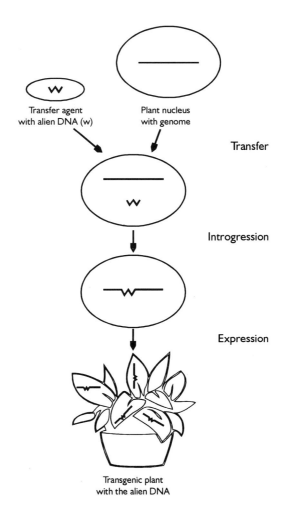

Figure 8.1.
Schematic illustration of a transformation process of a plant.

T1. Transfer via *Agrobacterium* spp.

Category	Genetic engineering technique.
Subcategory	Biological vector.
Description	An explant (small fragment of tissue) or a cell suspension is culti-vated in a media with *Agrobacterium tumefaciens* or *A. rhizogenes*, a detailed description is available (712). The bacterium strain harbours a plasmid with the alien genes, which is transferred into the nucleus of the plant. Depending on the transfer success, the alien genes will be incorporated into the genome of the host plant and expressed in the plant. See the section on detection of transformation, p. 196. The use of cointegrative vectors (the alien DNA is in a cis position to the *vir* region of the plasmid) or binary vectors systems is often used (e.g., 136, 968). The mode of incorporation is investigated (551) and the importance of tagged DNA is elucidated, as it may help to identify if the complete alien DNA molecule has been inserted in the genome of the plant. Transformation of plants can be achieved by cutting the plant and applying the plasmid suspension to the wound (422). The success rate increases, if the transfer is conducted under low pressure (72). Transgenic haploid and fertile diploid plants have been developed from pollen embryos, which may speed up the pro-duction of transgenic plants in breeding programs (800). Stability of inserted traits has been studied for kanamycin-resistant tobacco and Mendelian segregation of the resistance of the progeny, obtained by selfing, was found for 10 out of 16 transgenic plants (665). This success rate of the process is high.
Assumptions and restrictions	The method requires that the target plant can be infected by *Agro-bacterium* and hence the method mainly applies to dicots. However, for some monocots *Agrobacterium* can be used as a vector for DNA transfer (48), either by itself or in combination with other transfer methods (e.g., T3). The efficiency of the transformation varies with the strain of *Agrobacterium* and with the ecotype of the plant (13).
Test system	Laboratory experiment.
Advantages	This is an established method for gene transfer in dicot species which is able to transfer the alien DNA fragment to a functional position in the host genome (48). Due to the use of explants meristems or seeds as recipients for the delivery of the DNA, no complex regeneration is needed as in **transfer via electroporation** (T2) or **transfer via polyethylene glycol (PEG)** (T3).
Application	The method has been used for transferring an array of traits to a wide range of species. A review on plant species shows that *Agro-bacterium* mediated transfer is the most used method so far (258).
Evaluation	Sensitivity: 5 Requirements: 4 Time: 3 Cost: 3
References	13, 48, 57, 72, 136, 148, 258, 368, 422, 451, 452, 506, 525, 551, 598, 623, 629, 665, 694, 712, 794, 800, 825, 841, 939, 968, 1008, 1026.

T2. Transfer via electroporation

Category	Genetic engineering technique.
Subcategory	Electroporation.

Description

The method **electroporation** (electric field-mediated membrane permeabilization) is based on the fact that electric pulses can open the cell membrane and allow penetration of alien DNA. The mechanism underlying **electroporation** has been described (404). The method should be distinguished from electrofusion (electric field-induced cell-to-cell fusion, see 67, 994), which has been used for production of hybrids.

Protocols for the method are available (269, 713, 840). It has been demonstrated that $CaCl_2$ increases the efficiency of electroporation (269). The effect of various media have been investigated in *Arachis hypogaea* protoplasts (505). The importance of the vector sequence has been elucidated (146). Heat shock, in combination with **electroporation**, has been used resulting in a higher efficiency of the transformation (41). The method has been compared to the direct delivering or transfer method, **transfer via polyethylene glycol (PEG)** (T3) and was found to be more efficient (741). Uptake of macromolecules may be checked by **flow cytometry** (M15) (91).

Assumptions and restrictions

The method requires protoplasts, because **electroporation** is only efficient on cell membranes. These protoplasts, harbouring the alien DNA, have to be developed into callus and regenerated into a plant. However, this restriction may be circumvented by the fact, that other cells than protoplasts can be transformed. Two recent publications have described transformation of callus from sugarcane (*Saccharum officinarum*) (34) and transformation of germinating seed embryos from rice (*Oryza sativa*) (147).

Test system

Laboratory experiment.

Advantages

The advantage of **electroporation** is that it can be applied to protoplasts in angiosperms. The recent application of the method to callus and to germinating seed embryos may enhance its usefulness.

Application

The method has been used for transforming several monocots, but also dicots.

Evaluation

Sensitivity: 4	Requirements: 5
Time: 4	Cost: 3

References

9, 14, 34, 40, 41, 48, 67, 91, 130, 146, 147, 155, 220, 269, 404, 505, 571, 713, 741, 840, 994.

T3. Transfer via polyethylene glycol (PEG)

Category	Genetic engineering technique.
Subcategory	Chemical poration.
Description	The chemical compound polyethylene glycol changes the pore size of the cell membranes, which enhances the probability of an alien DNA molecule penetrating into the cell.
	Protocols describing the method are available (547, 840). Heat shock may enhance the uptake of DNA (48). The effect of the concentration of PEG has been elucidated (611).
Assumptions and restrictions	The method is mainly applied to protoplasts. This means that the protoplast has to be propagated into a callus and regenerated into a plant, which makes the method very time-consuming. However, microspores from maize have been applied to the method (253), but still the method is time-consuming.
Test system	Laboratory experiment.
Advantages	The method applies to both monocots and dicots.
Application	**Transfer via polyethylene glycol** has been used for transfer of DNA to monocots as well as dicots, including transfer of mRNA to protoplasts (282).
Evaluation	Sensitivity: 4 Requirements: 4 Time: 5 Cost: 4
References	48, 220, 253, 272, 282, 460, 547, 571, 611, 642, 741, 813, 840, 877.

T4. Transfer via biolistics

Category	Genetic engineering technique.
Subcategory	Microprojectile bombardment.
Description	The principle of transfer via **biolistics** is to shoot particles coated with DNA into selected tissues or cells by a particle gun. The gun may be driven by either air pressure or gunpowder (e.g., 89), or helium (609, 904). A micro-targeting device has been described (711, 802). The particles may consist of either tungsten or gold carrying the DNA and the size may vary.
Assumptions and restrictions	Any growing plant tissue may be used for this method, but the plant material has to be regenerated via a callus transformation. However, the particle gun is expensive, even if you have a licence from one of the patent holders. Tungsten particles are more toxic to the transformed plant tissue than particles of gold (787).
Test system	Laboratory experiment.
Advantages	The method has shown to be useful both with monocots and dicots, a review is available (157). By this method, it has been possible to transfer not only plasmid DNA, but also the much longer YAC (yeast artificial chromosome) DNA (212). An array of tissue may be used: Microspore (398), meristem (87, 786, 802), leaf (384), inflorescence tissue (56), protocorm (472) and scutellar tissue (73).
Application	The applicability of the method to microbes, plants and animals is reviewed (448). Even plastid transformation is possible with the method (1031).
Evaluation	Sensitivity: 4 Requirements: 4 Time: 4 Cost: 4
References	56, 73, 87, 89, 131, 135, 155, 157, 212, 306, 347, 384, 398, 448, 472, 609, 711, 786, 787, 802, 904, 955, 1031.

T5. Transfer via microinjection

Category	Genetic engineering technique.
Subcategory	Microinjection.
Description	The method transfers a very small solution of DNA into a selected cell by **injection** with a capillary tube or a pipette (850). The procedure has to be performed under a microscope. Improved injection may be obtained under pressure (666). It may be difficult to certify that the DNA really is delivered into the cell (23), but the use of dyes may help, e.g., 664.
	A protocol for the procedure has been described (712, 751).
	Injection of microspores (94, 410, 664), embryo (521, 627, 876), protoplast (625), and meristem (521) has been performed.
Assumptions and restrictions	Being a highly advanced manual procedure, only a limited number of cells can be treated per day. It requires expensive equipment such as micromanipulator, inverted microscope and very skilled personnel.
Test system	Laboratory experiment.
Advantages	The advantage of the method is that the particular fragment of DNA can be transferred exactly into the target cells which are to express the DNA.
Application	The method has been applied to monocots as well as dicots.

Evaluation

Sensitivity: 3	Requirements: 5
Time: 5	Cost: 5

References   23, 94, 410, 462, 521, 625, 626, 627, 664, 666, 712, 750, 751, 850, 876.

9. Inserted traits for transgenic plants

By: Kathrine Hauge Madsen and Gitte Silberg Poulsen

Introduction

The biotechnological development in recent years has resulted in an increasing number of transgenic crop species. Most are still under development, but a few have been approved for marketing. Data for approvals in the USA are easily obtained from the World Wide Web (URL//http:www.aphis.usda.gov/bbep/bp/), whereas data from the European Union are accessible from the national competent authorities.

In the following sections we aim to provide the reader with a quick overview of the many and diverse approaches in the development of new traits in plants by genetic engineering. For an overview of inserted traits see Table 9.1. The chapter does not give a complete overview of genes, methods or plant species, but is intended to be an introduction to the numerous and rapidly growing number of reports in this area.

Table 9.1. Traits that have been inserted in transgenic plants. Further information is given in the text.

Herbicide tolerance	Glyphosate tolerance Glufosinate tolerance Bromoxynil tolerance Tolerance to ALS-inhibitors
Insect tolerance	*Bt* toxins Protease inhibitors α-amylase inhibitors Lectin proteins
Virus tolerance	Coat proteins Satellite RNA's Replicase Antisense Defective interfering *Cis*-acting elements Movement protein Ribozymes
Tolerance to fungi	Phytoalexins Ribosome inactivation proteins Chitinases and glucanases
Tolerance to bacteria	Lysozymes Lytic peptides Toxins H_2O_2

Table 9.1., contd.	
Flower colours	Flavonoid pigments
Male sterility	Ribonuclease Glucanase
Altered metabolites	Aminoacids and proteins Lipids and plastics Carbohydrates Vaccines
Drought tolerance	Proline Betaine Fructans
Oxidative stress tolerance	Superoxide dismutase Glutathione reductase
Cold tolerance	Glycerol-3-phosphate
Salinity tolerance	Mannitol-phosphate dehydrogenase

9.1. Herbicide tolerance

Natural tolerance

There are three physiological mechanisms by which a plant can be naturally tolerant to a herbicide:

1) Reduced sensitivity at the molecular site of action, either by gene amplification or by an altered site of action.

2) Degradation of the herbicide by metabolic processes.

3) Avoidance of the herbicide through lack of uptake or by inactivation if the herbicide is absorbed.

The first two mechanisms have been employed by biotechnologists to generate herbicide tolerant crops (209).

Selectivity

Herbicides are compounds especially designed to kill or control green plants. Some herbicides have no or only minor effects on crops, whereas other plant species, e.g., weeds, are severely damaged or killed by the compound. These herbicides are denoted selective herbicides. Other herbicides are non-selective herbicides because they control or kill crops and weeds indiscriminately. Much effort has been put into developing crops with tolerance to the non-selective herbicides glyphosate and glufosinate. These two herbicides are usually considered to be less environmentally suspect and so far there have been no reports of weeds which have developed resistance to either of these herbicides.

Glyphosate tolerance

The non-selective herbicide glyphosate has broad spectrum herbicidal activity against a wide range of weeds and it has been used extensively for more than 20 years (533). Glyphosate is an inhibitor of EPSPS (5-enolpyruvylshikimate-3-phosphate synthase), an enzyme of the shikimate pathway that produces three essential amino acids, phenylalanine, tyrosine and tryptophane (26). Three approaches for creating glyphosate tolerance in crop plants by genetic engineering were investigated. One was overproduction (amplification) of the wild type EPSPS. This led to identification of the nucleotide sequence of the gene coding for EPSPS in *Petunia hybrida* L. (831). Another approach was isolation of a glyphosate tolerant mutant EPSPS, which proved to be the more efficient. In enteric bacteria the enzyme is encoded for by the *aro*A gene (878). One of these tolerance genes came from a *Salmonella typhimurium* where proline was replaced with serine at position 101 (166, 878). Another *aro*A gene came from an *Escherichia coli* bacteria where the EPSPS contained a single amino acid substitution at position 96, where glycine was replaced with alanine (443). This mutant EPSPS enzyme was improved with second site variants (61). Later EPSPS from *Agrobacterium* sp. strain CP4 showed higher tolerance to glyphosate. These genes (CP4-genes) have little homology to the previous plant analogous EPSPS mutant genes (61). A third approach, degradation genes, had been investigated and a gene from *Achromobacter* sp. strain LBAA, which codes for the glyphosate oxidoreductase (GOX) enzyme was chosen. This enzyme catalyses the cleavage of the C-N bond of glyphosate yielding aminomethyl-phosphonic acid (AMPA) and glyoxylate as reaction products (61). Glyphosate tolerance does not seem to be associated with a yield penalty in the crop (528, 883). At present, glyphosate tolerant soybean and cotton are approved for marketing in the USA and a glyphosate tolerant oilseed rape is approved in Canada. The glyphosate tolerant soybean products are now also beeing imported into the European Union, but the plants are not allowed to be grown.

Glufosinate tolerance

The non-selective and broad spectrum herbicide glufosinate-ammonium is a chemically synthesized phosphinothricin whereas bialaphos is the natural antibiotic produced by *Streptomyces hygroscopicus*. Both act as inhibitors of GS (glutamine synthase), an enzyme which is active in the assimilation of ammonia and the regulation of nitrogen metabolism in plants. Furthermore, it is the only enzyme in plants known to detoxify ammonia. Inhibition of GS causes a rapid accumulation of ammonia which leads to death of the plant cell (901). A gene (*bar* gene) that encodes for an enzyme that detoxifies glufosinate was found in a *Streptomyces hygroscopicus* strain. The gene codes for an enzyme which converts phosphinotricin into a non-herbicidal acetylated form which makes it possible for the bacteria to avoid its own toxin (919). A *pat* gene from *Streptomyces viridochromogenes* has significant sequence homology to the *bar* gene and it also inactivates phosphinotricin by acetylation (207). The tolerance gene has been inserted into numerous crop species (93) and there does not seem to be a yield penalty associated with the gene (100, 268, 318). Furthermore, it is a convenient dominant selectable marker both in tissue culture and in field trials for plants which are genetically engineered with genes encoding for other traits, e.g., the male sterile oilseed rape developed by Plant Genetic Systems (539, 540). This variety was approved for marketing in the EU in 1996. However, use of the oil in food and animal feed has yet to be approved. In Canada, a glufosinate tolerant oilseed rape (Innovator) has been approved, but the distribution of the variety in 1995 was limited to 0.3 percent of the Canadian area planted with oilseed rape (11).

Bromoxynil tolerance

Bromoxynil is a herbicide used for post-emergence control of dicotyledonous weeds in cereals. The herbicide is a potent inhibitor of photosynthetic electron transport at the photosystem II site, a mode of action which is common to several groups of herbicides. Furthermore, it acts as an uncoupler of oxidative and photosynthetic phosphorylation (247). A gene that detoxifies bromoxynil (BXN gene) was identified, isolated and cloned from a plasmid in the bacteria *Klebsiella ozaenae* (880, 881). This gene codes for a nitrilase that metabolizes bromoxynil into the primary metabolite 3,5-dibromo-4-hybroxybenzoic acid which is at least 100-fold less toxic to plant cells than bromoxynil. The metabolites found in the transgenic bromoxynil tolerant cotton were identical to metabolites found in the naturally tolerant cereals. The bromoxynil tolerant cotton gives high yields and growth does not seem to be negatively affected by the gene (879). A bromoxynil tolerant tobacco variety was approved for marketing in the EU in 1995 and a cotton variety was approved in the USA in 1994.

Tolerance to ALS-inhibitors

The ALS (acetolactate synthase) inhibitors have wide crop selectivity, high efficacy and are used in relatively low rates. They include the following chemical classes of herbicides: Sulfonylureas, imidazolinones, triazolopyrimidines and pyrimidinyl thiobenzoates. At present, herbicide tolerant crops have been developed for sulfonylureas and imidazolinones. ALS (acetolactate synthase) is an essential enzyme in the biosynthesis of the branched-chain amino acids: Valine, leucine and isoleucine. In all known reports of genetically engineered tolerance to ALS inhibitors, the tolerance mechanism has been due to an ALS that is less sensitive to ALS inhibitors (789). There have been numerous reports on natural resistance to ALS-inhibitors occurring in different weed species (789). These led the chemical company Dupont (the largest producer of sulfonylureas) to reduce the effort to create sulfonylurea tolerant crops, but some ALS tolerant crops have been developed (790). Plants with tolerance to ALS-inhibitors can be obtained in several ways: Selection of natural resistant plants, mutant selection and by transformation (790, 832). Genes that code for a mutant ALS enzyme were found in an *Arabidopsis thaliana* biotype where the enzyme differed from the wild type by substitution of a serine for proline in position 197 (*csr1-1* gene), in tobacco (*SurB* and *SurA*-gene) and oilseed rape (*ahas3r*) (790). Another gene was isolated from a mutant

Arabidopsis thaliana where serine was substituted with asparagine at position 653 which resulted in tolerance to imidazolinones but not to sulfonylurea (832). Tolerance towards ALS-inhibitors does not seem to be associated with any yield penalty (790, 832).

Tolerance to other herbicides   Tolerance to the herbicide 2,4-D with auxin activity has been introduced through selection in cell tissue culture and through transformation with a 2,4-D degrading monooxygenase gene *tfdA* from the bacteria *Alcaligenes eutrophus* (70, 173).

A gene that detoxifies the PSII-inhibitor phenmedipham was found in the bacteria *Arthrobacter oxidans* strain P52. This gene encodes for an enzyme that catalyses the hydrolic cleavage of phenmedipham (891).

Non-transgenic strategies   The first crop plant with tolerance to photosystem II (PSII) inhibiting herbicides was an oilseed rape variety. This variety was the product of traditional breeding techniques, where triazine tolerance had been transferred from *Brassica campestris* into *B. napus*. The tolerance was caused by a mutation in a chloroplast gene (*psbA*), which greatly reduced the affinity of the *s*-triazines to the active PSII site. It was, however, associated with a yield penalty because of a reduced photosynthetic capacity (322). A similar strategy has been chosen for other crops and herbicides where naturally occurring herbicide resistance can be found by selection of plants.

9.2. Pest tolerance

State of the art   Most of the reported transgenic plants tolerant to pests have been developed by inserting genes from the bacteria *Bacillus thuringiensis* (*Bt*) coding for endotoxins, which makes them tolerant to attacks by mainly *Lepidopteran* larvae. However, plants with other genes inserted are beginning to appear (review in 287). Also transgenic plants resistant to nematodes have been reported (46, 948).

Bt toxin   *Bacillus thuringiensis* (*Bt*) toxin has been used as a microbial insecticide (a biopesticide) for more than 30 years (248) mainly against attacks of *Lepidopteran* larvae. *B. thuringiensis* is a gram-positive, spore-forming bacterium in soil. The crystals formed during sporulation contain proteins (δ-endotoxins) that become toxic when eaten and dissolved in the midgut of specific insects (review in 68). The genes coding for the proteins have been isolated from *B. thuringiensis* (first time by 816) and have been transferred, most often in a modified form, to different crop plants, the first being tobacco (952). When the target insect feeds on plants with one of the *Bt* genes it ingests the δ-endotoxin with identical insecticidal effect as when *Bt* is applied as a biopesticide. The genes can be divided into different groups according to their specificity. The main groups produce δ-endotoxins which are toxic either to *Lepidoptera* (CryI), *Lepidoptera* and *Diptera* (CryII), *Coleoptera* (CryIII) or *Diptera* (CryIV)

(371) and finally *Lepidoptera* and *Coleoptera* (CryV) (494). Screening numerous *Bt* strains, toxins against other taxonomic groups have also been found (248). Insects have already developed natural resistance against *Bt* toxins when used as a microbial insecticide (900). The possible mechanism of resistance is a reduced binding of toxins to receptors at the binding site in the midgut of the insects. Transgenic plants, which contain the *Bt* toxin in the environment for longer periods, will probably increase the speed of resistance development among insects (899).

Protease inhibitors

A strategy to control insects, other than by killing them, might be through alteration of the plant in a way which affects the life conditions of the insects. One rather simple way is to use those mechanisms in plants which already restrict insect attacks. In a biotechnological context, mechanisms which are coded by a single gene are preferable. Protease inhibitors are among the possibilities (reviewed by 365, 788). These are antimetabolic proteins and the insect control mechanisms are increased mortality, decreased growth rate and a prolonged larvae development period due to malnutrition (285, 319). The mechanism of action is only partially understood (412). The direct toxic effects of Bt could be combined with the more indirect control caused by the protease inhibitors (526) or with other genes with insecticidal effects (101). Originally, the protease inhibitors could either be constitutively present in the plants or induced by wounding (e.g., insect attacks). The protease inhibitors can be grouped in different families (365, 788). Here we describe the cowpea trypsin inhibitor (CpTI), which belongs to the Bowman-Birk family and briefly the wound-induced families PI-I and PI-II. Protease inhibitors from other families have also been transferred to plants (e.g., 77, 287, 546). Genes coding for protease inhibitors can also be isolated from insects (416) and transferred to plants (917).

CpTI

Some cowpea *(Vigna unguiculata)* accessions have shown rather good tolerance against attacks of the cowpea bruchid beetle (*Callosobruchus maculatus* F.). This tolerance is a result of a 2- to 4-fold elevated level of the cowpea trypsin inhibitor (CpTI) compared with the susceptible accessions (286). The gene coding for CpTI has been identified and transferred to plants which lack this natural defence against insect attacks. The first reported successful transformation was to tobacco in order to obtain tolerance to tobacco budworms (*Heliothis virescens*). The CpTI-expressing plants had a significantly enhanced tolerance to *H. virescens* (366). In general it has been shown that CpTI has an effect on several species from the insect orders *Lepidoptera* and *Coleoptera* and also to the orthopteran *Locustus migratoria* (365).

PI-I, PI-II

PI-I and PI-II are two families of wound-induced protease inhibitors occurring in potato and tomato (319). The PI-II protease encodes a trypsin inhibitor with some chymotrypsin activity (287, 405). Transferred to tobacco with a constitutive promoter (CaMV 35S), this protease inhibitor gave some tolerance to attacks of tobacco hornworm (*Manduca sexta*), whereas plants expressing PI-II, which is a chymotrypsin inhibitor, did not control tobacco hornworm (405).

α-amylase inhibitors

A different approach to using known plant tolerance mechanisms against pests is the transfer of genes that code for α-amylase inhibitors. The insecticidal mechanism is inhibition of amylases in the midgut of the insect resulting in retarded growth. Such a gene protects common bean (*Phaseolus vulgaris*) from attacks by cowpea weevil (*Callosobruchus maculatus*) (382) and feeding studies have shown that α-amylase inhibits growth of two beetles (cowpea weevil and azuki bean weevil (*C. chinensis*)) of the family *Bruchidae* (391). The first successful transformation was in tobacco in 1990 (20). Amylase inhibitors are often expressed in the seeds. When the α-amylase inhibitor is transferred to peas (*Pisum sativum* L.) the transgenic peas show tolerance to cowpea weevils and azuki bean weevils, which often cause severe damage during seed storage (830). The transgenic peas are not only protected during storage. The amylase gene also blocks the development of pea weevil (*Bruchus pisorum*) larvae during growth (821).

Lectins

Lectins are carbohydrate-binding plant proteins that bind glycans and glucoproteins, glucolipids or polysaccharides with high affinity. Similar to protease inhibitors and amylase inhibitors, the most likely function of lectins is plant defence and the control mechanisms are assumed to be alike. When the plant tissue is eaten by predators lectins are released and come into contact with glucoproteins in the intestinal tracts of the predator, thereby possibly inhibiting the absorption of nutrients. Different plant families produce different lectins, which, during feeding studies, have been shown to have different defence properties against insects or fungi (review in 156). Two examples are presented: GNA, which has been cloned from snowdrop (*Galanthus nivalis* L.) (186) and pea lectin (P-Lec).

GNA

The binding site of GNA is exclusive for mannose. This gene has protected transgenic tobacco, potatoes and lettuce against sap-sucking insects of the order *Homoptera*. The size of peach potato aphid (*Myzus persicae*) populations and their growth rate were significantly reduced when feeding on transgenic tobacco with a GNA gene (367). Normally one would expect that only insects which feed by chewing the plant tissue would be influenced by an increased level of lectin, but even with a suboptimal promoter (CaMV35S) the expression of GNA in the ploem was high enough to retard growth of the aphids. Plants with a ploem specific promoter are under development (367).

P-Lec

The P-Lec gene has been cloned from pea (*Pisum sativum*) seeds (289) and has been transferred to tobacco with the result that mortality of tobacco budworm larvae (*Heliothis virescens*) was significantly higher than in control plants (101). Transgenic tobacco plants which had two tolerance genes (CpTI and P-Lec) showed a higher level of tolerance compared to plants with only one of the genes (101). In this case, the promoter was CaMV35S.

Until now, very few papers report direct success with engineered resistance against nematodes. Therefore, only the potential strategies are described. Nematodes are plant parasites and most of them penetrate roots and modify plant cells into nematode feeding sites. The possible ways to control the nematodes are when they penetrate the roots, in the feedings cells or by introducing an anti-nutritional factor (46). Another part of the strategy creating transgenic nematode resistant plants is the choice of promoter. One possibility is a constitutive promoter (e.g., CaMV35S), but this promoter is deregulated in the feeding cells (301) and is therefore not an optimal option. Other constitutive promoters are only expressed in roots (e.g., TobRB7) and they could be useful in strategies that focus on the invasion and establishment of the nematodes (667). This is also the strategy for using promoters of wound-responsive genes. The best promoter would probably be one that controls genes in the induced feeding cells, because this would ensure that the tolerance genes are only expressed when the plant is attacked by nematodes. Unfortunately, it has been difficult to isolate genes specific for the feeding sites (46). If feeding cell specific promoters are recognized, they could possibly be used in a sort of suicide approach - expression of a strong cell poison (e.g., barnase) under control of a feeding cell specific promoter. This would destroy the feeding cells, which are very important for development of the nematodes (328). The root specific constitutive promoter and the wound induced promoter could be combined with either proteinase inhibitors, lectins or antibodies (bind to sensory organs) or maybe collagenases, which attack structural proteins of the nematode during migration. Some promising results have been reported for proteinase inhibitors. Cow pea trypsin (CpTI) inhibitor expressed in potato resulted in reduced fecundity of *Meloidogyne* spp. and switched the population of *Globodera pallida* from most females to most males, which are less harmful (46, 948). Another example of the anti-feedant strategy is introduction of a gene coding for oryzacystatin I from rice. In hairy root cultures of tomato, this gene reduced growth of *G. pallida*, even though the suboptimal promoter CaMV35S was used (948). The effect of CpTI, as well as of cystatin, is inhibition of cysteine proteinases (461).

9.3. Pathogen tolerance

This topic covers the attempts to develop tolerance against diseases caused by either virus, fungi or bacteria. So far, virus tolerance has had a high success rate with genes coding for coat protein mediated tolerance. So far, no reports concerning associated penalty in crop yield or other side effects, due to insertion of these genes, have been presented.

9.3.1. Virus tolerance
Several genetic engineering strategies have been pursued to develop transgenic virus tolerant plants. Most emphasis has been given to

transformation with viral nucleotide sequences (pathogen derived resistance) (799), but recently, genes of non-viral origin have been inserted into plants to obtain virus resistance.

Strategies with pathogen derived tolerance

The transformation strategies with viral nucleotide sequences can be divided into the following groups (modified after 958):
1. Coat protein genes.
2. Sequences of satellites.
3. Replicase-mediated tolerance.
4. RNA complementary to infecting virus (antisense RNA).
5. Nucleotide sequence of defective interfering RNA.
6. Expression of *cis*-acting elements.
7. Movement protein mediated tolerance.

When reviewing the literature, it is striking that the precise mechanisms, by which the different strategies confer tolerance or resistance, are still theories or unknown. The most common theories are listed in the following paragraphs.

Coat protein

The first theories concerning the mechanism of coat protein mediated resistance were that the transgenic plant accumulating coat protein was tolerant, because the coat protein bound to the receptor and thereby prevented the association of virions with the receptor, which rendered them unable to initiate infection. Another initial theory was that the initiation of uncoating was triggered by a change in physiological conditions upon entry into the cell and this process was reversible. The equilibrium between uncoating and recoating of the end of the virion would be shifted in favour of recoating (259). However, conflicting results were found with potyviruses, where the tolerance was probably caused by the RNA transcript of the coat protein gene rather than the coat protein itself (508). The latest theories suggest that, when the coat protein gene is inserted into the plant the level of primary RNA transcripts of this gene exceeds a certain threshold level. As a consequence the cell starts to degrade the RNA transcript including similar RNA transcripts from the virus (509, 860). A gene coding for an untranslatable coat protein sequence provided protection against homologous and closely related isolates (682). This was interpreted to be caused by inhibition of viral replication or degradation. Numerous cases of tolerance in plants by insertion of viral coat protein genes has been demonstrated for different viral groups: Tobamo, potex, cucumo, tobra, carla, poty and alfalfa mosaic plus the luteovirus group (508). Coat protein mediated protection usually results in a delay of symptom appearance and the level of protection depends on the concentration of inoculum (682).

Satellites

Virus satellites (encoding their own coat protein) and satellite RNA's (encapsidated in the coat protein of the helper virus) depend on the helper virus for replication and are not related to the virus by sequence similarity. The mechanisms of tolerance are still unknown (958).

Replicase	Transgenic expression of genes encoding the viral RNA-dependent RNA polymerase is generally referred to as replicase-mediated tolerance. The mechanisms of replicase-mediated resistance are likely to be diverse. In some cases, the defective protein probably disrupts or competes with the viral replicase complex, thereby inhibiting replication (356).
Antisense	Inhibition of gene expression in eukaryotes may occur by one or both of the following mechanisms: Hybridization with mRNA may occur in the nucleus and prevent processing and/or transport of the target message or antisense RNA may hybridize with the sense message in the cytoplasm, causing blockage of translation of specific mRNAs (958). Antisense to inhibit viral infections has not been very successful in the past. One possible explanation is that viral proteins contain helicase activity which could unwind any antisense-viral RNA duplex formed in the cytoplasm (217).
Defective interfering	Defective interfering RNAs and DNAs are truncated forms of a wild type virus that require the parental (helper) virus for replication. They are derived from their parental virus by a series of sequence deletions and rearrangements. These probably reduce the accumulation of the wild type virus by competition for factors necessary for replication (356, 542).
Cis-acting elements	In principle *cis*-acting elements of viral genome expressed transgenically could serve as decoy molecules distracting the proteins away from the inoculated viral genome and thereby away from involvement in virus replication and spread (69).
Movement protein	There is evidence that unrelated viruses move between cells in a uniform manner. It is therefore expected that movement protein mediated resistance will have broad-spectrum effectiveness. The mechanism may be due to the defective movement protein interfering with the viral movement protein (69). A gene encoding a defective mutant of the tobacco mosaic virus was inserted into tobacco plants (168). Significant delays in the time were seen before decease symptoms appeared when the plants had been inoculated with five vira from different taxonomic groups.
Antiviral proteins	In recent years, reports about the use of genes coding for antibody against virus encoded proteins have been published (69).
Ribozymes	RNA enzymes (ribozymes) have been shown to cleave target RNA sequences and to inhibit RNA transcript. Ribozymes are a diverse group of enzymes involved in the processing of RNA. The most promising cleavage structure seems to be the hammerhead type (named after the secondary structure) that cleaves the target. Usage of this technique requires knowledge of sequence of the virus in order to construct the ribozyme (217).

9.3.2. Resistance to fungi

As for virus resistance, several approaches have been used in order to create tolerance against pathogenic fungi. This area of genetic engineering can become of considerable benefit to the plant breeders because resistance obtained by years of traditional breeding is often due to one gene and, within a few years, the fungus developes races which are tolerant to the gene. With genetic engineering techniques it may (with some limits for size of the gene construct) become possible to insert multiple genes in a cassette of genes which can be manipulated as a single Mendelian unit in classical breeding programmes (577). This will save the plant breeders time and provide crops with tolerance to the fungus for a longer period of time.

Phytoalexins

Phytoalexins are low molecular weight antimicrobial secondary plant products which are rapidly synthesized as a response to microbial infections. A positive correlation between phytoalexin synthesis and disease tolerance has been found in several host-pathogen interactions. Genes from grapevine coding for stilbene synthase, which is responsible for synthesis of the phytoalexin resveratrol, were inserted into tobacco, tomato and potato and increased the tolerance to common fungal diseases (257, 332).

Ribosome inactivating proteins

Ribosome inactivating proteins do not inactivate their "own" ribosomes but show activity towards ribosomes from distantly related species including fungi. Expression of barley genes coding for ribosome inactivating proteins in tobacco increased the tolerance against *Rhizoctonia solani* (477).

Chitinases and glucanases

These enzymes catalyse the hydrolysis of chitin and β-1,3-glucanase, both major components of the cell wall of many fungi (171 after 995). Chitinases and glucanases can act synergistically (228, 1040). Similar results were found for genes coding for ribosome-inactivating protein and chitinase inserted into tobacco (393). The first successful report on chitinase genes was insertion of a bean vacuolar gene into tobacco and *Brassica napus* which resulted in decreased symptom formation by *Rhizoctonia solani* (106, 107). Several example of increased tolerance to fungi after successful introduction of chitinase and glucanase genes into tobacco, tomato, rice, etc. can be found in literature (228, 507, 516, 631).

9.3.3 Resistance to bacteria

Proteins from insects may be used for protection against bacterial diseases in plants. Insects do not have an antigen-antibody reaction, but have unique self-defence mechanisms. Lysozyme, cecropin, attacin, diptericine and insect defencin are five types of insect proteins with antibacterial activity (1027).

Lysozymes

A transgenic potato with increased tolerance against bacteria was obtained by inserting a chimeric gene coding for barley α-amylase signal peptide and bacteriophage T4 lysozyme (210). A gene coding for hen egg white lysozyme was inserted into tobacco and the lysozymes that were extracted from the tobacco were able to inhibit the growth of some bacterial and fungal plant pathogens. Furthermore, the authors suggest that the bacteriophage T4 lysozyme provides more tolerance to gram-negative bacteria, whereas the hen egg white lysozyme is more active against gram-positive bacteria and some fungi (938).

Lytic peptides

A designed gene coding for the peptide Shiva-1 (46% homology to cecropin B) was inserted into tobacco. This resulted in delayed and decreased disease symptoms from bacteria and the authors suggest potentials for fungal resistance (399). A gene from *Hyalophora cecropia* coding for attacin was inserted into apple and the resulting transgenic plant was significantly more tolerant to the bacterium *Erwinia amylovora* which causes fire blight (632). However, e.g., cecropin should be inserted into plants at a limited level because it may cause a toxic effect, not only to bacterial cells but also to plant protoplasts (210).

Toxins

A tobacco with insertion of a gene encoding for the enzyme ornithine carbamoyltransferase was insensitive to phaseolotoxin, an important virulence component of *Pseudomonas syringae* (270). These plants were less prone to infection by this bacteria. A gene coding for a subunit of a cholera toxin was inserted into tobacco and decreased the susceptibility to *Pseudomonas tabaci* (610). The gene activated the signalling pathway, resulting in constitutive expression of pathogenesis related proteins.

H_2O_2

A gene from *Aspergillus niger* coding for glucose oxidase which generates H_2O_2 was inserted into potato. The transgenic potatoes exhibited strongly enhanced tolerance to the bacteria *Erwinia corotovora* subsp. *corotovora* and an enhanced tolerance to the fungi *Phytophthora infestans* (1020).

9.4. Changed flower characteristics

9.4.1. Flower colours

"Classical flower breeding by continuous crossing and selection has its limitations, for example, no one has succeeded in breeding a blue rose or an orange petunia" (591). Pigmentation in plants are mainly due to three different types of compounds: Flavonoids, carotenoids and betalains. Flavonoids are the most common flower pigments which provide the greatest range of flower colours. The anthocyanins are the most important flavonoid pigments. Many structural genes encoding flavonoid enzymes have been isolated and characterized in detail (263). Colours available in *Petunia hybrida* are predominantly based on the derivatives of the flavonoids cyanidins and delphinidins giving a colour range from white via salmon, pink and red to violet. An orange pelargonidin-producing petunia variety was

Orange petunia

obtained by expressing the dihydro-flavonol-4-reductase gene (*dfr* gene) from maize in petunia. However, the flowers were pale brick-red. By crossing the transformed line with non-transgenic breeding lines, it was possible to obtain petunia with orange flower colour in combination with a good general performance (574, 672). Development of a blue rose is more complicated because the presence of the following three factors is required: Delphinidins (3',5'-hydroxylated anthocyanins), flavonol co-pigments and a relatively high vacuolar pH, but so far no blue roses have been created (591). However, a violet carnation has recently been approved for marketing in Australia. This colour was obtained by inserting two genes coding for the enzymes flavonoid 3',5'-hydroxylase and dihydroflavonol reductase which resulted in production of dephinidin (292). The intensity of flower colours has been modified by various methods. Pure white and pure pink flowers have been produced by inserting sense and antisense genes into petunia, chrysanthemum, gerbera and rose (591). Other traits, such as changed plant size, flower architecture, inflorescence and flower deterioration are currently being developed with the use of genetic engineering, but so far there are no reports on the modification of floral fragrances (591).

9.4.2. Male sterility

Ribonuclease

Several male sterile crop plants have been produced by combining a promoter gene, *TA29* from tobacco that is expressed specifically in anther tapetal cells with a natural ribonuclease gene (barnase gene) from *Bacillus amyloliquefaciens*. The presence of this chimeric gene selectively destroys the tapetum during anther development (539). In order to restore male fertility, oilseed rape plants were transformed with the *TA29* promoter gene and another gene (barstar gene) from *Bacillus amyloliquefaciens*. Barstar is a protein that corresponds to and inhibits the ribonuclease produced by the barnase gene by forming a stable complex with barnase in the cytoplasm. The *TA29*-barstar gene is a dominant suppressor of cytotoxic *TA29*-barnase gene activity and restores male fertility. The *TA29*-barstar transformants were male fertile and produced normal flowers, therefore the *bar* gene (phosphinotricin tolerance) was used as a selectable marker (540).

Glucanase

Male sterility in transgenic tobacco has been introduced with a slightly different approach (941). An anther specific promoter (*Osg6B* gene) from rice was fused with a gene coding for endo-ß-1,3 glucanase from soybean which caused premature digestion of the callose wall that surrounds microsporocytes at the time of meiosis.

Break down of self-incompatibility

In order to break down self-incompatibility in *Brassica* species an antisense gene of the S-locus glycoprotein (SLG gene), isolated from self-incompatible *B. campestris*, was introduced into another self-incompatible *B. campestris* (839). In *Brassica* self-incompatibility is controlled by multiallelic genes at the *S*-locus. The transgenic plants set seeds in about 80% of the flowers, whereas the control plants set seeds in less than 3% of the flowers.

9.5 Changed metabolic content

Changes in metabolic content can be either alterations in compounds that are naturally produced in the plant (protein, oil, carbohydrates, etc.) or it can be production of compounds not normally present in the plant. This new approach is often denoted "molecular farming" or "plants as bioreactors or factories" and these transgenic plants seem to have great potential for production of carbohydrates, fatty acids, pharmaceutical polypeptides, industrial enzymes and biodegradable plastics (302).

Amino acids and proteins

Lysine and threonine are considered very important among the essential amino acids. They are synthesized from aspartate by a pathway that is primarily regulated by sensitivity of the two key enzymes, aspartate kinase and dihydrodipicolinate synthase, to feed-back inhibition from lysine and threonine. One approach for increasing the content of lysine and threonine has been to insert genes coding for feed-back insensitive enzymes. A tobacco with increased content of lysine and threonine was produced by inserting two *Escherichia coli* genes coding for insensitive enzymes (279). Similarly, an oilseed rape and soybean with increased content of lysine were produced by inserting a gene from *E. coli* coding for an insensitive aspartate kinase and a gene from *Corynebacterium* coding for dihydrodipicolinate synthase (245). Another approach, which elevated the levels of lysine, methionine and cysteine in rape seeds, was used by introduction of an antisense gene coding for cruciferin, which is the most abundant storage protein in rape seeds (458).

Lipids

Triacylglycerides (used for manufacturing of detergents and for specialized nutritional applications) were produced by inserting the thioesterase gene from the California Bay Tree into *Arabidopsis thaliana* (302). There is growing interest for production of polyhydroxy-alkanoate polymers that can be used for synthesis of biodegradable plastics. A polyhydroxyalkanoate polymer was produced by *Arabidopsis thaliana* after introduction of genes from *Alcaligenes eutrophus* (709).

Carbohydrates

The amount of amylose in potato starch was reduced by inserting an antisense gene coding for granule-bound starch synthase (473). A cyclodextrin (used in pharmaceuticals delivery or as food additives) producing potato was obtained by inserting a gene from *Klebsiella pneumoniae* coding for cyclodextrin glycosyltransferase. A fructan storing potato was produced by introduction of a gene from *Bacillus subtilis* coding for fructosyltransferase. A mannitol producing tobacco was obtained by inserting a gene (*mtlD*) from *Escherichia coli* coding for mannitol-q-phosphate dehydrogenase. Similarly a pinitol producing tobacco was produced with a gene from *Mesembryanthemum crystallinum* (ice plant) (302).

Pharmaceuticals

Plants can be used to produce vaccines. The hepatitis B surface antigen was expressed in tobacco and potato and had similar qualitative properties as the commercial vaccine (203, 915). A human interleukin-6 gene was inserted into tobacco in order to find a new host for the production of this compound (474). Interleukin-6 plays an important role in human defence against various pathogens. Four different genes producing protein chains for secretory immuno-globulin were inserted into separate tobacco plants and these plants were crossed (523). The resulting tobacco plant was able to express all four proteins which assembled into a functional high molecular weight secretory immunoglobulin. Secretory immunoglobulin is abundant in mucosal secretions as the first line of defence against infectious agents.

9.6. Stress tolerance

9.6.1. Drought tolerance

The mechanisms by which a plant can grow under conditions of limited water can be divided into two groups: If the plant cell is able to continue metabolism under reduced water potential, the mechanism is considered to be drought tolerance. If the plant is able to find and retain water in an arid environment, then this is con-sidered to be a drought-avoidance mechanism (640). Several strategies have been used to develop drought tolerant plants.

Proline

A gene coding for glutamic acid kinase (*ProB* gene) from yeast was inserted into tobacco and was able to elevate the content of free proline which probably prevents osmotic stress (1036). A gene from mothbean (*Vigna aconitifolia*) coding for D-pyrroline-5-carboxylate synthase, a bifunctional enzyme that catalyzes the conversion of glutamate to D-pyrroline-5-carboxylate, which is then reduced to proline, was inserted into tobacco and rice (441). These tobacco plants had higher osmotic potentials of leaf sap and enhanced root biomass and flower development under drought stress than the wildtype (442).

Betaine

A gene from barley coding for betaine aldehyde dehydrogenase was inserted into tobacco. However, the transformants did not exhibit increased resistance to osmotic stress, and the authors suggest that introduction of choline monooxygenase is also required to produce enough betaine to act as an osmoprotectant (392).

Fructans

A gene from *Bacillus subtilis* (*SacB* gene) coding for fructans was inserted into tobacco, which does not produce fructans naturally. Under drought stress, biomass increased significantly in the trans-genic plants compared to the wild type. This difference was most pronounced in the roots (705).

It has been stated that drought tolerance in plants is a complex trait composed of many genes and that far more sophisticated genetic engineering approaches are needed before real progress in drought tolerance can be observed (640).

9.6.2. Oxidative stress tolerance
Oxidative stress occurs in chloroplasts when light intensities are high and the oxygen supply is abundant, when carbon dioxide fixation is limited (e.g., because of low temperatures or drought stress) or when air pollutants are abundant. Under these circumstances reactive oxygen intermediates are created which may cause biological damage to the plant. To prevent damage, the chloroplasts contain antioxidant systems which include enzymes, e.g., superoxide dismutase (transforming superoxide radicals to peroxide) and ascorbate peroxidase (transforming peroxide to water) (104).

Superoxide dismutase

Genetic engineering has successfully been used to increase the activity of the antioxidant enzymes in tobacco and potato by inserting a Cu/Zn superoxide dismutase from pea (16, 326, 987) and a gene from tomato was inserted into potato (694). A chimeric gene from *Nicotiana plumbaginifolia* coding for a Mn superoxide dismutase has been successfully introduced into tobacco, alfalfa and cotton. The overexpression of the chloroplastic Mn superoxide dismutase seems to provide increased chilling or drought tolerance in these crops (16, 562).

Glutathione reductase

Glutathione reductase is one of the enzymes involved in regeneration of oxidized ascorbate. A chimeric gene from *Escherichia coli* (*gor* gene) has been inserted into tobacco and poplar (16, 32, 755) and a gene from pea that overexpresses glutathione has been introduced into tobacco (104). However, the concluding remarks in a review (755) state that enhancement of just one of the components of the complex antioxidant system is usually insufficient and they suggest that genes coding for superoxide dismutase and glutathione reductase should be introduced together.

9.6.3. Other stress factors
Cold tolerance

Somatic hybrids between tomato (*Lycopersicon esculentum*) and *L. peruvianum* showed higher growth rates under low temperature conditions than tomato, but less than *L. peruvianum* (796). Genes coding for glycerol-3-phosphate from *Arabidopsis thaliana* and squash were introduced into tobacco (605). *Arabidopsis* is a species with a high proportion of unsaturated fatty acids which causes tolerance to chilling (similar in cold tolerance to spinach), whereas squash has a low proportion. The degree of chilling sensitivity in the transformed tobacco plants was similar to either squash or *Arabidopsis* and the wildtype tobacco was intermediate.

Salinity tolerance

Increased tolerance to salinity (NaCl) was obtained by inserting a bacterial gene (*mtlD*) coding for mannitol-phosphate dehydrogenase into *Arabidopsis thaliana* and tobacco. This enzyme catalyses the production of the acyclic polyol mannitol. However, the mechanism through which mannitol acts is still unknown (918).

9.7. Outlook for the future

So far, genetic engineering of plants has focused on inserting traits into the major crops of the world, but also transgenic forest trees and ornamental plants are produced. Great effort has been put into herbicide tolerant crops partly because the spectrum of crops, which tolerate existing herbicides, is broadened and partly because new herbicides are exceedingly expensive to develop. This makes it profitable to adapt biology (crops) to chemistry (herbicides) rather than adapting chemistry to biology as in the past. Insect tolerance, by inserting genes coding for *Bt*-toxins, has also received attention, but naturally developed tolerance against these toxins has already been found in insects and it is difficult to predict whether or not these crops will become a success for agriculture in the long term. For the plant breeders, genetic engineering is predicted to be an efficient and time saving tool. The farmers will have new options to control pests and weeds and the industry may obtain new production methods, e.g., "plants as factories". However, it is important to study and assess the long-term consequences of growing these transgenic crops to prevent "shooting ourselves in the foot" by creating self-inflicted problems such as increased pesticide use, herbicide tolerant weeds, tolerant insects, hybrid plants with unknown metabolites, etc. By constructing models that simulate different scenarios of herbicide tolerant crops in different crop rotations, the effects of introducing herbicide tolerant crops on herbicide use may be evaluated (529). A similar approach could be adopted for evaluation of other transgenic traits.

10. Principles and procedures for ecological risk assessment of transgenic plants

By: Gösta Kjellsson

Introduction

The emphasis of this chapter is on the practical procedures and theoretical framework which are currently applied or have been suggested for ecological risk assessment of transgenic plants. In this context, reference will be made to the general concepts and principles for risk assessment which have been established for other fields of environmental problems. This means that the experience from handling risk from toxic compounds in nature (ecotoxicology), simulation of processes (modelling) and management of populations and genetic resources (conservation ecology) will be included. Also, the experience from introduced species and monitoring invasion and gene-flow will be discussed.

10.1. The theory of ecological risk assessment

Definitions of risk

The field of ecological risk assessment is based on experience with human health risk and was established in the 1980s as a tool for making environmental decisions (511, 896). A paradigm for evaluating environmental risk can be expressed as:

Risk = Occurrence x Hazard,

where "occurrence" is the probability of occurrence of an event causing harm (a similar concept, exposure, is often used) and "hazard" refers to the resulting undesired effect (amount of harm) with negative consequences to the environment (490, 947). A short definition of "risk" would be "the potential for damage as a result of a hazard being realized" (139). Similar to this, the central questions to ecological risk assessment of GMPs are:

1. Can the GMP or its offspring inflict a hazard on the environment.
2. Can the transgene material be transferred to other organisms which could then inflict a hazard on the environment.

General paradigm for risk assessment

The need for a framework for procedures in risk assessment of GMP is evident when analytical methods, such as described in this book, are put into context of decision-making. Suggestions for a general paradigm for risk assessment of GMP with flow of procedures and necessary steps is presented in Figure 10.1. The procedures are based on experience for predictive risk assessment of chemicals (896, 937) and adapted to GMPs. This paradigm and the main steps will be discussed and compared to the procedures that are currently employed for risk assessment of GMPs. What are the conceptual strengths, how do the problems translate into GMP terminology and what are the main differences to risk assessment of transgenic plants?

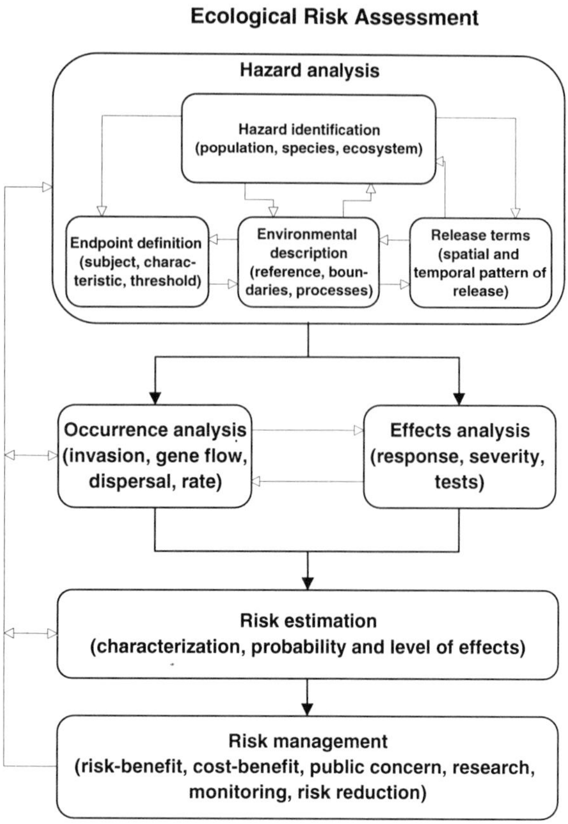

Ecological Risk Assessment

Figure 10.1
Stages in ecological risk assessment (based on: 896, 937).

10.1.1. Hazard analysis

The first stage in a predictive risk assessment is identification and analysis of the hazards involved including: Measurement variables (i.e., endpoint definition), description of the environment and the terms for the release (Figure 10.1). The main questions are: 1. What is at risk in the environment? and 2. How should the effects be defined?

Hazard identification

Hazard identification should be done in a strict manner by systematically listing and reviewing known hazards. To identify hazards for ecological effects of chemicals, various techniques for formal analyses of relationships between organisms have been used, e.g.: Identification of receptor organisms, scoring of intensity of occurrence and sensitivity, linkage trees to establish causal chains and modelling. Similar techniques could be applied to GMPs.

Hazards related to GMPs

Potential hazards to the ecosystem are not always easily identified when new organisms are introduced. The complexity of biological and environmental processes makes predictions difficult. However, experience gained from earlier cases and from related issues may help. The specific organism attributes and altered traits (e.g., genomic structure, origin, habitats, organism-interactions, etc.) indicate potential hazards (925). The main types of hazards, that have been identified as possible from the spread of GMPs into the environment, are shown in Table 10.1. The harmful effects caused by GMPs are described from species over population to the ecosystem level. The type of hazard may also relate to the specific type of environment. The loss of valuable biological resources for potential use (crops, chemical compounds) adds to the list of hazards (165). The major processes that are involved in creating the negative effects relate to gene-transfer within or between species and to changes in competitive and mutualistic interactions between species.

Table 10.1.
Potential ecological hazards from the release of GMPs into the environment.

The main types of conceivable effects to the environment and the type of interactions and processes that may cause the effects are shown.[1] The table is mainly based on information from: 139, 165, 177.

Harmful effects to the environment	Processes which can cause the effects
Cultivated ecosystems [2]	
Increased use of pesticides	Selection for resistance and transport to compatible plants
Creation of new agricultural weeds	Gene-flow and hybridization
Creation of new crop pests	Pathogen-plant interactions Herbivore-plant interactions [1]
Natural ecosystems [3]	
Invasion of new habitats	Pollen- and seed dispersal Disturbance [1] Competition [1]
Loss of genetic diversity in species	Gene-flow and hybridization Competition [1]
Reduction of population numbers	Gene-flow and hybridization Competition [1]
Harm to non-target species	Altered mutualistic relations
Loss of biodiversity	Competition [1] Environmental stress Added effects (gene, population, species)
Changes of nutrient cycles and geochemical processes	Interactions with abiotic environment [1] e.g., GMPs with N_2 fixing systems
Changes in primary production	Altered species composition [1]
Increased soil erosion	Added effects (interactions with environment, species composition)

[1] : Processes covered in vol. 1.
[2] : Only effects specific to cultivated ecosystems are shown.
[3] : Most effects are also relevant to cultivated ecosystems.

Endpoint definition

In ecotoxicology, the term "endpoint" refers to the assessment variables that are used for measuring effects. Endpoints are chosen on the basis of hazard definition and susceptibility to the hazardous agent (896). Besides, the biological relevance, the ease of measurement and a clear definition, are important issues. Examples of endpoints used in ecotoxicology tests are: Mortality levels (e.g., LD_{50}), effect concentrations (EC) and no observed effect concentration (NOEC) measured on life-history traits such as survival, growth and reproduction.

For risk assessment of GMPs relevant measurement variables (endpoints) could include: Fitness measures (e.g., survival, biomass accumulation and seed set), population rate of increase, genetic diversity and biodiversity in a receiver habitat, etc.

Environmental description

The description may include the use of reference environments, which are representative for the sites where a release would occur or habitats which are subject to invasion. It may be an abstraction (e.g., worst case) or an actual test site. The characteristics of the environment is described, such as ideally: Habitat-type, physical boundaries, biodiversity (plant and animal), vegetation density, natural or man-made disturbances, main interactive processes (herbivory, pollinator-plant, competitive relationships) and climatic and edaphic conditions. See also Section 10.3 and Chapter 7.3 in vol. 1.

Risk cascades

Interactive feedback loops between organisms will partly determine ecosystem stability. Disturbances to keystone species (see below) may cause a cascade of effects to a number of species and should be considered in environmental risk assessment (511).

Release terms - Spatial and temporal patterns

In ecotoxicology "source terms" are the temporal and spatial pattern of release of a chemical substance (e.g., pesticide spraying or emissions from a point source). A parallel concept "release terms" could easily be applied to risk assessment of GMPs and could cover: Location of release, number of releases and area covered, number of transplants, timing of the release, etc. When this information is included in official databases or perhaps a GIS-system, potentially unwanted events (e.g., hybridization, spread of transgenes) may be handled more effectively. In fact, in most countries, this information is already delivered by the industry when trials are made.

10.1.2. Analysis of occurrence

In risk assessment of chemicals, information of exposure to target and non-target species (plant, animal or man) is an essential element of risk assessment. The exposure analysis includes knowledge of the distribution paths and degradation of substances through ecosystems and the duration of processes (959).

Invasibility analysis

For GMPs, the concept of "occurrence" could refer to the product of relevant probability factors, e.g., the probability for gene transfer, hybridization and dispersal of plants into new areas in time (invasibility). Hence, it could include the probability of escape from a release and the subsequent fate of plants and inserted traits in the environment (300). Models for gene transfer, dispersal and invasion could become valuable to predict the probability of invasibility (see Section 10.4). The above factors would mainly be relevant for experimental trial releases where safety-measures to prevent gene dispersal are important. In risk assessment for commercial releases, the probability of escape is considered close to 1 and focus must be on probabilities for long-term establishment and negative effects in the environment (420).

10.1.3. Effects analysis

The effects from increasing levels of exposure to the hazardous agent are quantified during the effects analysis (e.g., as dose-response relationships). Studies of effects in ecological risk assessment have mainly been done at the species level, but tests of effects at the population and ecosystem level are needed and would be of perhaps greater use.

Individual level tests

Tests on individual organisms are easily designed for use in laboratory and greenhouse. Tests on toxicity of chemical compounds and of GMPs have been applied to many different organisms (e.g., algae, collemboles and mice) that are easily kept in culture. Data from single species tests may, with caution, be extrapolated to genus or higher levels by different statistical techniques. A selection of test methods relevant for risk analysis of GMPs for: Pollinator activity, pollen dispersal, cross and self-pollination, gene-flow, hybridization and plant competition, is shown in Table 10.2. For more detailed information see respective subcategories in Chapter 6 and in Chapter 5 in vol 1.

Table 10.2.
Levels of test systems for GMPs and main types of measures.

Examples of relevant methods for different measures relevant to risk analysis of GMPs are given. For a comprehensive list of methods, see Chapter 6 and Chapter 3 in vol. 1.

Level
    Measure
	Test methods
	    (examples)

Gene and genome
    Structural chromosome changes
	Genetic map construction (M23)
	 Karyotype analysis: Chromosome number, size
	    and form (M33)

Insert detection, Genetic marker [2] (Marker gene)	Morphological character analysis (M39) Polymerase chain reaction (PCR) (M59) Southern and Northern hybridization (M71)

Table 10.2. continued	
Individual [3]	
Cross-pollination, Self-pollination	Pollen deposition on stigmas: Pollen numbers and fertility (M51)
	Seed development analysis (M66)
	Spatial autocorrelation analysis (M72)
Mating system, Incompatibility	Diallel cross (M9)
	Experimental pollination (M13)
	Pollen germination tests (M52)
Pollinator activity	Pollinator attraction: Chemical cues (M55), Visual cues (M56)
	Pollinator foraging behaviour (M57)
Plant competition	Leaf area (M35) [1]
	Relative growth rate (M58) [1]
	Two-species mixtures (M81) [1]
Population [3]	
Population dynamics [1], Recruitment [1]	Germination test (M18, M19) [1]
	Half-life (M25) [1]
	Leslie matrix (M38) [1]
	Seed burial treatments (M65) [1]
	Effective population-size N_e (M11)
Pollen dispersal	Paternity analysis (M45)
	Pollen counts (M50)
	Pollen traps (M53)
	Pollen viability tests (M54)
Gene-flow, Hybridization, Introgression	Amplified restriction fragment polymorphism (AFLP) (M1)
	DNA sequencing (M10)
	F-statistics (M19)
	Haplotype statistics (M26)
	Meiotic analysis: Chromosome pairing and recombination (M37)
	Morphological character analysis (M39)
	Protein electrophoresis: Isozyme analysis (M60)
	RAPD (M61)
	RFLP (M64)
Genetic stability, Genetic diversity, Genetic drift, Outcrossing	Fitness measurement (M14)
	Genetic distance (M22)
	Genetic neighbourhood-size (M24)
	Selfing and outcrossing rate (M69)
Ecosystem	
Invasion, Invasibility	Biogeographical assay and monitoring (M4)
	Local adaptation analysis (M35)
	Reproduction in alien plants (M62)
	Transplantation experiment (M76)
Biodiversity [1], Community structure [1]	Diversity index (M11) [1]
	Sampling procedures: Vegetation analysis (M64) [1]
	Spatial pattern analysis (M76) [1]

[1] : Measures and methods covered in volume 1.
[2] : For specific suggestions of methods, see Table 6.1.
[3] : The listing of measures to individual or population is tentative.

Population level tests

Tests at the population level is necessary in order to include the genetic variation in susceptibility and response to hazardous agents. In addition, tests on development and fluctuations of population size (including modelling, see Section 10.4) give valuable parameters for evaluation of effects. The different issues that are important to risk analyses of GMPs include: Population dynamics, recruitment, mating system, genetic diversity, genetic stability, genetic drift and introgression. A selection of test methods are given in Table 10.2. For extensive references see "Population" in Chapter 6 and Chapter 5 in vol. 1.

The threshold for an organism to successfully invade a community is for the population to achieve a net reproductive rate > 1. The GMP may have obtained a selective advantage which could contribute to the invasiveness of the organism. The reproductive rate represents a summary measure of the species competitive ability, stress-tolerance and reproductive success in a given community.

Ecosystem level tests

Tests performed in actual ecosystems will provide the most direct and complete set of information on effects for risk assessment, but may not always be environmentally desirable unless strict precautions are taken. Valuable information on species interactions and ecosystem response is available, but data can be difficult to analyze due to a high degree of complexity and to spatial and temporal (year-to-year) variation. A higher degree of control is obtained in micro and mesocosm test-systems where small portions of actual ecosystems are contained and test compounds or test species added. Mesocosms have recently been adapted to tests of plant performance (e.g., GMPs) in terrestrial ecosystems (446, 888). Controlled Environmental Facilities, CEFs, have been used to study environmental change of terrestrial ecosystems and to simulate effects of the loss of biodiversity on ecosystem processes (486). Contained tests on GMPs in CEFs are possible, although expensive to produce. Field trials comparing GMPs with non-GMPs, which are sowed or transplanted into different types of habitats, give optimal information but may require such safety measures that they become difficult to perform.

Hierarchical test systems

A number of tests at different levels can be integrated in a hierarchical test procedure in a decision tree. For GMPs, the tests at the individual plant level could, e.g., consider the effects of inserted traits on competive ability, stress tolerance, reproductive output, etc., in relation to the unmodified plant. Suggestions for appropriate tests for effect analysis of GMPs at different levels are given in Table 10.2. An outline of a suggested integrated test system for GMPs with three levels is shown in Figure 10.2. At the first level (tier 1), analysis is based on available data or literature information on the organism (e.g., relatives, inserted trait, life-history, etc.). At tier 2 and 3, the analysis is based on test data from laboratory, greenhouse and field. Information at tier 1 is mainly qualitative, while it mainly is quantitative at tier 2 and 3. The complexity of the problems involved and the relevance to risk assessment increases from tier 1 to tier 3.

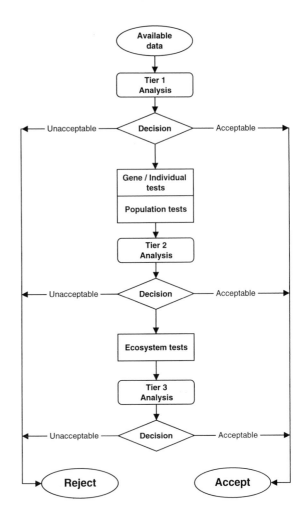

Figure 10.2.
Procedures in ecological risk analysis (adapted from: 490, 896). See text for
further information.

A decision of "accept" or "reject" is made at each level, if the results
of the analysis are clear cut. If information is insufficient or a
decision cannot be made, the analysis proceeds to the next tier. An
important issue is how to decide upon the relevant decision criteria
in a test procedure. This is especially important when quantitative
measures are used. In tests where GMPs are compared to non-
transgenic relatives, a "safe" requirement would be a statistically
non-significant deviation from expectations under ecologically
relevant conditions. Null hypothesis: The measurement values of
GMPs and non-GMPs are equal, or the GMPs perform less well
under the same experimental conditions.

Long-term effects

A general consensus of the importance to include information on long-term effects to risk assessment of GMPs exists (140), but has to be followed up with more studies addressing these questions. The definition in terms of years can be difficult, although for practical reasons 5 to 10 years are commonly used (177, 447). However, this should depend on the life-cycle of the GMP and the succession rate of the background environment. A monitoring program for spread of GMPs, transfer of transgenes to wild relatives and possible ecosystem changes, will be required (683). The release of GMPs in centres of origin (e.g., potatoes in South-America) should be considered especially problematic both on account of direct effects (weedy hybrids) and for long-term protection of original gene pools.

Effect levels

Identification of the mechanisms behind the effects to the ecosystem may be detected after a hazard has occurred (retrospective risk assessment). It has been suggested that this identification should be done at three different levels (139) by:

a. Characterization of the ecological interactions that caused the effects.

b. Characterization of the phenotypic or physiological traits of the organisms that caused the interactions.

c. Characterization of the genetic traits that led to phenotypic or physiological traits of the organism.

The reverse order of identification would apply to normal, prognostic, risk assessment. A number of biological cases have been presented to support this view (139).

10.1.4. Risk estimation
Estimates of risks can be derived from data on occurrence probabilities and data from tests on effects of the release.

Characterization of risks

Results from the analyses of occurrence and effects must be integrated in order to characterize the risks and make estimates of effect levels. This means describing and presentation (i.e., in form of graphics or models) of relationships between, e.g., occurrence and effects or between time and effects. Hence, occurrence probabilities for different scenarios (gene-flow, invasion) and level of effects should be given.

Probability and level of effects

The view followed in this chapter is that risk can only be stated meaningfully in probabilistic terms (62). This is certainly an optimal goal but not always possible to achieve when analysis of complex situations are made (e.g., ecosystem effects). However, models which incorporate the main parameters may be employed for estimation of hazard levels including stochastic elements.

Review data and test data   The different types of review and test data that are required for the
ecological risk assessment of GMPs have been summarized in Table
10.3. The data are obtained from industry, research and specific
environmental risk analysis of current cases and form the basis for
authorities to make decisions for field trials and commercial release.

Table 10.3
Requirements for review data and test data for ecological risk
assessment of GMPs.

Information which is useful as background for assessment has been
included. The table is partly based on: 945.

Receiver organism (plant)
    genetic composition
    taxonomy
    evolutionary history
    life-history traits
    competitive ability (weediness)
    pollination and gene-transfer
    reproduction
    naturalization and habitat choice

Transgene (Inserted trait)
    type of trait (see Chapter 9)

Biotechnological method and stability
    precision of technique (see Chapter 8)
    nature of biological vectors used (- // -)
    copy number and number of sites

Transgenic plant (GMP)
    expression of inserted gene (phenotype)
    toxicity of transgene products or metabolites
    altered life-history traits

Deliberate release conditions
    site characteristics
    trial design
    confinement measures

Fate of transgene
    gene-transfer and hybridization
    seed dispersal and survival (seed bank)
    establishment of population

Ecological effects.
    see Table 10.1

10.1.5. Risk management

Risk management involves the decision-making which attempts to minimize the undesired effects of a hazard. It involves political issues as well as practical precautions and restoration measures. Valuation of the ecological resources at risk depend on different considerations (scientific, public recreation, aesthetic values, etc.) which must be balanced. Analytic procedures as cost-benefit may be included. Proper communication between risk managers and risk analysts (scientists) is necessary.

Good developmental principles

Risk management of GMPs for field releases should be concerned with two main issues: 1. Application of relevant safety conditions based on the knowledge of organism and inserted trait; and 2. Effectuation of monitoring procedures during and after the trials. When doing basic research, during development of transgenic plants, minimum risk procedures should be followed. OECD has, based on experience and field tests, defined some general principles called "Good Developmental Principles (GDP)" in order to allow field research with negligible risks (645, 914). The GDP principles concern safety factors caused by the characteristics of the organism and the research site and the design of good experimental conditions.

Information on releases

The Information Resource for the release of Organisms into the Environment, IRRO (http://www.bdt.org.br/irro/irro.html) under the UN environmental program is gathering global information (438). The databases BIOTRACK from OECD with summary information on GMO releases, etc. by governments of member countries (http://cs1-hq.oecd.org/ehs/service.htm) and BIOCAT (Release of control organisms for pests) are especially valuable. Information on release permits in U.S.A. may be obtained from Department of Agriculture, Animal and Plant Health Inspection Service, APHIS (http://www.aphis.usda.gov/bbep/bp/).

Biosafety

The term "biosafety" has been defined as "the policies and procedures adopted to ensure the environmentally safe application of modern biotechnology" (463). Biosafety regulations may include a framework of existing or new statutes which may be risk or technology based. Also a case-by-case approach and step-by-step analysis are included (see below). International harmonization of regulations is a central issue, especially for safe use of GMPs in developing countries. An international register on biosafety under the United Nations Environmental Programme (UNEP) is easily available (http://irpct.unep.ch/biodiv). Also a Biosafety Information Network and Advisory Service (BINAS) has been established by UNIDO (http://binas.unido.org/binas/). Work on a biosafety protocol for GMOs has been initiated by UNEP in an attempt to rationalize international regulations and enhance collaboration (603).

10.2.   Established and new concepts for ecological risk assessment of GMPs

Case-by-case procedure

In the European Union and in most other countries a thorough risk assessment of every new GMP is required and done by governmental authorities before field release (e.g., field trial). Up to now this case-by-case procedure has been applied even if donor plant and inserted trait are both well known from former cases (see "familiarity", below). Background information on GMPs being assessed is usually requested and obtained from producers and additionally from agricultural, environmental and food quality authorities. The EU Directive (90/220/EEC) stipulates the terms for information which is given as "Summary Notification Information Format", SNIF. The information concerns: The characteristics of the host, the characteristics of the genetic insert, the characteristics of the GMP and a description of its use in relation to the environment for release. A simplified Annex (94/15/EC) for releases of transgenic crop plants has since been added by the EU. SNIFs on present cases are distributed via the Commission to all members of the EU for comments or request for supplementary information within 30 days. The criteria for risk assessment are under constant debate and movement towards more focus on long-term and secondary impacts and ecosystem response has been made (e.g., 290). Several efforts are currently being made to harmonize international guidelines for use of agricultural biotechnology (944). Recommendations for safety measures for biotechnology have been elaborated through several symposia and publications by the OECD (e.g., 644, 646, overview in 914).

Step-by-step procedure

In the step-by-step procedure, the testing of the GMPs is done in four stages of decreased safety measures: 1. Contained research (laboratory and greenhouse), 2. Confined trials (small-scale), 3. Unconfined trials (large-scale) and 4. Commercial release (372, 644). The possible adverse effects on the environment are assessed at each stage. Commercial tests are mainly concerned with the agronomic performance under cultivation. Plant gene-flow and invasive ability in native environments are not assessed in a direct manner and conclusions with respect to environmental concerns may be of limited use (177).

Familiarity and simplified procedures

The experience gained from previous releases should ideally be incorporated in regulatory risk assessment without lowering high safety standards. Hence, there is a need for simplified procedures based on familiarity with GMPs and inserted traits. The concept of familiarity has been defined according to the NRC-framework (341, 634) and can be evaluated in three steps:

1.   Is the plant a product of classical genetic methods or genetically modified (GMP)?
2.   Is the GMP phenotypically equivalent to a product of a classical method?

3.   Is the plant modified only by a marker gene or DNA sequence that will have no agricultural or environmental effects?

If the answer is "YES" to all questions, the GMP is regarded as familiar and no special measures (e.g., confinement in field tests) need to be taken, whereas a "NO" requires more research and data. The question about phenotypic response may be technically simple to answer if only one environment is involved, but more realistic measures may require several different types of environmental backgrounds. Knowledge about environmental effects is usually incomplete or not available. Examples on how specific traits (e.g., toxic substances) may affect organisms at different levels (weediness, resistance and non-target effects) have been described (80). In the OECD Environment Directorate, activities to establish "Minimal Data Requirements" for GMP releases are presently made, based on current practices in member countries (647).

Index procedures

The use of an index approach to risk classification of GMPs, as suggested by Dutch authors (570, 964), is currently being adapted to Swiss conditions (25). The GMPs are classified for three categories (chances for gene dispersal and hybridization, chances for seed dispersal and frequency of distribution of wild relatives) into five risk levels (low to high) based on available information from research and field observations. Potentially risky cases are identified and useful summary information is presented, which may be used further in risk assessment.

Gene migration rates

It has been argued that rates of spread of genes in natural, non-transgenic, populations may be used as valuable background information for design of mitigation procedures for GMPs (300). Inevitably, transgenes will escape from cultivation (418), as domesticated plants have done for centuries whether directly or through hybridization. Historical examples of cultigens running wild in Europe include, a.o.: *Brassica napus, Cichorium intybus, Daucus carota, Medicago sativa, Pyrus communis, Prunus serotina* and *Panicum miliaceum* (65). However, knowledge on gene-transfer and rates of migration between species or populations may provide valuable information when handling potential risk situations. An analogy is the control of spread of diseases (i.e., epidemic invasion of pathogens) where field data and models are used for prognostic purposes. Hence, the possibility to model the spread of genes by gene migration rates exists (see Section 10.4). The problems arising from release of GMPs should most often be considered as irreversible (177).

Significance of the transgene insert

A main issue is the combination of inserted traits, the genetical background and the ecological conditions under which these traits are expressed. The different traits that have been inserted in GMPs are described in Chapter 9 (see Table 9.1). The use of morphological, cytological, metabolic and protein traits for markers and the changes of genome structure are described in the subcategory marker (see

Table 6.1). The genetical background for the inserted traits may significantly affect the expression and the structural stability of the construct (185). The genetic factors and the specific trait are important to risk assessment and to the need for specific tests.

Levels of test systems and procedures

Decision making in risk assessment is made easier by employing simple decision trees or complex flowcharts where most decisions come down to simple YES or NO answers (848). When test systems are developed, it will be important to proceed in a stepwise fashion in a decision tree to test for effects at different levels from: 1. Gene, 2. Individual, 3. Population to 4. Ecosystem. Some main types of measures for each level are suggested in Table 10.2, where examples on relevant methods are given (see Chapter 6 for detailed information). A step-wise procedure for risk assessment of transgenic crop weediness and gene-flow to wild relatives has been proposed (768).

10.3. Risk assessment and the type of environment

The interactions between the GMP and the receptor environment will determine both the competitive ability, the success of invasion and effects to the environment. Hence, both the nature of the invader and the particular environment has to be considered.

Habitat invasibility

It has long been known that different habitat-types show different susceptibility to invasion of alien species. The risk of invasion mainly depend on: Successional stage, humidity level and frequency of disturbances. The relationship between species diversity of a habitat and its invasibility is, however, not clear. See Chapter 6, p. 55 and 7.3 in vol. 1 for more information. Methods to measure habitat invasibility exist (see Table 10.2). In addition, a possible procedure would be to rank European habitats in order of their invasibility (177).

Weediness

The ability of a transgenic crop or an introduced species to become a weed in cultivated soil (weediness) may be considered a separate problem (744, 768). Occurrence of new weeds or increased weediness in an existing species may result in crop reduction and lead to environmentally undesirable measures (e.g., an increased use of herbicides, etc.; see also Table 10.1). Furthermore, the possible spread of the species to natural habitats has to be considered (see below). However, the concept of weediness, as such, is probably not that useful. Weediness, much similar to "competitive ability", is not an inherent trait of the species, but depends on the type of habitat, the vegetation composition and the edaphic conditions (177). Plant species that may become, or have become, a problem to plant production in cultivated soil, i.e., weeds, are usually annual, but otherwise show large differences. Hence, weediness cannot be predicted from plant characters (e.g., Bakers characters) and opinions on the weediness of actual species often differ (1002, 1003). Increased weediness may be caused by hybridization between crop and wild relatives (403).

Regional aspects

An important general concern of risk assessment would be to include the regional aspects to assessment procedures. Differences in climate, soil properties and vegetation are significant between, e.g., lowland NW Europe, mountainous Central Europe and Mediterranean Europe. Hence, the problems and questions asked concerning possible trangene invasion and impact on habitat types may change. Furthermore, the awareness of government authorities and the public may differ strongly between countries as may the approaches of the industry.

Invasion of exotic species

Much controversy has been raised about the invasion of exotic species and the question if and how this subject can be applied to risk estimation for GMPs (21, 749, 1002). It has been suggested that invasion of new genotypes within native species may cause less environmental damage than invasion of exotic species (844). However, potential hazards will depend on the inserted trait and its ecological effects. Invasion of foreign species may have severe effects on plant cover and the natural vegetation present. Examples include the invasion of foreign species into seashore environments where they may become dominant (keystone) such as *Rosa rugosa* in NV Europe and *Carpobrotus* spp. in Mediterranean Europe. It seems to be very difficult to predict which invasive species will become pests in various habitats, based solely on plant characters (1002). A general opinion is that a broad niche facilitates invasion. Invaders possessing traits such as high growth rate, large biomass accumulation and high reproductive capacity, represent potential hazards, but specific test procedures are clearly needed to make an assessment.

Keystone species

Identification of keystone species, which are important determinants of community structure, in different types of habitats, would be highly desirable from an ecological standpoint, although it might be difficult (177). Generally, the impact of a keystone species on the ecosystem should be greater than expected from its abundance. Hence, a keystone plant may be identified from the ratio of its biomass and its effect on the ecosystem (889).

10.4. The role of modelling in ecological risk assessment

Mathematical models can be used to predict the outcome of situations, either where trials are impossible to perform due to safety consideration or where it is impractical to cover a whole range of situations. A sensitivity analysis can be used to identify key-parameters in the model, which can point to areas where additional information and research is needed (62, 117). Requirements for accuracy and the specific parameters (endpoints) needed are important to determine before models are chosen. Model types that are used to predict ecosystem changes precisely, usually require a large amount of different data and the complexity may inhibit model performance. However, spatial models of vegetation dynamics (occurrence of gaps, etc.) could prove valuable for prediction of risk of invasion. Relatively simple models that simulate interactive processes between 2 or 3 organisms (e.g., competition, population dynamics or herbivory) in different environments can be used to extrapolate data from test systems to a range of conditions. As always, the statistical analyses of data requires careful considerations. Especially for ecologically based models, it is important to secure that the models are validated by field data.

A few specific uses of modelling for GMPs can be mentioned. Mathematical models on resource competition may be useful for identifying transgenic plants that have the potential for altering community structure (28). Models may also be used to estimate risk of gene spread from field trials (417). Many of the methods indicated under the category "Population" in this book are based on genetic models.

Experience from
conservation ecology

In conservation ecology, the main questions concern the abundance and change of populations over a period of time and the risks of extinction. Some of the problems are similar to those concerning invasion and establishment of GMPs in natural environments, although from an opposite viewpoint (risk of extinction). Stochastic models for population size over a period, including density-dependence and age structure (see M38 in vol. 1), are much used (117). Models for spatial structure, dynamics of metapopulations, for migration and inbreeding are also employed. It has been argued that the DNA technology may be used for conservation genetical purposes, although it may also have negative implications for the genetic diversity of natural populations (495).

The role of modelling in risk assessment must not be overstressed into making confident predictions about success of invasion, etc., perhaps based on limited data (see 419). The principle of *caveat emptor* must, as always, guide us through the perils of future promises.

Reviews

62, 117, 139, 463, 490, 896, 945.

11. References

1. *Ågren, J., Schemske, D.W. (1993)*: Outcrossing rate and inbreeding depression in two annual monoecious herbs, *Begonia hirsuta* and *B. semiovata*. Evolution 47: 125-135.

2. *Ågren, J., Schemske, D.W. (1995)*: Sex allocation in the monoecious herb *Begonia semiovata*. Evolution 49: 121-130.

3. *Abbo, S., Miller, T.E., King, I.P. (1993)*: Primer-induced in situ hybridization to plant chromosomes. Genome 36: 815-817.

4. *Abbott, R.J., Irwin, J.A. (1988)*: Pollinator movements and the polymorphism for outcrossing rate at the ray floret locus in groundsel, *Senecio vulgaris* L. Heredity 60: 295-298.

5. *Adam, K.D., Köhler, W.H. (1996)*: Evolutionary genetic considerations on the goals and risks in releasing transgenic crops. In: Tomiuk, J., Wöhrmannn, K., Sentker, A., (eds.) Transgenic organisms: Biological and social implications. Birkhäuser, Basel, 59-79.

6. *Adam-Blondon, A.-F., Sevignac, M., Dron, M., Bannerot, H. (1994)*: A genetic map of common bean to localize specific resistance genes against anthracnose. Genome 37: 915-924.

7. *Adams, M.D., Fields, C., Venter, J.C. (1994)*: Automated DNA sequencing and analysis. Academic Press, London, 368 pp.

8. *Adams, R.P. (1982)*: A comparison of multivariate methods for the detection of hybridization. Taxon 31: 646-661.

9. *Adang, M.J., Brody, M.S., Cardineau, G., Eagan, N., Roush, R.T., Shewmaker, C.K., Jones, A., Oakes, J.V., McBride, K.E. (1993)*: The reconstruction and expression of a *Bacillus thuringiensis crylIIa* gene in protoplasts and potato plants. Plant Mol. Biol. 21: 1131-1145.

10. *Affre, L., Thompson, J.D., Debussche, M. (1995)*: The reproductive biology of the Mediterranean endemic *Cyclamen balearicum* Willk. (Primulaceae). Bot. J. Linn. Soc. 118: 309-330.

11. *AgrEvo (1996)*: The liberty link system: Profitable canola made simple. Connecting industry with farmers A1-A8.

12. *Aizen, M.A. (1993)*: Self pollination shortens flower lifespan in *Portulaca umbraticola* H.B.K. (Portulacaceae). Int. J. Plant Sci. 154: 412-415.

13. *Akama, K., Shiraishi, H., Ohta, S., Nakamura, K., Okada, K., Shimura, Y. (1992)*: Efficient transformation of *Arabidopsis thaliana*: Comparison of the efficiencies with various organs, plant ecotypes and *Agrobacterium* strains. Plant Cell Rep. 12: 7-11.

14. *Akella, V., Lurquin, P.F. (1993)*: Expression in cowpea seedlings of chimeric transgenes after electroporation into seed-derived embryos. Plant Cell Rep. 12: 110-117.

15. *Ali, M., Copeland, L.O., Elias, S.G., Kelly, J.D. (1995)*: Relationship between genetic distance and heterosis for yield and morphological traits in winter canola (*Brassica napus* L.). Theor. Appl. Genet. 91: 118-121.

16. *Allen, R.D. (1995)*: Dissection of oxidative stress tolerance using transgenic plants. Plant Physiol. 107: 1049-1054.

17.   *Allendorf, F.W. (1983)*: Isolation, gene flow, and genetic differentiation among populations. In: Schonewald-Cox, C.M., Chambers, S.M., MacBride, B., Thomas, W.L., (eds.) Genetics and conservation: A reference for managing wild animal and plant populations. The Benjamin/Cummings Publishing Company, London, 51-65.

18.   *Alpert, P., Lumaret, R., Giusto, F. di (1993)*: Population structure inferred from allozyme analysis in the clonal herb *Fragaria chiloensis* (Rosaceae). Amer. J. Bot. 80: 1002-1006.

19.   *Alston, R.E. (1965)*: Flavonoid chemistry of *Baptisia*: A current evaluation of chemical methods in the analysis of interspecific hybridizations. Taxon 14: 268-274.

20.   *Altabella, T., Chrispeels, M.J. (1990)*: Tobacco plants transformed with the bean αai gene express an inhibitor of insect α-amylase in their seeds. Plant Physiol. 93: 805-810.

21.   *Altmann, M. (1993)*: Gene technology and biodiversity. Experientia 49: 187-189.

22.   *Altmann, M., Ammann, K. (1992)*: Gentechnologie im gesellschaftlichen Spannungsfeld: Züchtung transgener Kulturpflanzen. GAIA 4: 204-213.

23.   *Aly, M.A.M., Owens, L.D. (1987)*: A simple system for plant-cell microinjection and culture. Plant Cell Tissue Organ Cult. 10: 159-174.

24.   *Ammann, K., Felber, F., Keller, J., Jacot, Y., Küpfer, P., Rufener Al Mazyad, P., Savova, D. (1994)*: Dynamic biogeography and natural hybridization of selected weedy species in Switzerland. In: Jacot, Y., Ammann, K., Pythoud, F., (eds.) Gene transfer: Are wild species in danger? Proceedings of the "Le Louverain" Symposium, Nov. 9th, 1993. Federal Office of Environment, Forests and Landscape (FOEFL), Bern, 36-39.

25.   *Ammann, K., Jacot, Y., Rufener Al Mazyad, P. (1996)*: Field release of transgenic crops in Switzerland, an ecological risk assessment. In: Schulte, E., Käppeli, O., (eds.) Gentechnisch veränderte krankheits- und schädlingsresistente Nutzpflanzen. Eine Option für die Landwirtschaft? Schwerpunktprogramm Biotechnologie des Schweizerischen National-fonds, Bern, 101-157.

26.   *Amrhein, N., Deus, B., Gehrke, P., Steinrücken, H.C. (1980)*: The site of the inhibition of the shikimate pathway by glyphosate. II. Interference of glyphosate with chorismate formation *in vivo* and *in vitro*. Plant Physiol. 66: 830-834.

27.   *Anderson, E. (1949)*: Introgressive hybridization. Biological Research Series, John Wiley & Sons, New York, 109 pp.

28.   *Andow, D.A. (1994)*: Community response to transgenic plant release: Using mathematical theory to predict effects of transgenic plants. Mol. Ecol. 3: 65-70.

29.   *Antonius, K., Nybom, H. (1994)*: DNA fingerprinting reveals significant amounts of genetic variation in a wild raspberry *Rubus idaeus* population. Mol. Ecol. 3: 177-180.

30.   *Antonovics, J., Clay, K., Schmitt, J. (1987)*: The measurement of small-scale environmental heterogeneity using clonal transplants of *Anthoxanthum odoratum* and *Danthonia spicata*. Oecologia 71: 601-607.

31.   *Antos, J.A., Allen, G.A. (1994)*: Biomass allocation among reproductive structures in the dioecious shrub *Oemleria cerasiformis* - a functional interpretation. J. Ecol. 82: 21-29.

32.   *Aono, M., Kubo, A., Saji, H., Tanaka, K., Kondo, N. (1993)*: Enhanced tolerance to photo-oxidative stress of transgenic *Nicotiana tabacum* with high chloroplastic glutathione reductase activity. Plant Cell Physiol. 34: 129-135.

33. *Aradhya, K.M., Mueller-Dombois, D., Ranker, T.A. (1993)*: Genetic structure and differentiation in *Metrosideros polymorpha* (Myrtaceae) along altitudinal gradients in Maui, Hawaii. Genet. Res. 61: 159-170.

34. *Arencibia, A., Molina, P.R., Delariva, G., Selmanhousein, G. (1995)*: Production of transgenic sugarcane (*Saccharum officinarum* L.) plants by intact cell electroporation. Plant Cell Rep. 14: 305-309.

35. *Armstrong, J.S., Gibbs, A.J., Peakall, R., Weiller, G. (1994)*: The RAPDistance Package. Obtainable via the World Wide Web.

36. *Arnold, M.L., Bennett, B.D. (1993)*: Natural hybridization in Louisiana irises: Genetic variation and ecological determinants. In: Harrison, R.G., (ed.) Hybrid zones and the evolutionary process. Oxford University Press, New York, 115-139.

37. *Arnold, M.L., Hodges, S.A. (1995)*: Are natural hybrids fit or unfit relative to their parents? Trend. Ecol. Evolut. 10: 67-71.

38. *Arnold, M.L., Robinson, J.J., Buckner, C.M., Bennet, B.D. (1992)*: Pollen dispersal and interspecific gene flow in Louisiana irises. Heredity 68: 399-404.

39. *Arroyo, J., Dafni, A. (1995)*: Variations in habitat, season, flower traits and pollinators in dimorphic *Narcissus tazetta* L. (Amaryllidaceae) in Israel. New Phytol. 129: 135-145.

40. *Asano, Y., Otsuki, Y., Ugaki, M. (1991)*: Electroporation-mediated and silicon carbide fiber-mediated DNA delivery in *Agrostis alba* L. (Redtop). Plant Sci. 79: 247-252.

41. *Asano, Y., Ugaki, M. (1994)*: Transgenic plants of *Agrostis alba* obtained by electroporation-mediated direct gene transfer into protoplasts. Plant Cell Rep. 13: 243-246.

42. *Ashman, T.-L. (1992)*: Indirect costs of seed production within and between seasons in a gynodioecious species. Oecologia 92: 266-272.

43. *Ashman, T.-L., Stanton, M. (1991)*: Seasonal variation in pollination dynamics of sexually dimorphic *Sidalcea oregana* ssp. *spicata* (Malvaceae). Ecology 72: 993-1003.

44. *Asmussen, C.B. (1993)*: Pollination biology of the sea pea, *Lathyrus japonicus*: Floral characters and activity and flight patterns of bumblebees. Flora 188: 227-237.

45. *Asmussen, M.A., Basten, C.J. (1994)*: Sampling theory for cytonuclear disequilibria. Genetics 138: 1351-1363.

46. *Atkinson, H.J., Urwin, P.E., Hansen, E., McPherson, M.J. (1995)*: Designs for engineered resistance to root-parasitic nematodes. Trends Biotech. 13: 369-374.

47. *Avise, J.C. (1994)*: Molecular markers, natural history and evolution. Chapman & Hall, New York, 511 pp.

48. *Ayres, N.M., Park, W.D. (1994)*: Genetic transformation of rice. Crit. Rev. Plant Sci. 13: 219-239.

49. *Bachem, C.W.B., Hoeven, R.S. van der, Bruijn, S.M. de, Vreugdenhil, D., Zabeau, M., Visser, R.G.F. (1996)*: Visualization of differential gene expression using a novel method of RNA fingerprinting based on AFLP: Analysis of gene expression during potato tuber development. Plant J. 9: 745-753.

50. *Bachmann, K. (1994)*: Tansley review No. 63. Molecular markers in plant ecology. New Phytol. 126: 403-418.

51.   *Baker, H.G., Baker, I. (1975)*: Studies of nectar-constituents and pollinator-plant coevolution. In: Gilbert, L.E., Raven, P.H., (eds.) Coevolution of plants and animals. University of Texas Press, Austin, 100-140.

52.   *Baker, H.G., Baker, I. (1979)*: Sugar ratios in nectars. Phytochem. Bull. 12: 43-45.

53.   *Baker, I. (1979)*: Methods for the determination of volumes and sugar concentrations from nectar spots on paper. Phytochem. Bull. 12: 40-42.

54.   *Baker, I., Baker, H.G. (1976a)*: Analyses of amino acids in flower nectars of hybrids and their parents, with phylogenetic implications. New Phytol. 76: 87-98.

55.   *Baker, I., Baker, H.G. (1976b)*: Analysis of amino acids in nectar. Phytochem. Bull. 9: 4-6.

56.   *Barcelo, P., Hagel, C., Becker, D., Martin, A., Lorz, H. (1994)*: Transgenic cereal (*Tritordeum*) plants obtained at high efficiency by microprojectile bombardment of inflorescence tissue. Plant J. 5: 583-592.

57.   *Barker, H., Reavy, B., Kumar, A., Webster, K.D., Mayo, M.A. (1992)*: Restricted virus multiplication in potatoes transformed with the coat protein gene of potato leafroll luteovirus: Similarities with a type of host gene-mediated resistance. Ann. Appl. Biol. 120: 55-64.

58.   *Barrett, S.C.H., Harder, L.D. (1996)*: Ecology and evolution of plant mating. Trend. Ecol. Evolut. 11: 73-79.

59.   *Barrett, S.C.H., Richardson, B.J. (1986)*: Genetic attributes of invading species. In: Groves, R.H., Burdon, J.J., (eds.) Ecology of biological invasions. Cambridge University Press, Cambridge, 21-33.

60.   *Barrett, S.C.H., Shore, J.S. (1989)*: Isozyme variation in colonizing plants. In: Soltis, D.E., Soltis, P.S., (eds.) Isozymes in plant biology. Chapman & Hall, London, 106-126.

61.   *Barry, G., Kishore, G., Padgette, S., Taylor, M., Kolacz, K., Weldon, M., Re, D., Eichholtz, D., Fincher, K., Hallas, L. (1992)*: Inhibitors of amino acid biosynthesis: Strategies for imparting glyphosate tolerance to crop plants. In: Singh, B.K., Flores, H.E., Shannon, J.C., (eds.) Biosynthesis and molecular regulation of amino acids in plants. American Society of Plant Physiologists, 139-145.

62.   *Bartell, S.M., Gardner, R.H., O'Neil, R.V. (1992)*: Ecological risk estimation. Lewis Publishers, Boca Raton, 252 pp.

63.   *Barton, N.H., Hewitt, G.M. (1985)*: Analysis of hybrid zones. Annu. Rev. Ecol. Syst. 16: 113-148.

64.   *Barton, N.H., Slatkin, M. (1986)*: A quasi-equilibrium theory of the distribution of rare alleles in a subdivided population. Heredity 56: 409-415.

65.   *Bartsch, D., Sukopp, H., Sukopp, U. (1993)*: Introduction of plants with special regard to cultigens running wild. In: Wöhrmann, K., Tomiuk, J., (eds.) Transgenic organisms: Risk assessment of deliberate release. Birkhäuser Verlag, Basel, 135-151.

66.   *Bates, G.W. (1992)*: Molecular analysis of nuclear genes in somatic hybrids. Physiol. Plant. 85: 308-314.

67.   *Bates, G.W., Hasenkampf, C.A. (1985)*: Culture of plant somatic hybrids following electrical fusion. Theor. Appl. Genet. 70: 227-233.

68. *Bauer, L.S. (1995)*: Resistance: A threat to the insecticidal crystal proteins of *Bacillus thuringiensis*. Fla. Entomol. 78: 414-443.

69. *Baulcombe, D. (1994)*: Novel strategies for engineering virus resistance in plants. Curr. Opin. Biotechnol. 5: 117-124.

70. *Bayley, C., Trolinder, N., Ray, C., Morgan, M., Quisenberry, J.E., Ow, D.W. (1992)*: Engineering 2,4-D resistance into cotton. Theor. Appl. Genet. 83: 645-649.

71. *Bazzaz, F.A. (1986)*: Life history of colonizing plants: Some demographic, genetic, and physiological features. In: Mooney, H.A., Drake, J.A., (eds.) The ecology of biological invasions of North America and Hawaii. Springer Verlag, Berlin, 96-110.

72. *Bechtold, N., Ellis, J., Pelletier, G. (1993): In planta Agrobacterium* mediated gene transfer by infiltration of adult *Arabidopsis thaliana* plants.C.R. Acad. Sci. Ser. III - Vie 316: 1194-1199.

73. *Becker, D., Brettschneider, R., Lorz, H. (1994)*: Fertile transgenic wheat from microprojectile bombardment of scutellar tissue. Plant J. 5: 299-307.

74. *Becker, H.C., Engqvist, G.M., Karlsson, B. (1995)*: Comparison of rapeseed cultivars and resynthesized lines based on allozyme and RFLP markers. Theor. Appl. Genet. 91: 62-67.

75. *Belhassen, E., Atlan, A., Couvet, D., Gouyon, P.H., Quétier, F. (1993)*: Mitochondrial genome of *Thymus vulgaris* L. (Labiatae) is highly polymorphic between and among natural populations. Heredity 71: 462-472.

76. *Belzer, N.F., Ownbey, M. (1971)*: Chromatographic comparison of *Tragopogon* species and hybrids. Amer. J. Bot. 58: 791-802.

77. *Benchekroun, A., Michaud, D., Binh, N.Q., Overney, S., Desjardins, Y., Yelle, S. (1995)*: Synthesis of active oryzacystatin I in transgenic potato plants. Plant Cell Rep. 14: 585-588.

78. *Benhamou, N. (1995)*: Immunocytochemistry of plant defense mechanisms induced upon microbial attack. Microsc. Res. Technique 31: 63-78.

79. *Berg, E.E., Hamrick, J.L. (1995)*: Fine-scale genetic structure of a Turkey oak forest. Evolution 49: 110-120.

80. *Bergmans, H.E.N. (1993)*: The development of risk evaluation for the deliberate release of genetically modified organisms: An outline of the international discussion. In: Vuijk, D.H., Dekkers, J.J., van der Plas, H.C., (eds.) Developing agricultural biotechnology in the Netherlands. Pudoc Scientific Publishers, Wageningen, 297-302.

81. *Bergström, G., Dobson, H.E.M., Groth, I. (1995)*: Spatial fragrance patterns within the flowers of *Ranunculus acris (Ranunculaceae)*. Plant Syst. Evol. 195: 221-242.

82. *Bernatzky, R., Tanksley, S.D. (1989)*: Restriction fragments as molecular markers for germplasm evaluation and utilisation. In: Brown, A.H.D., Marshall, D.R., Frankel, O.H., Williams, J.T., (eds.) The use of plant genetic resources. Cambridge University Press, Cambridge, 353-362.

83. *Bertin, R.I., Newman, C.M. (1993)*: Dichogamy in angiosperms. Bot. Rev. 59: 112-152.

84. *Bertin, R.I., Sholes, O.D.V. (1993)*: Weather, pollination and the phenology of *Geranium maculatum*. Amer. Midland Naturalist 129: 52-66.

85. *Best, L.S., Bierzychudek, P. (1982)*: Pollinator foraging on foxglove (*Digitalis purpurea* L): A test of a new model. Evolution 36: 70-79.

86.   *Bevan, M.W., Harrison, B.D., Leaver, C.J. (1994)*: The production and uses of genetically transformed plants. Chapman & Hall, London, 111 pp.

87.   *Bilang, R., Zhang, S., Leduc, N., Iglesias, V.A., Gisel, A., Simmonds, J., Potrykus, I., Sautter, C. (1993)*: Transient gene expression in vegetative shoot apical meristems of wheat after ballistic microtargeting. Plant J. 4: 735-744.

88.   *Binelli, G., Gianfranceschi, L., Pè, M.E., Taramino, G., Busso, C., Stenhouse, J., Ottaviano, E. (1992)*: Similarity of maize and sorghum genomes as revealed by maize RFLP probes. Theor. Appl. Genet. 84: 10-16.

89.   *Birch, R.G., Franks, T. (1991)*: Development and optimization of microprojectile systems for plant genetic transformation. Aust. J. Plant Physiol. 18: 453-469.

90.   *Blaakmeer, A., Geervliet, J.B.F., Loon, J.J.A. van, Posthumus, M.A., Beek, T.A. van, Groot, A. de (1994)*: Comparative headspace analysis of cabbage plants damaged by two species of *Pieris* caterpillars: Consequences for in-flight host location by *Cotesia* parasitoids. Entomol. Exp. Appl. 73: 175-182.

91.   *Blackhall, N.W., Finch, R.P., Power, J.B., Cocking, E.C., Davey, M.R. (1995)*: Flow cytometric quantification of electroporation mediated uptake of macromolecules into plant protoplasts. Protoplasma 186: 50-56.

92.   *Block, M. de (1993)*: The cell biology of plant transformation: Current state, problems, prospects and the implications for the plant breeding. Euphytica 71: 1-14.

93.   *Block, M. de, Botterman, J., Vandewiele, M., Dockx, J., Thoen, C., Gosselé, V., Movva, N.R., Thompson, C., Montagu, M. van, Leemans, J. (1987)*: Engineering herbicide resistance in plants by expression of a detoxifying enzyme. EMBO J. 6: 2513-2518.

94.   *Bolik, M., Koop, H.U. (1991)*: Identification of embryogenic microspores of barley (*Hordeum vulgare* L.) by individual selection and culture and their potential for transformation by microinjection. Protoplasma 162: 61-68.

95.   *Bolten, A.B., Feinsinger, P., Baker, H.G., Baker, I. (1979)*: On the calculation of sugar concentration in flower nectar. Oecologia 41: 301-304.

96.   *Borghi, B., Perenzin, M. (1994)*: Diallel analysis to predict heterosis and combining ability for grain yield, yield components and bread-making quality in bread wheat (*T. aestivum*). Theor. Appl. Genet. 89: 975-981.

97.   *Borg-Karlson, A.K., Valterová, I., Nilsson, L.A. (1994)*: Volatile compounds from flowers of six species in the family *Apiaceae*: Bouquets for different pollinators. Phytochemistry 35: 111-119.

98.   *Bos, M., Harmens, H., Vrieling, K. (1986)*: Gene flow in *Plantago*. I. Gene flow and neighbourhood size in *P. lanceolata*. Heredity 56: 43-54.

99.   *Boshier, D.H., Chase, M.R., Bawa, K.S. (1995)*: Population genetics of *Cordia alliodora* (Boraginaceae), a neotropical tree. 3. Gene flow, neighborhood, and population substructure. Amer. J. Bot. 82: 484-490.

100. *Botterman, J., D'Halluin, K., Block, M. de, Greef, W. de, Leemans, J. (1991)*: Engineering of glufosinate resistance and evaluation under field conditions. In: Caseley, J.C., Cussans, G.W., Atkin, R.K., (eds.) Herbicide resistance in weeds and crops. ButterworthHeinemann, Oxford, 355-365.

101. *Boulter, D., Edwards, G.A., Gatehouse, A.M.R., Gatehouse, J.A., Hilder, V.A. (1990)*: Additive protective effects of different plant-derived insect resistance genes in transgenic tobacco plants. Crop Prot. 9: 351-354.

102. *Brailsford, R.W., Voesenek, L.A.C.J., Blom, C.W.P.M., Smith, A.R., Hall, M.A., Jackson, M.B. (1993)*: Enhanced ethylene production by primary roots of *Zea mays* L. in response to sub-ambient partial pressures of oxygen. Plant Cell Environ. 16: 1071-1080.

103. *Brauner, S., Crawford, D.J., Stuessy, T.F. (1992)*: Ribosomal DNA and RAPD variation in the rare plant family Lactoridaceae. Amer. J. Bot. 79: 1436-1439.

104. *Broadbent, P., Creissen, G., Wellburn, F.A.M., Mullineaux, P.M., Wellburn, A.R. (1994)*: Biochemical effects of tropospheric ozone in transgenic plants. Biochem. Soc. Trans. 22: 1020-1025.

105. *Brochmann, C. (1993)*: Reproductive strategies of diploid and polyploid populations of arctic *Draba* (Brassicaceae). Plant Syst. Evol. 185: 55-83.

106. *Broglie, K., Chet, I., Holliday, M., Cressman, R., Biddle, P., Knowlton, S., Mauvais, C.J., Broglie, R. (1991)*: Transgenic plants with enhanced resistance to the fungal pathogen *Rhizoctonia solani*. Science 254: 1194-1197.

107. *Broglie, R., Broglie, K., Roby, D., Chet, I. (1993)*: Production of transgenic plants with enhanced resistance to microbial pathogens. In: Kung, S.-D., Wu, R., (eds.) Transgenic plants. Engineering and utilization. Academic Press, San Diego, 1: 265-276.

108. *Brown, A.H.D. (1975)*: Sample sizes required to detect linkage disequilibrium between two or three loci. Theor. Pop. Biol. 8: 184-201.

109. *Brown, A.H.D., Allard, R.W. (1970)*: Estimation of the mating system in open-pollinated maize populations using isozyme polymorphisms. Genetics 66: 133-145.

110. *Brown, A.H.D., Feldman, M.W. (1981)*: Population structure of multilocus associations. Proc. Nat. Acad. Sci. USA 73: 5913-5916.

111. *Brown, A.H.D., Grant, J.E., Pullen, R. (1986)*: Outcrossing and paternity in *Glycine argyrea* by paired fruit analysis. Biol. J. Linn. Soc. 29: 283-294.

112. *Brown, B.A., Clegg, M.T. (1984)*: Influence of flower color polymorphism on genetic transmission in a natural population of the common morning glory, *Ipomoea purpurea*. Evolution 38: 796-803.

113. *Broyles, S.B., Schnabel, A., Wyatt, R. (1994)*: Evidence for long-distance pollen dispersal in milkweeds (*Asclepias exaltata*). Evolution 48: 1032-1040.

114. *Boyles, S.B., Wyatt, R. (1993)*: Allozyme diversity and genetic structure in Southern Appalachian populations of poke milkweed, *Asclepias exaltata*. Syst. Bot. 18: 18-30.

115. *Bults, G., Horwitz, B.A., Malkin, S., Cahen, D. (1982)*: Photoacustic measurements of photosynthetic activities in whole leaves. Photochemistry and gas exchange. Biochim. Biophys. Acta 679: 452-465.

116. *Burd, M., Head, G. (1992)*: Phenological aspects of male and female function in hermaphroditic plants. Amer. Naturalist 140: 305-324.

117. *Burgman, M.A., Ferson, S., Akçakaya, H.R. (1993)*: Risk assessment in conservation biology. Chapman & Hall, London, 314 pp.

118. *Burkhardt, D. (1964)*: Colour discrimination in insects. Adv. Insect Physiol. 2: 131-173.

119. *Búrquez, A., Corbet, S.A. (1991)*: Do flowers reabsorb nectar? Funct. Ecol. 5: 369-379.

120. *Burson, B.L., Voigt, P.W., Sherman, R.A., Dewald, C.L. (1990)*: Apomixis and sexuality in eastern gamagrass. Crop Sci. 30: 86-89.

121. *Bustos, M.M., Kalkan, F.A., VandenBosch, K.A., Hall, T.C. (1991)*: Differential accumulation of four phaseolin glycoforms in transgenic tobacco. Plant Mol. Biol. 16: 381-395.

122. *Butcher, P.A., Bell, J.C., Moran, G.F. (1992)*: Patterns of genetic diversity and nature of the breeding system in *Melaleuca alternifolia* (Myrtaceae). Aust. J. Bot. 40: 365-375.

123. *Byrne, M., Moran, G.F. (1994)*: Population divergence in the chloroplast genome of *Eucalyptus nitens*. Heredity 73: 18-28.

124. *Caballero, A., Hill, W.G. (1992)*: Effective size of nonrandom mating populations. Genetics 130: 909-916.

125. *Campbell, D.R. (1985)*: Pollinator sharing and seed set of *Stellaria pubera*: Competition for pollination. Ecology 66: 544-553.

126. *Campbell, D.R. (1991)*: Comparing pollen dispersal and gene flow in a natural population. Evolution 45: 1965-1968.

127. *Campbell, D.R. (1992)*: Variation in sex allocation and floral morphology in *Ipomopsis aggregata* (Polemoniaceae). Amer. J. Bot. 79: 516-521.

128. *Campbell, D.R., Dooley, J.L. (1992)*: The spatial scale of genetic differentiation in a hummingbird-pollinated plant: Comparison with models of isolation by distance. Amer. Naturalist 139: 735-748.

129. *Campbell, D.R., Waser, N.M. (1989)*: Variation in pollen flow within and among populations of *Ipomopsis aggregata*. Evolution 43: 1444-1455.

130. *Campbell, M.A., Kinlaw, C.S., Neale, D.B. (1992)*: Expression of luciferase and β-glucuronidase in *Pinus radiata* suspension cells using electroporation and particle bombardment. Can. J. Forest Res. 22: 2014-2018.

131. *Cao, J., Duan, X.L., McElroy, D., Wu, R. (1992)*: Regeneration of herbicide resistant transgenic rice plants following microprojectile-mediated transformation of suspension culture cells. Plant Cell Rep. 11: 586-591.

132. *Carney, S.E., Cruzan, M.B., Arnold, M.L. (1994)*: Reproductive interactions between hybridizing irises: Analyses of pollen-tube growth and fertilization success. Amer. J. Bot. 81: 1169-1175.

133. *Carr, D.E. (1991)*: Sexual dimorphism and fruit production in a dioecious understory tree, *Ilex opaca* Ait. Oecologia 85: 381-388.

134. *Carré, S., Badenhausser, I., Taséi, J.N., Guen, J. le, Mesquida, J. (1994)*: Pollen deposition by *Bombus terrestris* L., between male-fertile and male-sterile plants in *Vicia faba* L. Apidologie 25: 338-349.

135. *Casas, A.M., Kononowicz, A.K., Zehr, U.B., Tomes, D.T., Axtell, J.D., Butler, L.G., Bressan, R.A., Hasegawa, P.M. (1993)*: Transgenic sorghum plants via microprojectile bombardment. Proc. Nat. Acad. Sci. USA 90: 11212-11216.

136. *Casse-Delbart, F. (1992)*: Methods and uses of genetic transformation of superior plants. C. R. Soc. Biol. 186: 515-527.

137. *Castiglione, S., Wang, G., Damiani, G., Bandi, C., Bisoffi, S., Sala, F. (1993)*: RAPD fingerprints for identification and for taxonomic studies of elite poplar (*Populus* spp.) clones. Theor. Appl. Genet. 87: 54-59.

138. *Castillo, R.F. del (1994)*: Factors influencing the genetic structure of *Phacelia dubia*, a species with a seed bank and large fluctuations in population size. Heredity 72: 446-458.

139. *CDPE (1992)*: Ecological impact of genetically modified organisms. A survey of literature, guidelines and legislation. Steering committee for the conservation and management of the environment and natural habitats. Council of Europe, Strasbourg, 89 pp.

140. *CEP (1995)*: Pan-European conference on the potential long-term ecological impact of genetically modified organisms. Proceedings Strasbourg, 24-26 November 1993. Council of Europe Press, Strasbourg, 319 pp.

141. *Chaboudez, P. (1994)*: Patterns of clonal variation in skeleton weed (*Chondrilla juncea*), an apomictic species. Aust. J. Bot. 42: 283-295.

142. *Charlesworth, D. (1995)*: Multi-allelic self-incompatibility polymorphisms in plants. Bioessays 17: 31-38.

143. *Charlesworth, D., Charlesworth, B. (1987)*: Inbreeding depression and its evolutionary consequences. Ann. Rev. Ecol. Syst. 18: 237-268.

144. *Chase, M.W., Hills, H.H. (1991)*: Silica gel: An ideal material for field preservation of leaf samples for DNA studies. Taxon 40: 215-220.

145. *Chase, M.W., Soltis, D.E., Olmstead, R.G., Morgan, D., Les, D.H., Mishler, B.D., Duvall, M.R., Price, R.A., Hills, H.G., Qiu, Y.L., Kron, K.A., Rettig, J.H., Conti, E., Palmer, J.D., Manhart, J.R., Sytsma, K., Michaels, H.J., Kress, W.J., Karol, K.G., Clark, W.D., Hedren, M., Gaut, B.S., Jansen, R.K., Kim, K.J., Wimpee, C.F., Smith, J.F., Furnier, G.R., Strauss, S.H., Xiang, Q.Y., Plunkett, G.M., Soltis, P.S., Swensen, S.M., Williams, S.E., Gadek, P.A., Quinn, C.J., Eguiarte, L.E., Golenberg, E., Learn, G.H., Graham, S.W., Barrett, S.C.H., Dayanandan, S., Albert, V.A. (1993)*: Phylogenetics of seed plants. Ann. Mo. Bot. Gard. 80: 528-580.

146. *Chaudhury, A., Maheshwari, S.C., Tyagi, A.K. (1993)*: Transient expression of electroporated gene in leaf protoplasts of *Indica* rice and influence of template topology and vector sequences. Physiol. Plant. 89: 842-846.

147. *Chaudhury, A., Maheshwari, S.C., Tyagi, A.K. (1995)*: Transient expression of *gus* gene in intact seed embryos of *Indica* rice after electroporation-mediated gene delivery. Plant Cell Rep. 14: 215-220.

148. *Chee, P.P., Slightom, J.L. (1991)*: Transfer and expression of cucumber mosaic virus coat protein gene in the genome of *Cucumis sativus*. J. Amer. Hort. Sci. 116: 1098-1102.

149. *Chen, B.Y., Cheng, B.F., Jørgensen, R.B., Heneen, W.K. (1996)*: Production and cytogenetics of *Brassica campestris-alboglabra* chromosome addition lines. Theor. Appl. Genet. (in press).

150. *Chen, B.Y., Heneen, W.K. (1992)*: Inheritance of seed colour in *Brassica campestris* L. and breeding for yellow-seeded *B. napus* L. Euphytica 59: 157-163.

151. *Chen, J.L., Beversdorf, W.D. (1992)*: Cryopreservation of isolated microspores of spring rapeseed (*Brassica napus* L.) for *in vitro* embryo production. Plant Cell Tissue Organ Cult. 31: 141-149.

152. *Cheng, B.F., Heneen, W.K. (1995)*: Satellited chromosomes, nucleolus organizer regions and nucleoli of *Brassica campestris* L., *B. nigra* (L.) Koch, and *Sinapis arvensis* L. Hereditas 122: 113-118.

153. *Cheng, F.S., Weeden, N.F., Brown, S.K. (1996)*: Identification of co-dominant RAPD markers tightly linked to fruit skin color in apple. Theor. Appl. Genet. 93: 222-227.

154. *Chesser, R.K., Rhodes Jr., O.E., Sugg, D.W., Schnabel, A. (1993)*: Effective sizes for subdivided populations. Genetics 135: 1221-1232.

155. *Chowdhury, M.K.U., Vasil, I.K. (1992)*: Stably transformed herbicide resistant callus of sugarcane via microprojectile bombardment of cell suspension cultures and electro-poration of protoplasts. Plant Cell Rep. 11: 494-498.

156. *Chrispeels, M.J., Raikhel, N.V. (1991)*: Lectins, lectin genes, and their role in plant defense. Plant Cell 3: 1-9.

157. *Christou, P. (1992)*: Genetic transformation of crop plants using microprojectile bombardment. Plant J. 2: 275-281.

158. *Cipollini, M.L., Whigham, D.F. (1994)*: Sexual dimorphism and cost of reproduction in the dioecious shrub *Lindera benzoin* (Lauraceae). Amer. J. Bot. 81: 65-75.

159. *Clark, G. (1981)*: Staining procedures. Williams & Wilkins, Baltimore, 135 pp.

160. *Clark, G.M. (1995)*: Relationship between developmental stability and fitness: Application for conservation biology. Conservation Biology 9: 18-24.

161. *Clough, G.H., Hamm, P.B. (1995)*: Coat protein transgenic resistance to watermelon mosaic and zucchini yellows mosaic virus in squash and cantaloupe. Plant Dis. 79: 1107-1109.

162. *Cluster, P.D., Allard, R.W. (1995)*: Evolution of ribosomal DNA (rDNA) genetic structure in colonial Californian populations of *Avena barbata*. Genetics 139: 941-954.

163. *Coates, D.J., Sokolowski, R.E.S. (1992)*: The mating system and patterns of genetic variation in *Banksia cuneata* A.S. George (Proteaceae). Heredity 69: 11-20.

164. *Cockerham, C.C. (1973)*: Analysis of gene frequencies. Genetics 74: 679-700.

165. *Colwell, R.K. (1994)*: Potential ecological and evolutionary problems of introducing transgenic crops into the environment. In: Krattiger, A.F., Rosemarin, A., (eds.) Biosafety for sustainable agriculture: Sharing biotechnology regulatory experiences of the western hemisphere. ISAAA & SEI, Ithaca, 33-46.

166. *Comai, L., Facciotti, D., Hiatt, W.R., Thompson, G., Rose, R.E., Stalker, D.M. (1985)*: Expression in plants of a mutant *aro*A gene from *Salmonella typhimurium* confers tolerance to glyphosate. Nature 317: 741-745.

167. *Conner, A.J., Dale, P.J. (1996)*: Reconsideration of pollen dispersal data from field trials of transgenic potatoes. Theor. Appl. Genet. 92: 505-508.

168. *Cooper, B., Lapidot, M., Heick, J.A., Dodds, J.A., Beachy, R.N. (1995)*: A defective movement protein of TMV in transgenic plants confers resistance to multiple viruses whereas the functional analog increases susceptibility. Virology 206: 307-313.

169. *Corbet, S.A. (1978a)*: Bee visits and the nectar of *Echium vulgare*. In: Richards, A.J., (ed.) The pollination of flowers by insects. Academic Press, London, 21-30.

170. *Corbet, S.A. (1978b)*: Bee visits and the nectar of *Echium vulgare* L. and *Sinapsis alba* L. Ecol. Entomol. 3: 25-37.

171. *Cornelissen, B.J.C., Melchers, L.S. (1993)*: Strategies for control of fungal diseases with transgenic plants. Plant Physiol. 101: 709-712.

172. *Corriveau, J.L., Coleman, A.W. (1988)*: Rapid screening method to detect potential biparental inheritance of plastid DNA and results for over 200 angiosperm species. Amer. J. Bot. 75: 1443-1458.

173. *Coupland, D. (1994)*: Resistance to the auxin analog herbicide. In: Powles, S.B., Holtum, J.A.M., (eds.) Herbicide resistance in plants. Biology and biochemistry. Lewis Publishers, Boca Raton, 171-215.

174. *Courtney, S.P., Hill, C.J., Westerman, A. (1982)*: Pollen carried for long periods by butterflies. Oikos 38: 260-263.

175. *Crawford, D.J., Brauner, S., Cosner, M.B., Stuessy, T.F. (1993)*: Use of RAPD markers to document the origin of the intergeneric hybrid x *Margyracaena skottsbergii* (Rosaceae) on the Juan Fernandez islands. Amer. J. Bot. 80: 89-92.

176. *Crawford, T.J. (1984)*: The estimation of neighbourhood parameters for plant populations. Heredity 52: 273-283.

177. *Crawley, M.J. (1995)*: Long term ecological impacts of the release of genetically modified organisms. In: CEP, Pan-European conference on the potential long-term ecological impact of genetically modified organisms. Proceedings Strasbourg, 24-26 November 1993. Council of Europe Press, Strasbourg, 43-66.

178. *Creelman, R.A., Mullet, J.E. (1995)*: Jasmonic acid distribution and action in plants: Regulation during development and response to biotic and abiotic stress. Proc. Nat. Acad. Sci. USA 92: 4114-4119 .

179. *Croy, R.R.D. (1993)*: Plant molecular biology: Labfax. BIOS Scientific Publishers, Oxford, 382 pp.

180. *Cruden, R.W. (1977)*: Pollen-ovule ratios: A conservative indicator of breeding systems in flowering plants. Evolution 31: 32-46.

181. *Cruden, R.W., Hermann, S.M. (1983)*: Studying nectar? Some observations on the art. In: Bentley, B., Elias, T., (eds.) The biology of nectaries. Columbia University Press, New York, 223-241.

182. *Cruzan, M.B. (1990)*: Variation in pollen size, fertilization ability, and postfertilization siring ability in *Erythronium grandiflorum*. Evolution 44: 843-856.

183. *Dafni, A. (1992)*: Pollination ecology. A practical approach. IRL Press at Oxford University Press, Oxford, 250 pp.

184. *Dale, P.J., Irwin, J.A., Scheffler, J.A. (1993)*: The experimental and commercial release of transgenic crop plants. Plant Breed. 111: 1-22.

185. *Dale, P.J., Scheffler, J.A. (1994)*: Release of transgenic oil crops. In: Murphy, D.J., (ed.) Designer oil crops: Breeding, processing and biotechnology. VCH Verlagsgesellschaft MBH, Weinheim, 283-296.

186.   *Damme, E.J.M. van, Clercq, N. de, Claessens, F., Hemschoote, K., Peeters, B., Peumans, W.J. (1991)*: Molecular cloning and characterization of multiple isoforms of the snowdrop (*Galanthus nivalis* L.) lectin. Planta 186: 35-43.

187.   *Darmency, H. (1994)*: The impact of hybrids between genetically modified crop plants and their related species: Introgression and weediness. Mol. Ecol. 3: 37-40.

188.   *Davis, L.E., Stephenson, A.G., Winsor, J.A. (1987)*: Pollen competition improves performance and reproductive output of the common zucchini squash under field conditions. J. Amer. Soc. Hort. Sci. 112: 712-716.

189.   *Deboo, G.B., Albertsen, M.C., Taylor, L.P. (1995)*: Flavanone 3-hydroxylase transcripts and flavonol accumulation are temporally coordinate in maize anthers. Plant J. 7: 703-713.

190.   *Dehio, C., Grossmann, K., Schell, J., Schmülling, T. (1993)*: Phenotype and hormonal status of transgenic tobacco plants overexpressing the *rolA* gene of *Agrobacterium rhizogenes* T-DNA. Plant Mol. Biol. 23: 1199-1210.

191.   *Delph, L.F. (1993)*: Factors affecting intraplant variation in flowering and fruiting in the gynodioecious species *Hebe subalpina*. J. Ecol. 81: 287-296.

192.   *Delph, L.F., Lu, Y., Jayne, L.D. (1993)*: Patterns of resource allocation in a dioecious *Carex* (Cyperaceae). Amer. J. Bot. 80: 607-615.

193.   *Delph, L.F., Meagher, T.R. (1995)*: Sexual dimorphism masks life history trade-offs in the dioecious plant *Silene latifolia*. Ecology 76: 775-785.

194.   *DePamphilis, C.W., Wyatt, R. (1990)*: Electrophoretic confirmation of interspecific hybridization in *Aesculus* (Hippocastanaceae) and the genetic structure of a broad hybrid zone. Evolution 44: 1295-1317.

195.   *Depeiges, A., Goubely, C., Lenoir, A., Cocherel, S., Picard, G., Raynal, M., Grellet, F., Delseny, M. (1995)*: Identification of the most repeated motifs in *Arabidopsis thaliana* microsatellite loci. Theor. Appl. Genet. 91: 160-168.

196.   *Derks, F.H.M., Hakkert, J.C., Verbeek, W.H.J., Colijn-Hooymans, C.M. (1992)*: Genome composition of asymmetric hybrids in relation to the phylogenetic distance between the parents: Nucleus-chloroplast interaction. Theor. Appl. Genet. 84: 930-940.

197.   *Devlin, B., Ellstrand, N.C. (1990)*: The development and application of a refined method for estimating gene flow from angiosperm paternity analysis. Evolution 44: 248-259.

198.   *Devlin, B., Roeder, K., Ellstrand, N.C. (1988)*: Fractional paternity assignment: Theoretical development and comparison to other methods. Theor. Appl. Genet. 76: 369-380.

199.   *Deynze, A.E. van, Landry, B.S., Pauls, K.P. (1995)*: The identification of restriction fragment length polymorphisms linked to seed colour genes in *Brassica napus*. Genome 38: 534-542.

200.   *Dietrich, J.T., Kaminek, M., Blevins, D.G., Reinbott, T.M., Morris, R.O. (1995)*: Changes in cytokinins and cytokinin oxidase activity in developing maize kernels and the effects of exogenous cytokinin on kernel development. Plant Physiol. Biochem. 33: 327-336.

201.   *Dole, J., Ritland, K. (1993)*: Inbreeding depression in two *Mimulus* taxa measured by multigenerational changes in the inbreeding coefficient. Evolution 47: 361-373.

202.   *Domansky, N., Ehsani, P., Salmanian, A.H., Medvedeva, T. (1995)*: Organ-specific expression of hepatitis B surface antigen in potato. Biotechnol. Lett. 17: 863-866.

203. *Doughty, K.J., Porter, A.J.R., Morton, A.M., Kiddle, G., Bock, C.H., Wallsgrove, R. (1991)*: Variation in the glucosinolate content of oilseed rape (*Brassica napus* L.) leaves. II. Response to infection by *Alternaria brassicae* (Berk.) Sacc. Ann. Appl. Biol. 118: 469-477.

204. *Dow, B.D., Ashley, M.V., Howe, H.F. (1995)*: Characterization of highly variable (GA/CT)$_n$ microsatellites in the bur oak, *Quercus macrocarpa*. Theor. Appl. Genet. 91: 137-141.

205. *Downie, S.R., Katz-Downie, D.S. (1996)*: A molecular phylogeny of *Apiaceae* subfamily *Apioideae*: Evidence from nuclear ribosomal DNA internal transcribed spacer sequences. Amer. J. Bot. 83: 234-251.

206. *Doyle, J.J., Dickson, E.E. (1987)*: Preservation of plant samples for DNA restriction endonuclease analysis. Taxon 36: 715-722.

207. *Dröge, W., Broer, I., Pühler, A. (1992)*: Transgenic plants containing the phosphino-thricin-N-acetyltransferase gene metabolize the herbicide L-phosphinothricin (glufosinate) differently from untransformed plants. Planta 187: 142-151.

208. *Dudash, M.R., Ritland, K. (1991)*: Multiple paternity and self-fertilization in relation to floral age in *Mimulus guttatus* (Scrophulariaceae). Amer. J. Bot. 78: 1746-1753.

209. *Duke, S.O., Lawrence, C.A., Hess, D.F., Holt, J.S. (1991)*: Herbicide-resistant crops. Cast: Council for Agricultural Science and Technology 1991-1: 1-25.

210. *Düring, K. (1994)*: Strategies towards introducing resistance to bacterial pathogens in transgenic potatoes. In: Belknap, W.R., Vayda, M.E., Park, W.D., (eds.) The molecular and cellular biology of the potato / 2nd ed. CAB International, Wallingford, 221-232.

211. *Eck, H.J. van, Voort, J.R. van der, Draaistra, J., Zandvoort, P.M. van, Enckevort, E. van, Segers, B., Peleman, J., Jacobsen, E., Helder, J., Bakker, J. (1995)*: The inheritance and chromosomal localization of AFLP markers in a non-inbred potato offspring. Mol. Breed. 1: 397-410.

212. *Eck, J.M. van, Blowers, A.D., Earle, E.D. (1995)*: Stable transformation of tomato cell cultures after bombardment with plasmid and YAC DNA. Plant Cell Rep. 14: 299-304.

213. *Eckert, C.G., Barrett, S.C.H. (1993)*: The inheritance of tristyly in *Decodon verticillatus* (Lythraceae). Heredity 71: 473-480.

214. *Eckert, C.G., Barrett, S.C.H. (1994)*: Inbreeding depression in partially self-fertilizing *Decodon verticillatus* (Lythraceae): Population genetic and experimental analyses. Evolution 48: 952-964.

215. *Eckhard, A. (1988)*: Kryokonservierung als Methode der Langzeitlagerung von in-vitro-Material. (Cryopreservation as a method of long-term storage of in-vitro material). Arch. Gartenbau 36: 323-332.

216. *Eckhart, V.M. (1992)*: The genetics of gender and the effects of gender on floral characters in gynodioecious *Phacelia linearis* (Hydrophyllaceae). Amer. J. Bot. 79: 792-800.

217. *Edington, B.V., Nelson, R.S. (1992)*: Utilization of ribozymes in plants. Plant viral resistance. In: Erickson, R.P., Izant, J.G., (eds.) Gene regulation: Biology of antisense RNA and DNA. Raven Press, New York, 209-221.

218. *Eguiarte, L.E., Búrquez, A., Rodríguez, J., Martínez-Ramos, M., Sarukhán, J., Piñero, D. (1993)*: Direct and indirect estimates of neighborhood and effective population size in a tropical palm, *Astrocaryum mexicanum*. Evolution 47: 75-87.

219. *Ehlers, B.K., Olesen, J.M. (1997)*: The fruit-wasp route to toxic nectar in *Epipactis* orchids? Flora (in press).

220. *Eimert, K., Siegemund, F. (1992)*: Transformation of cauliflower (*Brassica oleracea* L. var. *botrytis*) - an experimental survey. Plant Mol. Biol. 19: 485-490.

221. *Ellingboe, J., Gyllensten, U.B. (1992)*: DNA sequencing. Eaton Publishing Company, Natick, 170 pp.

222. *Ellstrand, N.C. (1984)*: Multiple paternity within the fruits of the wild radish, *Raphanus sativus*. Amer. Naturalist 123: 819-828.

223. *Ellstrand, N.C. (1992)*: Gene flow by pollen: Implications for plant conservation genetics. Oikos 63: 77-86.

224. *Ellstrand, N.C., Marshall, D.L. (1985a)*: Interpopulational gene flow by pollen in wild radish, *Raphanus sativus*. Amer. Naturalist 126: 606-616.

225. *Ellstrand, N.C., Marshall, D.L. (1985b)*: The impact of domestication on distribution of allozyme variation within and among cultivars of radish, *Raphanus sativus* L. Theor. Appl. Genet. 69: 393-398.

226. *Ellstrand, N.C., Marshall, D.L. (1986)*: Patterns of multiple paternity in population of *Raphanus sativus*. Evolution 40: 837-842.

227. *Elton, C.S. (1958)*: The ecology of invasions by plants and animals. Methuen, London, 181 pp.

228. *Elzen, P.J.M. van den, Jongedijk, E., Melchers, L.S., Cornelissen, B.J.C. (1993)*: Virus and fungal resistance: From laboratory to field. Phil. Trans. Roy. Soc. London B 342: 271-278.

229. *Emms, S.K. (1993)*: Andromonoecy in *Zigadenus paniculatus* (Liliaceae): Spatial and temporal patterns of sex allocation. Amer. J. Bot. 80: 914-923.

230. *Endler, J.A. (1973)*: Gene flow and population differentiation. Science 179: 243-250.

231. *Endler, J.A. (1980)*: Geographic variation, speciation, and clines. Princeton University Press, Princeton, 246 pp.

232. *English-Loeb, G.M., Karban, R. (1992)*: Consequences of variation in flowering phenology for seed head herbivory and reproductive success in *Erigeron glaucus* (Compositae). Oecologia 89: 588-595.

233. *Ennos, R.A. (1994)*: Estimating the relative rates of pollen and seed migration among plant populations. Heredity 72: 250-259.

234. *Ennos, R.A., Clegg, M.T. (1982)*: Effect of population substructuring on estimates of outcrossing rates in natural plant populations. Heredity 48: 283-292.

235. *Epperson, B.K. (1995)*: Spatial distributions of genotypes under isolation by distance. Genetics 140: 1431-1440.

236. *Eriksson, O., Bremer, B. (1993)*: Genet dynamics of the clonal plant *Rubus saxatilis*. J. Ecol. 81: 533-542.

237. *Erlich, H.A. (1989)*: PCR technology: Principles and applications for DNA amplification. Stockton Press, New York, 246 pp.

238. *Espinasse, A., Foueillassar, J., Kimber, G. (1995)*: Cytogenetical analysis of hybrids between sunflower and four wild relatives. Euphytica 82: 65-72.

239. *Excoffier, L., Slatkin, M. (1995)*: Maximum-likelihood estimation of molecular haplotype frequencies in a diploid population. Mol. Biol. Evol. 12: 921-927.

240. *Excoffier, L., Smouse, P.E., Quattro, J.M. (1992)*: Analysis of molecular variance inferred from metric distances among DNA haplotypes: Application to human mitochondrial DNA restriction data. Genetics 131: 479-491.

241. *Fægri, K., Iversen, J. (1989)*: Textbook of pollen analysis. John Wiley & Sons, Chichester, 328 pp.

242. *Fægri, K., Pijl, L. van der (1979)*: The principles of pollination ecology. Pergamon Press, Oxford, 3. edition, 244 pp.

243. *Fahselt, D., Ownbey, M. (1968)*: Chromatographic comparison of *Dicentra* species and hybrids. Amer. J. Bot. 55: 334-345.

244. *Fairley, D., Batchelder, G.L. (1986)*: A study of oak-pollen production and phenology in Northern California: Prediction of annual variation in pollen counts based on geographic and meterologic factors. J. Allerg. Clin. Immunol. 78: 300-307.

245. *Falco, S.C., Guida, T., Locke, M., Mauvais, J., Sanders, C., Ward, R.T., Webber, P. (1995)*: Transgenic canola and soybean seeds with increased lysine. Bio/Technology 13: 577-582.

246. *Falconer, D.S. (1981)*: Introduction to quantitative genetics. Longman Scientific & Technical, 464 pp.

247. *Fedtke, C. (1982)*: Auxin-inhibitor herbicides. In: Fedtke, C., (ed.) Biochemistry and physiology of herbicide action. Springer Verlag, Berlin, 177-184.

248. *Feitelson, J.S., Payne, J., Kim, L. (1992)*: *Bacillus thuringiensis*: Insects and beyond. Bio/Technology 10: 271-275.

249. *Felber, F. (1988)*: Phénologie de la floraison de populations diploïdes et tétraploïdes d'*Anthoxanthum alpinum* et d'*Anthoxanthum odoratum*. Can. J. Bot. 66: 2258-2264.

250. *Feldman, M.W., Christiansen, F.B.* The effect of population subdivision on two loci without selection. Genet. Res. 24: 151-162.

251. *Feliner, G.N. (1991)*: Breeding systems and related floral traits in several *Erysimum* (Cruciferae). Can. J. Bot. 69: 2515-2521.

252. *Felsenstein, J. (1995)*: PHYLIP (Phylogeny inference package) version 3.572. Distributed over the World Wide Web, Seattle.

253. *Fennell, A., Hauptmann, R. (1992)*: Electroporation and PEG delivery of DNA into maize microspores. Plant Cell Rep. 11: 567-570.

254. *Fenster, C.B. (1991)*: Gene flow in *Chamaecrista fasciculata* (Leguminosae). I. Gene dispersal. Evolution 45: 398-409.

255. *Fenster, C.B., Ritland, K. (1992)*: Chloroplast DNA and isozyme diversity in two *Mimulus* species (Scrophulariaceae) with contrasting mating systems. Amer. J. Bot. 79: 1440-1447.

256. *Fenster, C.B., Ritland, K. (1994)*: Evidence for natural selection on mating system in *Mimulus* (Scrophulariaceae). Int. J. Plant Sci. 155: 588-596.

257. *Fischer, R., Hain, R. (1994)*: Plant disease resistance resulting from the expression of foreign phytoalexins. Curr. Opin. Biotechnol. 5: 125-130.

258. *Fisk, H.J., Dandekar, A.M. (1993)*: The introduction and expression of transgenes in plants. Sci. Hort., Amsterdam 55: 5-36.

259. *Fitchen, J.H., Beachy, R.N. (1993)*: Genetically engineered protection against viruses in transgenic plants. Annu. Rev. Microbiol. 47: 739-763.

260. *Floate, K.D., Whitham, T.G. (1995)*: Insects as traits in plant systematics: Their use in discriminating between hybrid cottonwoods. Can. J. Bot. 73: 1-13.

261. *Folkertsma, R.T., Rouppe van der Voort, J.N.A.M., Groot, K.E. de, Zandvoort, P.M. van, Schots, A., Gommers, F.J., Helder, J., Bakker, J. (1996)*: Gene pool similarities of potato cyst nematode populations assessed by AFLP analysis. Mol. Plant Microbe Interaction. 9: 47-54.

262. *Forfang, S., Olesen, J.M., Báez, M. (1997)*: Stress-induced male sterility and mixed mating in an island plant. Plant Syst. Evol. (in press).

263. *Forkmann, G. (1993)*: Control of pigmentation in natural and transgenic plants. Curr. Opin. Biotechnol. 4: 159-165.

264. *Fowler, N.L., Antonovics, J. (1981)*: Small-scale variability in the demography of transplants of two herbaceous species. Ecology 62: 1450-1457.

265. *Fox, M.D., Fox, B.J. (1986)*: The susceptibility of natural communities to invasion. In: Groves, R.H., Burdon, J.J., (eds.) The ecology of biological invasions. Cambridge University Press, Sydney, 57-66.

266. *Franklin, F.C.H., Lawrence, M.J., Franklin-Tong, V.E. (1995)*: Cell and molecular biology of self-incompatibility in flowering plants. International Review of Cytology - A Survey of Cell Biology 158: 1-64.

267. *Fredrikson, M. (1991)*: An embryological study of *Platanthera bifolia* (Orchidaceae). Plant Syst. Evol. 174: 213-220.

268. *Fredshavn, J.R., Poulsen, G.S., Huybrechts, I., Rudelsheim, P. (1995)*: Competitiveness of transgenic oilseed rape. Transgenic Res. 4: 142-148.

269. *Fromm, M., Callis, J., Taylor, L.P., Walbot, V. (1987)*: Electroporation of DNA and RNA into plant protoplasts. Meth. Enzymol. 153: 351-366.

270. *Fuente-Martinez, J.M. de la, Mosqueda-Cano, G., Alvarez-Morales, A., Herrera-Estrella, L. (1993)*: Pathogen-derived strategy to produce transgenic plants resistant to the bacterial toxin phaseolotoxin. In: Nester, E.W., Verma, D.P.S., (eds.) Advances in molecular genetics of plant-microbe interactions. Kluwer Academic Publishers, Dordrecht, 579-586.

271. *Fujimoto, H., Itoh, K., Yamamoto, M., Kyozuka, J., Shimamoto, K. (1993)*: Insect resistant rice generated by introduction of a modified δ-endotoxin gene of *Bacillus thuringiensis*. Bio/Technology 11: 1151-1155.

272. *Funatsuki, H., Kuroda, H., Kihara, M., Lazzeri, P.A., Muller, E., Lorz, H., Kishinami, I. (1995)*: Fertile transgenic barley generated by direct DNA transfer to protoplasts. Theor. Appl. Genet. 91: 707-712.

273. *Fütterer, J., Gisel, A., Iglesias, V., Klöti, A., Kost, B., Mittelsten Scheid, O., Neuhaus, G., Neuhaus-Url, G., Schrott, M., Shillito, R., Spangenberg, G., Wang, Z.Y. (1995)*: Standard molecular techniques for the analysis of transgenic plants. In: Potrykus, I., Spangenberg, G., (eds.) Gene transfer to plants. Springer Verlag, Berlin, 215-263.

274. *Gabel, B., Thiéry, D., Suchy, V., Marionpoll, F., Hradsky, P., Farkas, P. (1992)*: Floral volatiles of *Tanacetum vulgare* L. attractive to *Lobesia botrana* Den. et Schiff. females. J. Chem. Ecol. 18: 693-701.

275. *Galen, C. (1992)*: Pollen dispersal dynamics in an alpine wildflower, *Polemonium viscosum*. Evolution 46: 1043-1051.

276. *Galen, C., Kevan, P.G. (1983)*: Bumblebee foraging and floral scent dimorphism: *Bombus kirbyellus* Curtis (Hymenoptera: Apidae) and *Polemonium viscosum* Nutt. (Polemoniaceae). Am. J. Bot. 63: 488-491.

277. *Galen, C., Plowright, R.C. (1985)*: Contrasting movement patterns of nectar-collecting and pollen-collecting bumble bees (*Bombus terricola*) on fireweed (*Chamaenerion angustifolium*) inflorescences. Ecol. Entomol. 10: 9-17.

278. *Galen, C., Shore, J.S., Deyoe, H. (1991)*: Ecotypic divergence in alpine *Polemonium viscosum*: Genetic structure, quantitative variation, and local adaptation. Evolution 45: 1218-1228.

279. *Galili, G., Karchi, H., Shaul, O., Perl, A., Cahana, A., Tzchori, I.B.-T., Zhu, X.Z., Galili, S. (1994)*: Production of transgenic plants containing elevated levels of lysine and threonine. Biochem. Soc. Trans. 22: 921-925.

280. *Gallacher, D.J., Lee, D.J., Berding, N. (1995)*: Use of isozyme phenotypes for rapid discrimination among sugarcane clones. Aust. J. Agr. Res. 46: 601-609.

281. *Gallardo, F., Miginiac-Maslow, M., Sangwan, R.S., Decottignies, P., Keryer, E., Dubois, F., Bismuth, E., Galvez, S., Sangwan-Norreel, B., Gadal, P., Crétin, C. (1995)*: Monocotyledonous C_4 NADP(+)-malate dehydrogenase is efficiently synthesized, targeted to chloroplasts and processed to an active form in transgenic plants of the C_3 dicotyledon tobacco. Planta 197: 324-332.

282. *Gallie, D.R. (1993)*: Introduction of mRNA to plant protoplasts using polyethylene glycol. Plant Cell Rep. 13: 119-122.

283. *Gange, A.C. (1995)*: Aphid performance in an alder (*Alnus*) hybrid zone. Ecology 76: 2074-2083.

284. *Garnier-Gere, P., Dillmann, C. (1992)*: A computer program for testing pairwise linkage disequilibria in subdivided populations. J. Hered. 83: 239.

285. *Gatehouse, A.M.R., Boulter, D. (1983)*: Assessment of the antimetabolic effects of trypsin inhibitors from cowpea (*Vigna unguiculata*) and other legumes on development of the bruchid beetle *Callosobruchus maculatus*. J. Sci. Food Agric. 34: 345-350.

286. *Gatehouse, A.M.R., Gatehouse, J.A., Dobie, P., Kilminster, A.M., Boulter, D. (1979)*: Biochemical basis of insect resistance in *Vigna unguiculata*. J. Sci. Food Agric. 30: 948-958.

287. *Gatehouse, A.M.R., Hilder, V.A., Powell, K.S., Wang, M., Davison, G.M., Gatehouse, L.N., Down, R.E., Edmonds, H.S., Boulter, D., Newell, C.A., Merryweather, A., Hamilton, W.D.O., Gatehouse, J.A. (1994)*: Insect-resistant transgenic plants: Choosing the gene to do the job. Biochem. Soc. Trans. 22: 944-949.

288. *Gatehouse, A.M.R., Shi, Y., Powell, K.S., Brough, C., Hilder, V.A., Hamilton, W.D.O., Newell, C.A., Merryweather, A., Boulter, D., Gatehouse, J.A. (1993)*: Approaches to insect resistance using transgenic plants. Phil. Trans. Roy. Soc. London B 342: 279-286.

289. *Gatehouse, J.A., Bown, D., Evans, I.M., Gatehouse, L.N., Jobes, D., Preston, P., Croy, R.D. (1987)*: Sequence of the seed lectin gene from pea (*Pisum sativum* L.). Nucl. Acid. Res. 15: 7642.

290. *Gaugitsch, H., Torgersen, H. (1995)*: Streamlining regulations, keeping high safety standards: Revised criteria for the assessment of releases of genetically modified organisms (GMOs) into the environment. Ambio 24: 47-50.

291. *Gehring, J.L., Linhart, Y.B. (1992)*: Population structure and genetic differentiation in native and introduced populations of *Deschampsia caespitosa* (Poaceae) in the Colorado Alpine. Amer. J. Bot. 79: 1337-1343.

292. *Genetic Manipulation Advisory Committee (1995)*: Gmac Public Information Sheets on Planned Release Proposals. 1-176.

293. *Giese, H., Holm-Jensen, A.G., Jensen, H.P., Jensen, J. (1993)*: Localization of the Laevigatum powdery mildew resistance gene to barley chromosome 2 by the use of RFLP markers. Theor. Appl. Genet. 85: 897-900.

294. *Gilbert, C., Breen, P.J. (1987)*: Low pollen production as a cause of fruit malformation in strawberry. J. Amer. Soc. Hort. Sci. 112: 56-60.

295. *Gilbert, F.S., Haines, N., Dickson, K. (1991)*: Empty flowers. Funct. Ecol. 5: 29-39.

296. *Gillan, R., Cole, M.D., Linacre, A., Thorpe, J.W., Watson, N.D. (1995)*: Comparison of *Cannabis sativa* by random amplification of polymorphic DNA (RAPD) and HPLC of cannabinoids: A preliminary study. Sci. Justice 35: 169-177.

297. *Gillespie, L.H., Henwood, M.J. (1994)*: Temporal changes of floral nectar-sugar composition in *Polyscias sambucifolia* (Sieb. ex DC.) Harms (Araliaceae). Ann. Bot. 74: 227-231.

298. *Gilliam, M., Moffett, J.O., Kauffeld, N.M. (1983)*: Examination of floral nectar of citrus, cotton, and Arizona desert plants for microbes. Apidologie 14: 299-302.

299. *Ginsberg, H.S. (1983)*: Foraging ecology of bees in an old field. Ecology 64: 165-175.

300. *Gliddon, C. (1995)*: The potential role of monitoring in risk assessment and in fundamental ecological research. In: CEP, Pan-European conference on the potential long-term ecological impact of genetically modified organisms. Proceedings Strasbourg, 24-26 November 1993. Council of Europe Press, Strasbourg, 253-263.

301. *Goddijn, O.J.M., Lindsey, K., Lee, F.M. van der, Klap, J.C., Sijmons, P.C. (1993)*: Differential gene expression in nematode-induced feeding structures of transgenic plants harbouring promoter-*gus*A fusion constructs. Plant J. 4: 863-873.

302. *Goddijn, O.J.M., Pen, J. (1995)*: Plants as bioreactors. Trends Biotech. 13: 379-387.

303. *Godt, M.J.W., Hamrick, J.L. (1991)*: Estimates of outcrossing rates in *Lathyrus latifolius* populations. Genome 34: 988-992.

304. *Goldstein, D.B., Linares, A.R., Cavalli-Sforza, L.L., Feldman, M.W. (1995)*: An evaluation of genetic distances for use with microsatellite loci. Genetics 139: 463-471.

305. *Golenberg, E.M. (1989)*: Migration patterns and the development of multilocus associations in a selfing annual, *Triticum dicoccoides*. Evolution 43: 595-606.

306. *Gonsalves, C., Xue, B., Yepes, M., Fuchs, M., Ling, K.S., Namba, S., Chee, P., Slightom, J.L., Gonsalves, D. (1994)*: Transferring cucumber mosaic virus-white leaf strain coat protein gene into *Cucumis melo* L. and evaluating transgenic plants for protection against infections. J. Amer. Hort. Sci. 119: 345-355.

307. *Goodnight, K.F., Queller, D.C. (1994)*: Relatedness, version 4.2b. Goodnight software, Houston.

308. *Gosden, J., Hanratty, D. (1993)*: PCR *in situ*: A rapid alternative to *in situ* hybridization for mapping short, low copy number sequences without isotopes. BioTechniques 15: 78-80.

309. *Gosden, J., Hanratty, D., Starling, J., Fantes, J., Mitchell, A., Porteous, D. (1991)*: Oligonucleotide-primed in situ DNA synthesis (PRINS): A method for chromosome mapping, banding, and investigation of sequence organization. Cytogenet. Cell Genet. 57: 100-104.

310. *Gottlieb, L.D. (1981)*: Electrophoretic evidence and plant populations. In: Reinhold, L., Harborne, J.B., Swain, T., (eds.) Progress in phytochemistry. Pergamon Press, Oxford, 7: 1-46.

311. *Gottsberger, G., Arnold, T., Linskens, H.F. (1990)*: Variation in floral nectar amino acids with aging of flowers, pollen contamination, and flower damage. Isr. J. Bot. 39: 167-176.

312. *Goudet, J. (1995)*: FSTAT (Version 1.2): A computer program to calculate F-statistics. J. Hered. 86: 485-486.

313. *Govindaraju, D.R. (1988)*: Relationship between dispersal ability and levels of gene flow in plants. Oikos 52: 31-35.

314. *Govindaraju, D.R. (1989)*: Variation in gene flow levels among predominantly self-pollinated plants. J. Evol. Biol. 2: 173-181.

315. *Grant, V. (1980)*: Gene flow and the homogeneity of species populations. Biol. Zbl. 99: 157-169.

316. *Grant, W.F. (1994)*: The present status of higher plant bioassays for the detection of environmental mutagens. Mutat. Res. - Fundam. Mol. Mech. Mut. 310: 175-185.

317. *Grazzini, R., Hesk, D., Yerger, E., Coxfoster, D., Medford, J., Craig, R., Mumma, R.O. (1995)*: Distribution of anacardic acids associated with small pest resistance among cultivars of *Pelargonium X hortorum*. J. Amer. Hort. Sci. 120: 343-346.

318. *Greef, W. de, Delon, R., Block, M. de, Leemans, J., Botterman, J. (1989)*: Evaluation of herbicide resistance in transgenic crops under field conditions. Bio/Technology 7: 61-64.

319. *Green, T.R., Ryan, C.A. (1972)*: Wound-induced proteinase inhibitor in plant leaves: A possible defense mechanism against insects. Science 175: 776-777.

320. *Grime, J.P. (1977)*: Evidence for the existence of three primary strategies in plants and its relevance to ecological and evolutionary theory. Amer. Naturalist 111: 1169-1194.

321. *Grime, J.P., Hodgson, J.G., Hunt, R. (1988)*: Comparative plant ecology. Unwin Hyman, London, 742 pp.

322. *Gronwald, J.W. (1994)*: Resistance to photosystem II inhibiting herbicides. In: Powles, S.B., Holtum, J.A.M., (eds.) Herbicide resistance in plants. Biology and biochemistry. Lewis Publishers, Boca Raton, 27-60.

323. *Groose, R.W., Bingham, E.T. (1991)*: Gametophytic heterosis for in vitro pollen traits in alfalfa. Crop Sci. 31: 1510-1513.

324. *Guitián, J., Guitián, P., Sánchez, J.M. (1993)*: Reproductive biology of two *Prunus* species (Rosaceae) in the Northwest Iberian peninsula. Plant Syst. Evol. 185: 153-165.

325. *Günther, P., Pestemer, W. (1990)*: Risk assessment for selected xenobiotics by bioassay methods with higher plants. Environ. Manage. 14: 381-388.

326. *Gupta, A.S., Heinen, J.L., Holaday, A.S., Burke, J.J., Allen, R.D. (1993)*: Increased resistance to oxidation stress in transgenic plants that overexpress chloroplastic Cu/Zn superoxide dismutase. Proc. Nat. Acad. Sci. USA 90: 1629-1633.

327. *Gupta, V., Sita, G.L., Shaila, M.S., Jagannathan, V. (1993)*: Genetic transformation of *Brassica nigra* by *Agrobacterium* based vector and direct plasmid uptake. Plant Cell Rep. 12: 418-421.

328. *Gurr, S.J., McPherson, M.J., Scollan, C., Atkinson, H.J., Bowles, D.J. (1991)*: Gene expression in nematode-infected plant roots. Mol. Gen. Genet. 226: 361-366.

329. *Gustafsson, L., Gustafsson, P. (1994)*: Low genetic variation in Swedish populations of the rare species *Vicia pisiformis* (*Fabaceae*) revealed with RFLP (rDNA) and RAPD. Plant Syst. Evol. 189: 133-148.

330. *Guth, C.J., Weller, S.G. (1986)*: Pollination, fertilization and ovule abortion in *Oxalis magnifica*. Amer. J. Bot. 73: 246-253.

331. *Hagemann, R. (1992)*: Plastid genetics in higher plants. In: Herrmann, R.G., (ed.) Cell organelles. Springer Verlag, Wien, 65-96.

332. *Hain, R., Reif, H.J., Langebartels, R., Schreier, P.H., Stöcker, R.H., Thomzik, J.E., Stenzel, K., Kindl, H., Schmelzer, E. (1992)*: Foreign phytoalexin expression in plants results in increased disease resistance. In: Brighton Crop Prot. Conference: Pests and diseases - 1992. British Crop Protection Council, Farnham, 757-766.

333. *Hainsworth, F.R., Wolf, L.L. (1972)*: Energetics of nectar extraction in a small, high altitude tropical hummingbird *Selasphorus flammula*. J. Comp. Physiol. 80: 377-387.

334. *Hall, L.N., Tucker, G.A., Smith, C.J.S., Watson, C.F., Seymour, G.B., Bundick, Y., Boniwell, J.M., Fletcher, J.D., Ray, J.A., Schuch, W., Bird, C.R., Grierson, D. (1993)*: Antisense inhibition of pectin esterase gene expression in transgenic tomatoes. Plant J. 3: 121-129.

335. *Hamann, A., Zink, D., Nagl, W. (1995)*: Microsatellite fingerprinting in the genus *Phaseolus*. Genome 38: 507-515.

336. *Hamilton, M.B., Mitchell-Olds, T. (1994)*: The mating system and relative performance of selfed and outcrossed progeny in *Arabis fecunda* (Brassicaceae). Amer. J. Bot. 81: 1252-1256.

337. *Hamrick, J.L. (1987)*: Gene flow and distribution of genetic variation in plant populations. In: Urbanska, K.M., (ed.) Differentiation patterns in higher plants. Academic Press, London, 53-67.

338. *Hamrick, J.L. (1989)*: Isozymes and the analysis of genetic structure in plant populations. In: Soltis, D.E., Soltis, P.S., (eds.) Isozymes in plant biology. Chapman & Hall, London, 87-105.

339. *Hamrick, J.L., Godt, M.J.W. (1990)*: Allozyme diversity in plant species. In: Brown, A.H.D., Clegg, M.T., Kahler, A.L., Weir, B.S., (eds.) Plant population genetics, breeding and genetic resources. Sinauer Associates, Sunderland, 43-64.

340. *Hamrick, J.L., Godt, M.J.W., Sherman-Broyles, S.L. (1995)*: Gene flow among plant populations: Evidence from genetic markers. In: Hoch, P.C., Stephenson, A.G., (eds.) Experimental and molecular approaches to plant biosystematics. Missouri Botanical Garden, St. Louis, 215-232.

341. *Hanneman, R.E. (1994)*: The testing and release of transgenic potatoes in the North American center of diversity. In: Krattiger, A.F., Rosemarin, A., (eds.) Biosafety for sustainable agriculture: Sharing biotechnology regulatory experiences of the western hemisphere. ISAAA & SEI, Ithaca, 47-67.

342. *Harborne, J.B. (1993)*: Introduction to ecological biochemistry. Academic Press, London, 4. edition, 318 pp.

343. *Harding, R.M., Boyce, A.J., Martinson, J.J., Flint, J., Clegg, J.B. (1993)*: A computer simulation study of VNTR population genetics: Constrained recombination rules out the infinite alleles model. Genetics 135: 911-922.

344. *Harris, S.A. (1993)*: DNA analysis of tropical plant species: An assessment of different drying methods. Plant Syst. Evol. 188: 57-64.

345. *Harris, S.A., Ingram, R. (1991)*: Chloroplast DNA and biosystematics: The effects of intra-specific diversity and plastid transmission. Taxon 40: 393-412.

346. *Hartl, D.L., Clark, A.G. (1989)*: Principles of population genetics. Sinauer Associates, Sunderland, 682 pp.

347. *Hartman, C.L., Lee, L., Day, P.R., Tumer, N.E. (1994)*: Herbicide resistant turfgrass (*Agrostis palustris* Huds.) by biolistic transformation. Bio/Technology 12: 919-923.

348. *Harvey, C.F., Fraser, L.G., Pavis, S.E., Considine, J.A. (1987)*: Floral biology of two species of *Actinidia* (Actinidiaceae). 1. The stigma, pollination, and fertilization. Bot. Gaz. 148: 426-432.

349. *Hauser, T.P., Loeschcke, V. (1995)*: Inbreeding depression in *Lychnis flos-cuculi* (Caryophyllaceae): Effects of different levels of inbreeding. J. Evol. Biol. 8: 589-600.

350. *Hauser, T.P., Loeschcke, V. (1996)*: Drought stress and inbreeding depression in *Lychnis flos-cuculi* (Caryophyllaceae). Evolution 50: 1119-1126.

351. *Heinrich, B. (1976)*: The foraging specializations of individual bumblebees. Ecol. Monogr. 46: 105-128.

352. *Heinrich, B. (1977)*: Pollination energetics: An ecosystem approach. In: Mattson, W.J., (ed.) The role of arthropods in forest ecosystems. Springer Verlag, New York, 41-46.

353. *Heinrich, B. (1979)*: "Majoring" and "minoring" by foraging bumblebees, *Bombus vagans*: An experimental analysis. Ecology 60: 245-255.

354. *Heinrich, B., Raven, P.H. (1972)*: Energetics and pollination ecology. Science 176: 597-602.

355. *Helenurm, K., Barrett, S.C.H. (1987)*: The reproductive biology of boreal forest herbs. II. Phenology of flowering and fruiting. Can. J. Bot. 65: 2047-2056.

356. Henry, C.M., Barker, I., Pratt, M., Pemberton, A.W., Farmer, M.J., Cotten, J., Ebbels, D., Coates, D., Stratford, R. (1995): Risks associated with the use of genetically modified virus tolerant plants. MAFF 1-107.

357. Herrera, C.M. (1992): Individual flowering time and maternal fecundity in a summer-flowering Mediterranean shrub: Making the right prediction for the wrong reason. Acta Oecol. 13: 13-24.

358. Herrera, C.M. (1993): Selection on floral morphology and environmental determinants of fecundity in a hawk moth pollinated violet. Ecol. Monogr. 63: 251-275.

359. Herrera, J. (1986): Flowering and fruiting phenology in the coastal shrublands of Doñana, south Spain. Vegetatio 68: 91-98.

360. Hess, D. (1983): Die Blüte. Eine Einführung in Struktur und Funktion, Ökologie und Evolution der Blüten. Mit Anleitung zu einfachen Versuchen. Ulmer, Stuttgart, 458 pp.

361. Hessing, M.B. (1988): Geitonogamous pollination and its consequences in Geranium caespitosum. Amer. J. Bot. 75: 1324-1333.

362. Heywood, J.S. (1991): Spatial analysis of genetic variation in plant populations. Annu. Rev. Ecol. Syst. 22: 335-355.

363. Heywood, J.S. (1993): Biparental inbreeding depression in the self-incompatible annual plant Gaillardia pulchella (Asteraceae). Amer. J. Bot. 80: 545-550.

364. Hiesey, W.M., Nobs, M.A. (1970): Genetic and transplant studies on contrasting species and ecological races of the Achillea millefolium complex. Bot. Gaz. 131: 245-259.

365. Hilder, V.A., Gatehouse, A.M.R., Boulter, D. (1993): Transgenic plants conferring insect tolerance protease inhibitor approach. In: Kung, S.-D., Wu, R., (eds.) Transgenic plants. Engineering and utilization. Academic Press, San Diego, vol. 1: 317-338.

366. Hilder, V.A., Gatehouse, A.M.R., Sheerman, S.E., Barker, R.F., Boulter, D. (1987): A novel mechanism of insect resistance engineered into tobacco. Nature 300: 160-163.

367. Hilder, V.A., Powell, K.S., Gatehouse, A.M.R., Gatehouse, J.A., Gatehouse, L.N., Shi, Y., Hamilton, W.D.O., Merryweather, A., Newell, C.A., Timans, J.C., Peumans, W.J., Damme, E. van, Boulter, D. (1995): Expression of snowdrop lectin in transgenic tobacco plants results in added protection against aphids. Transgenic Res. 4: 18-25.

368. Hill, K.K., Jarvis-Eagan, N., Halk, E.L., Krahn, K.J., Liao, L.W., Mathewson, R.S., Merlo, D.J., Nelson, S.E., Rashka, K.E., Loesch-Fries, L.S. (1991): The development of virus-resistant alfalfa, Medicago sativa L. Bio/Technology 9: 373-377.

369. Hill, W.G. (1974): Estimation of linkage disequilibrium in randomly mating populations. Heredity 33: 229-239.

370. Hippe, S., Düring, K., Kreuzaler, F. (1989): In situ localization of a foreign protein in transgenic plants by immunoelectron microscopy following high pressure freezing. Freeze substitution and low temperature embedding. Eur. J. Cell Biol. 50: 230-234.

371. Höfte, H., Whiteley, H.R. (1989): Insecticidal crystal proteins of Bacillus thuringiensis. Microbiol. Rev. 53: 242-255.

372. Hollebone, J.E., Duke, L. (1994): Canadian approaches to biotechnology regulation. In: Krattiger, A.F., Rosemarin, A., (eds.) Biosafety for sustainable agriculture: Sharing biotechnology regulatory experiences of the western hemisphere. ISAAA & SEI, Ithaca, 79-99.

373. *Holmström, K.O., Welin, B., Mandal, A., Kristiansdottir, I., Teeri, T.H., Lamark, T., Strøm, A.R., Palva, E.T. (1994)*: Production of the *Escherichia coli* betaine-aldehyde dehydrogenase, an enzyme required for the synthesis of the osmoprotectant glycine betaine, in transgenic plants. Plant J. 6: 749-758.

374. *Holsinger, K.E. (1991)*: Mass-action models of plant mating systems: The evolutionary stability of mixed mating systems. Amer. Naturalist 138: 606-622.

375. *Holsinger, K.E., Mason-Gamer, R.J. (1996)*: Hierarchical analysis of nucleotide diversity in geographically structured populations. Genetics 142: 629-639.

376. *Holub, E.B., Beynon, L.J., Crute, I.R. (1994)*: Phenotypic and genotypic characterization of interactions between isolates of *Peronospora parasitica* and accessions of *Arabidopsis thaliana*. Mol. Plant Microbe Interaction. 7: 223-239.

377. *Hossaert-McKey, M., Gibernau, M., Frey, J.E. (1994)*: Chemosensory attraction of fig wasps to substances produced by receptive figs. Entomol. Exp. Appl. 70: 185-191.

378. *Houten, W.H.J. van, Heusden, A.W. van, Voort, J.R. van der, Raijmann, L., Bachmann, K. (1991)*: Hypervariable DNA fingerprint loci in *Microseris pygmaea* (Asteraceae, Lactuceae). Bot. Acta 104: 252-256.

379. *Hsieh, H.-L., Tong, C.-G., Thomas, C., Roux, S.J. (1996)*: Light-modulated abundance of an mRNA encoding a calmodulin-regulated chromatin-associated NTPase in pea. Plant Mol. Biol. 30: 135-147.

380. *Hu, J., Quiros, C.F. (1991)*: Molecular and cytological evidence of deletions in alien chromosomes for two monosomic addition lines of *Brassica campestris-oleracea*. Theor. Appl. Genet. 81: 221-226.

381. *Huang, H.C., Kochert, G. (1994)*: Comparative RFLP mapping of an allotetraploid wild rice species (*Oryza latifolia*) and cultivated rice (*O. sativa*). Plant Mol. Biol. 25: 633-648.

382. *Huesing, J.E., Shade, R.E., Chrispeels, M.J., Murdock, L.L. (1991)*: α-Amylase inhibitor, not phytohemagglutinin, explains resistance of common bean seeds to cowpea weevil. Plant Physiol. 96: 993-996.

383. *Hughes, C.E., Harris, S.A. (1994)*: The characterization and identification of a naturally occurring hybrid in the genus *Leucaena* (Leguminosae, Mimosoideae). Plant Syst. Evol. 192: 177-197.

384. *Hunold, R., Bronner, R., Hahne, G. (1994)*: Early events in microprojectile bombardment: Cell viability and particle location. Plant J. 5: 593-604.

385. *Husband, B.C., Barrett, S.C.H. (1991)*: Colonization history and population genetic structure of *Eichhornia paniculata* in Jamaica. Heredity 66: 287-296.

386. *Husband, B.C., Barrett, S.C.H. (1995)*: Estimating effective population size: A reply to Nunney. Evolution 49: 392-394.

387. *Husband, B.C., Schemske, D.W. (1995)*: Magnitude and timing of inbreeding depression in a diploid population of *Epilobium angustifolium* (Onagraceae). Heredity 75: 206-215.

388. *Inouye, D.W. (1978)*: Resource partitioning in bumblebees: Experimental studies of foraging behavior. Ecology 59: 672-678.

389. *Inouye, D.W., Favre, N.D., Lanum, J.A., Levine, D.M., Meyers, J.B., Roberts, M.S., Tsao, F.C., Wang, Y.Y. (1980)*: The effects of nonsugar nectar constituents on nectar energy content. Ecology 61: 992-996.

390. *Irwin, J.A., Abbott, R.J. (1992)*: Morphometric and isozyme evidence for the hybrid origin of a new tetraploid radiate groundsel in York, England. Heredity 69: 431-439.

391. *Ishimoto, M., Kitamura, K. (1989)*: Growth inhibitory effects of an α-amylase inhibitor from the kidney bean, *Phaseolus vulgaris* (L.) on three species of bruchids (Coleoptera: Bruchidae). Appl. Ent. Zool. 24: 281-286.

392. *Ishitani, M., Nakamura, T., Han, S.Y., Takabe, T. (1995)*: Expression of the betaine aldehyde dehydrogenase gene in barley in response to osmotic stress and abscisic acid. Plant Mol. Biol. 27: 307-315.

393. *Jach, G., Görnhardt, B., Mundy, J., Logemann, J., Pinsdorf, P., Leah, R., Schell, J., Maas, C. (1995)*: Enhanced quantitative resistance against fungal disease by combinatorial expression of different barley antifungal proteins in transgenic tobacco. Plant J. 8: 97-109.

394. *Jacobs, J.M.E., Eck, H.J. van, Arens, P., Verkerk-Bakker, B., Hekkert, B. te L., Bastiaanssen, H.J.M., El-Kharbotly, A., Pereira, A., Jacobsen, E., Stiekema, W.J. (1995)*: A genetic map of potato (*Solanum tuberosum*) integrating molecular markers, including transposons, and classical markers. Theor. Appl. Genet. 91: 289-300.

395. *Jacobsen, E., Jong, J.H. de, Kamstra, S.A., Berg, P.M. M.M. van den, Ramanna, M.S. (1995)*: Genomic *in situ* hybridization (GISH) and RFLP analysis for the identification of alien chromosomes in the backcross progeny of potato (+) tomato fusion hybrids. Heredity 74: 250-257.

396. *Jalikop, S.H., Kumar, P.S. (1990)*: Use of a gene marker to study the mode of pollination in pomegranate (*Punica granatum* L.). J. Hort. Sci. 65: 221-223.

397. *Jansen, M.A.K., Shaaltiel, Y., Kazzes, D., Canaani, O., Malkin, S., Gressel, J. (1989)*: Increased tolerance to photoinhibitory light in paraquat-resistant *Conyza bonariensis* measured by photoacoustic spectroscopy and $^{14}CO_2$-fixation. Plant Physiol. 91: 1174-1178.

398. *Jardinaud, M.F., Souvré, A., Alibert, G., Beckert, M. (1995)*: *uidA* Gene transfer and expression in maize microspores using the biolistic method. Protoplasma 187: 138-143.

399. *Jaynes, J.M. (1993)*: Use of genes encoding novel lytic peptides and proteins that enhance microbial disease resistance in plants. Acta Hortic. 336: 33-38.

400. *Jennersten, O. (1991)*: Cost of reproduction in *Viscaria vulgaris* (Caryophyllaceae): A field experiment. Oikos 61: 197-204.

401. *Jiang, J., Gill, B.S. (1994)*: Nonisotopic in situ hybridization and plant genome mapping: The first 10 years. Genome 37: 717-725.

402. *Jørgensen, R.B., Andersen, B. (1994)*: Spontaneous hybridization between oilseed rape (*Brassica napus*) and weedy *B. campestris* (Brassicaceae): A risk of growing genetically modified oilseed rape. Amer. J. Bot. 81: 1620-1626.

403. *Jørgensen, R.B., Hauser, T., Mikkelsen, T.R., Østergård, H. (1996)*: Transfer of engineered genes from crop to wild plants. Trends Plant Sci. 1: 356-358.

404. *Joersbo, M., Brunstedt, J. (1991)*: Electroporation: Mechanism and transient expression, stable transformation and biological effects in plant protoplasts. Physiol. Plant. 81: 256-264.

405. *Johnson, R., Narvaez, J., An, G., Ryan, C. (1989)*: Expression of proteinase inhibitors I and II in transgenic tobacco plants: Effects on natural defense against *Manduca sexta* larvae. Proc. Natl. Acad. Sci. USA 86: 9871-9875.

406. *Johnston, M.O. (1992)*: Effects of cross and self-fertilization on progeny fitness in *Lobelia cardinalis* and *L. siphilitica*. Evolution 46: 688-702.

407. *Johnston, M.O., Schoen, D.J. (1994)*: On the measurement of inbreeding depression. Evolution 48: 1735-1741.

408. *Jones, C.E., Buchmann, S.L. (1974)*: Ultraviolet floral patterns as functional orientation cues in hymenopterous pollination systems. Anim. Behav. 22: 481-485.

409. *Jones Jr., S.B. (1972)*: Hybridization of *Vernonia acaulis* and *V. noveboracensis* (Compositae) in the Piedmont of North Carolina. Castanea 37: 244-253.

410. *Jones-Villeneuve, E., Huang, B., Prudhomme, I., Bird, S., Kemble, R., Hattori, J., Miki, B. (1995)*: Assessment of microinjection for introducing DNA into uninuclear microspores of rapeseed. Plant Cell Tissue Organ Cult. 40: 97-100.

411. *Jong, T.J. de, Waser, N.M., Price, M.V., Ring, R.M. (1992)*: Plant size, geitonogamy and seed set in *Ipomopsis aggregata*. Oecologia 89: 310-315.

412. *Jongsma, M.A., Bakker, P.L., Peters, J., Bosch, D., Stiekema, W.J. (1995)*: Adaptation of *Spodoptera exigua* larvae to plant proteinase inhibitors by induction of gut proteinase activity insensitive to inhibition. Proc. Nat. Acad. Sci. USA 92: 8041-8045.

413. *Jonsson, O., Rosquist, G., Widén, B. (1991)*: Operation of dichogamy and herkogamy in five taxa of *Pulsatilla*. Holarc. Ecol. 14: 260-271.

414. *Jorgensen, R.A., Cluster, P.D. (1988)*: Modes and tempos in the evolution of nuclear ribosomal DNA: New characters for evolutionary studies and new markers for genetic and population studies. Ann. Mo. Bot. Gard. 75: 1238-1247.

415. *Kadmon, R., Shmida, A. (1992)*: Departure rules used by bees foraging for nectar: A field test. Evol. Ecol. 6: 142-151.

416. *Kanost, M.R., Prasad, S.V., Wells, M.A. (1989)*: Primary structure of a member of the serpin superfamily of proteinase inhibitors from an insect, *Manduca sexta*. J. Biol. Chem. 264: 965-972.

417. *Kareiva, P., Manasse, R. (1990)*: Using models to integrate data from field trials and estimate risks of gene escape and gene spread. In: MacKenzie, D.R., Henry, S.R., (eds.) Biological monitoring of genetically engineered plants and microbes. Agricultural Research Institute, Bethesda, Maryland, 31-42.

418. *Kareiva, P., Morris, W., Jacobi, C.M. (1994)*: Studying and managing the risk of cross-fertilization between transgenic crops and wild relatives. Mol. Ecol. 3: 15-21.

419. *Kareiva, P., Parker, I.M., Pascal, M. (1996)*: Can we use experiments and models in predicting the invasiveness of genetically engineered organisms? Ecology 77: 1670-1675.

420. *Kareiva, P., Stark, J. (1994)*: Environmental risks in agricultural biotechnology. Chem. Ind. (London) 2: 52-55.

421. *Karron, J.D., Thumser, N.N., Tucker, R., Hessenauer, A.J. (1995)*: The influence of population density on outcrossing rates in *Mimulus ringens*. Heredity 75: 175-180.

422.   *Katavic, V., Haughn, G.W., Reed, D., Martin, M., Kunst, L. (1994):* In planta transformation of *Arabidopsis thaliana.* Mol. Gen. Genet. 245: 363-370.

423.   *Kato, N., Esaka, M. (1996):* cDNA cloning and gene expression of ascorbate oxidase in tobacco. Plant Mol. Biol. 30: 833-837.

424.   *Kawata, M. (1995):* Effective population size in a continuously distributed population. Evolution 49: 1046-1054.

425.   *Kay, Q.O.N. (1984):* Variation, polymorphism and gene-flow within species. In: Heywood, V.H., Moore, D.M., (eds.) Current concepts in plant taxonomy. Academic Press, London, 25: 181-199.

426.   *Kearns, C.A., Inouye, D.W. (1993):* Techniques for pollination biologists. University Press of Colorado, Nowot, 583 pp.

427.   *Keeler, K.H., Kwankin, B., Barnes, P.W., Galbraith, D.W. (1987):* Polyploid polymorphism in *Andropogon gerardii.* Genome 29: 374-379.

428.   *Kellogg, E.A. (1987):* Apomixis in the *Poa secunda* complex. Amer. J. Bot. 74: 1431-1437.

429.   *Kephart, S.R., Wyatt, R., Parrella, D. (1988):* Hybridization in North American *Asclepias.* I. Morphological evidence. Syst. Bot. 13: 456-473.

430.   *Kerlan, M.C., Chevre, A.M., Eber, F. (1993):* Interspecific hybrids between a transgenic rapeseed (*Brassica napus*) and related species: Cytogenetical characterization and detection of the transgene. Genome 36: 1099-1106.

431.   *Kevan, P.G. (1983):* Floral colors through the insect eye: What they are and what they mean. In: Jones, C.E., Little, R.J., (eds.) Handbook of experimental pollination biology. Scientic & Academic Editions, Van Nostrand Reinhold Comp., New York, 3-30.

432.   *Kevan, P.G., Grainger, N.D., Mulligan, G.A., Robertson, A.R. (1973):* A gray-scale for measuring reflectance and color in the insect and human visual spectra. Ecology 54: 924-926.

433.   *Khasa, P.D., Cheliak, W.M., Bousquet, J. (1993):* Effects of buffer system pH and tissue storage on starch gel electrophoresis of allozymes in three tropical tree species. Ann. Sci. Forest. 50: 37-56.

434.   *Khatun, S., Flowers, T.J. (1995):* Effects of salinity on seed set in rice. Plant Cell Environ. 18: 61-67.

435.   *Kikkert, J.R., Hébert-Soulé, D., Wallace, P.G., Striem, M.J., Reisch, B.I. (1996):* Transgenic plantlets of "Chancellor" grapevine (*Vitis* sp.) from biolistic transformation of embryonic cell suspensions. Plant Cell Rep. 15: 311-316.

436.   *King, L.M. (1993):* Origins of genotypic variation in North American dandelions inferred from ribosomal DNA and chloroplast DNA restriction enzyme analysis. Evolution 47: 136-151.

437.   *Kirk, J.T.O., Tilney-Bassett, R.A.E. (1978):* The plastids. Their chemistry, structure, growth and inheritance. Elsevier/North-Holland Biomedical Press, Amsterdam, 960 pp.

438.   *Kirsop, B.H. (1993):* Development of the information resource for the release of organisms into the environment. Ann. New York Acad. Sci. 700: 173-176.

439.   *Kisaka, H., Kameya, T. (1994):* Production of somatic hybrids between *Daucus carota* L. and *Nicotiana tabacum.* Theor. Appl. Genet. 88: 75-80.

440. *Kisaka, H., Lee, H., Kisaka, M., Kanno, A., Kang, K., Kameya, T. (1994)*: Production and analysis of asymmetric hybrid plants between monocotyledon (*Oryza sativa* L.) and dicotyledon (*Daucus carota* L.). Theor. Appl. Genet. 89: 365-371.

441. *Kishor, P.B.K., Hong, Z.-L., Miao, G.-H., Hu, C.-A.A., Verma, D.P.S. (1995a)*: Overexpression of D1-pyrroline-5-carboxylate synthetase increases proline production and confers osmotolerance in transgenic plants. Rice Biotechnology Quarterly 23: 21-22.

442. *Kishor, P.B.K., Hong, Z.-L., Miao, G.-H., Hu, C.-A.A., Verma, D.P.S. (1995b)*: Overexpression of Delta(1)-pyrroline-5-carboxylate synthetase increases proline production and confers osmotolerance in transgenic plants. Plant Physiol. 108: 1387-1394.

443. *Kishore, G.M., Padgette, S.R., Fraley, R.T. (1992)*: History of herbicide-tolerant crops, methods of development and current state of the art - Emphasis on glyphosate tolerance. Weed Technol. 6: 626-634.

444. *Kiss, G.B., Csanádi, G., Kálmán, K., Kaló, P., Ökrész, L. (1993)*: Construction of a basic genetic map for alfalfa using RFLP, RAPD, isozyme and morphological markers. Mol. Gen. Genet. 238: 129-137.

445. *Kita, F., Hongo, A., Zou, H., Cheng, J., Zhao, Z. (1994)*: Cytology and pollen grain fertility in populations of *Astragalus adsurgens* Pall. indigenous to Yunwu Mountain in the Loess Plateau, north-west China. Euphytica 72: 225-230.

446. *Kjellsson, G. (1997)*: Prediction of plant fate in natural habitats from a greenhouse mesocosm test system. J. Appl. Ecol. (in press).

447. *Kjellsson, G., Simonsen, V. (1994)*: Methods for risk assessment of transgenic plants. I. Competition, establishment and ecosystem effects. Birkhäuser Verlag, Basel, 214 pp.

448. *Klein, T.M., Arentzen, R., Lewis, P.A., Fitzpatrick-McElligott, S. (1992)*: Transformation of microbes, plants and animals by particle bombardment. Bio/Technology 10: 286-291.

449. *Kleiner, K.W., Ellis, D.D., McCown, B.H., Raffa, K.F. (1995)*: Field evaluation of transgenic poplar expressing a *Bacillus thuringiensis* cry1A(a) d-Endotoxin gene against forest tent caterpillar (Lepidoptera: Lasiocampidae) and gypsy moth (Lepidoptera: Lymantriidae) following winter dormancy. Environ. Entomol. 24: 1358-1364.

450. *Klinkhamer, P.G.L., Jong, T.J. de, Nell, H.W. (1994)*: Limiting factors for seed production and phenotypic gender in the gynodioecious species *Echium vulgare* (Boraginaceae). Oikos 71: 469-478.

451. *Klopfenstein, N.B., McNabb, H.S., Hart, E.R., Hall, R.B., Hanna, R.D., Heuchelin, S.A., Allen, K.K., Shi, N.-Q., Thornburg, R.W. (1993)*: Transformation of *Populus* hybrids to study and improve pest resistance. Silvae Genet. 42: 86-90.

452. *Klopfenstein, N.B., Shi, N.-Q., Kernan, A., McNabb Jr., H.S., Hall, R.B., Hart, E.R., Thornburg, R.W. (1991)*: Transgenic *Populus* hybrid express a wound-inducible potato proteinase inhibitor II - CAT gene fusion. Can. J. Forest Res. 21: 1321-1328.

453. *Knox, J.S., Gutowski, M.J., Marshall, D.C., Rand, O.G. (1995)*: Tests of the genetic bases of character differences between *Helenium virginicum* and *H. autumnale* (Asteraceae) using common gardens and transplant studies. Syst. Bot. 20: 120-131.

454. *Knox, R.B., Zee, S.Y., Blomstedt, C., Singh, M.B. (1993)*: Male gametes and fertilization in angiosperms. New Phytol. 125: 679-694.

455. *Kochert, G. (1994)*: RFLP technology. In: Phillips, R.L., Vasil, I.K., (eds.) DNA-based markers in plants. Kluwer Academic Publishers, Dordrecht, 8-38.

456. *Kochmer, J.P., Handel, S.N. (1986)*: Constraints and competition in the evolution of flowering phenology. Ecol. Monogr. 56: 303-325.

457. *Kohn, J.R., Casper, B.B. (1992)*: Pollen-mediated gene flow in *Cucurbita foetidissima* (Cucurbitaceae). Amer. J. Bot. 79: 57-62.

458. *Kohno-Murase, J., Murase, M., Ichikawa, H., Imamura, J. (1995)*: Improvement in the quality of seed storage protein by transformation of *Brassica napus* with an antisense gene for cruciferin. Theor. Appl. Genet. 91: 627-631.

459. *Konishi, T., Linde-Laursen, I. (1988)*: Spontaneous chromosomal rearrangements in cultivated and wild barleys. Theor. Appl. Genet. 75: 237-243.

460. *Koop, H.U., Eigel, L., Spörlein, B. (1992)*: Protoplasts in organelle research: Transfer and transformation of plastids. Physiol. Plant. 85: 339-344.

461. *Koritsas, V.M., Atkinson, H.J. (1994)*: Proteinases of females of the phytoparasite *Globodera pallida* (potato cyst nematode). Parasitology 109: 357-365.

462. *Kost, B., Galli, A., Potrykus, I., Neuhaus, G. (1995)*: High efficiency transient and stable transformation by optimized DNA microinjection into *Nicotiana tabacum* protoplasts. J. Exp. Bot. 46: 1157-1167.

463. *Krattiger, A.F., Lesser, W.H. (1994)*: Biosafety - an environmental impact assessment tool - and the role of the convention on biological diversity. In: Krattiger, A.F., McNeely, J.A., Lesser, W.H., Miller, K.R., St. Hill, Y., Senanayake, R., (eds.) Widening perspectives on biodiversity. International Union for Conservation of Nature and Natural Resources (IUCN), Gland, 353-366.

464. *Krattiger, A.F., Rosemarin, A. (1994)*: Biosafety for sustainable agriculture. Sharing biotechnology regulatory experiences of the western hemisphere. ISAAA/SEI, Stockholm, 278 pp.

465. *Krauss, S.L. (1994a)*: Restricted gene flow within the morphologically complex species *Persoonia mollis* (Proteaceae): Contrasting evidence from the mating system and pollen dispersal. Heredity 73: 142-154.

466. *Krauss, S.L. (1994b)*: Preferential outcrossing in the complex species *Persoonia mollis* R. Br. (Proteaceae). Oecologia 97: 256-264.

467. *Krebs, S.L., Hancock, J.F. (1991)*: Embryonic genetic load in the highbush blueberry, *Vaccinium corymbosum* (Ericaceae). Amer. J. Bot. 78: 1427-1437.

468. *Kron, P., Stewart, S.C., Back, A. (1993)*: Self-compatibility, autonomous self-pollination, and insect-mediated pollination in the clonal species *Iris versicolor*. Can. J. Bot. 71: 1503-1509.

469. *Krueger, S.K., Knapp, S.J. (1991)*: Mating systems of *Cuphea laminuligera* and *Cuphea lutea*. Theor. Appl. Genet. 82: 221-226.

470. *Krutovskii, K.V., Bergmann, F. (1995)*: Introgressive hybridization and phylogenetic relationships between Norway, *Picea abies* (L.) Karst., and Siberian, *Picea obovata* Ledeb., spruce species studied by isozyme loci. Heredity 74: 464-480.

471. *Kuang, A.X., Musgrave, M.E., Matthews, S.W., Cummins, D.B., Tucker, S.C. (1995)*: Pollen and ovule development in *Arabidopsis thaliana* under spaceflight conditions. Amer. J. Bot. 82: 585-595.

472. *Kuehnle, A.R., Sugii, N. (1992)*: Transformation of *Dendrobium* orchid using particle bombardment of protocorms. Plant Cell Rep. 11: 484-488.

473. *Kuipers, A.G.J., Soppe, W.J.J., Jacobsen, E., Visser, R.G.F. (1994)*: Field evaluation of transgenic potato plants expressing an antisense granule-bound starch synthase gene: Increase of the antisense effect during tuber growth. Plant Mol. Biol. 26: 1759-1773.

474. *Kwon, S.Y., Yang, Y., Hong, C.B., Pyun, K.H. (1995)*: Expression of active human interleukin-6 in transgenic tobacco. Mol. Cells 5: 486-492.

475. *Kwon, T., Sasahara, T., Abe, T. (1995)*: Lysine accumulation in transgenic tobacco expressing dihydrodipicolinate synthase of *Escherichia coli*. J. Plant Physiol. 146: 615-621.

476. *Lachenaud, P. (1995)*: Variations in the number of beans per pod in *Theobroma cacao* L. in the Ivory Coast. II. Pollen germination, fruit setting and ovule development. J. Hort. Sci. 70: 1-6.

477. *Lamb, C.J., Ryals, J.A., Ward, E.R., Dixon, R.A. (1992)*: Emerging strategies for enhancing crop resistance to microbial pathogens. Bio/Technology 10: 1436-1445.

478. *Lamb, E.M., Davis, D.W., Andow, D.A. (1994)*: Mid-parent heterosis and combining ability of European corn borer resistance in maize. Euphytica 72: 65-72.

479. *Lamont, B.B., Klinkhamer, P.G.L., Witkowski, E.T.F. (1993)*: Population fragmentation may reduce fertility to zero in *Banksia goodii* - a demonstration of the Allee effect. Oecologia 94: 446-450.

480. *Lande, R., Schemske, D.W. (1985)*: The evolution of self-fertilization and inbreeding depression in plants. I. Genetic models. Evolution 39: 24-40.

481. *Langenheim, J.H. (1984)*: The roles of plant secondary chemicals in wet tropical ecosystems. In: Medina, E., Mooney, H.A., Vázquez-Yánes, C., (eds.) Physiological ecology of plants of the wet tropics. Dr.W. Junk Publishers, The Hague, 189-208.

482. *Lapidot, M., Gafny, R., Ding, B., Wolf, S., Lucas, W.J., Beachy, R.N. (1993)*: A dysfunctional movement protein of tobacco mosaic virus that partially modifies the plasmodesmata and limits virus spread in transgenic plants. Plant J. 4: 959-970.

483. *Lau, T.-C., Stephenson, A.G. (1993)*: Effects of soil nitrogen on pollen production, pollen grain size, and pollen performance in *Cucurbita pepo* (Cucurbitaceae). Amer. J. Bot. 80: 763-768.

484. *Lau, T.-C., Stephenson, A.G. (1994)*: Effects of soil phosphorus on pollen production, pollen size, pollen phosphorus content, and the ability to sire seeds in *Cucurbita pepo* (Cucurbitaceae). Sex. Plant Reprod. 7: 215-220.

485. *Lawrence, J.B. (1990)*: A fluorescense in situ hybridization approach for gene mapping and the study of nuclear organisation. In: Davies, K.E., Tilghman, S.M., (eds.) Genome analysis. Cold Spring Harbor Laboratory Press, Cold Spring Harbor, vol. 1: 1-38.

486. *Lawton, J.H. (1995)*: Ecological experiments with model systems. Science 269: 328-331.

487. *Leclerc-Potvin, C., Ritland, K. (1994)*: Modes of self-fertilization in *Mimulus guttatus* (Scrophulariaceae): A field experiment. Amer. J. Bot. 81: 199-205.

488. *Lee, Y.H., Balyan, H.S., Wang, B.J., Fedak, G. (1993)*: Cytogenetic analysis of three *Hordeum* x *Elymus* hybrids. Euphytica 72: 115-119.

489. *Lee, Y.O., Kanno, A., Kameya, T. (1996)*: The physical map of the chloroplast DNA from *Asparagus officinalis* L. Theor. Appl. Genet. 92: 10-14.

490. *Leeuwen, C.J. van (1995)*: General introduction. In: Leeuwen, C.J. van, Hermens, J.M.L., (eds.) Risk assessment of chemicals: An introduction. Kluwer Academic Publishers, Dordrecht, 1-17.

491. *Leeuwen, C.J. van, Hermens, J.L.M. (1995)*: Risk assessment of chemicals: An introduction. Kluwer Academic Publishers, Dordrecht, 374 pp.

492. *Lefort-Buson, M. (1986)*: Hétérosis chez le colza oléagineux (*Brassica napus* L.): Analyse génétique et prédiction. Thesis. Université de Paris Sud, Orsay, 228 pp.

493. *Leitch, A.R., Schwarzacher, T., Jackson, D., Leitch, I.J. (1994)*: In situ hybridization: A practical guide. BIOS Scientific Publishers, Oxford, 118 pp.

494. *Lereclus, D., Delécluse, A., Lecadet, M.-M. (1993)*: Diversity of *Bacillus thuringiensis* toxins and genes. In: Entwistle, P.F., Cory, J.S., Bailey, M.J., Higgs, S., (eds.) B*acillus thuringiensis*, an environmental biopesticide: Theory and practice. John Wiley & Sons, Chichester, 38-69.

495. *Levin, B.R. (1992)*: DNA technology and the release of genetically engineered organisms: Some implications for the conservation of genetic resources. In: Sandlund, O.T., Hindar, K., Brown, A.H.D., (eds.) Conservation of biodiversity for sustainable development. Studies in conservation biology, Scandinavian University Press, Oslo, vol. 1: 245-259.

496. *Levin, D.A. (1966)*: Chromatographic evidence of hybridization and evolution in *Phlox maculata*. Amer. J. Bot. 53: 238-245.

497. *Levin, D.A. (1967a)*: An analysis of hybridization in *Liatris*. Brittonia 19: 248-260.

498. *Levin, D.A. (1967b)*: Hybridization between annual species of *Phlox*: Population structure. Amer. J. Bot. 54: 1122-1130.

499. *Levin, D.A. (1981)*: Dispersal versus gene flow in plants. Ann. Mo. Bot. Gard. 68: 233-253.

500. *Levin, D.A. (1984)*: Inbreeding depression and proximity-dependent crossing success in *Phlox drummondii*. Evolution 38: 116-127.

501. *Levin, D.A. (1988)*: The paternity pools of plants. Amer. Naturalist 132: 309-317.

502. *Levin, D.A. (1993)*: S-gene polymorphism in *Phlox drummondii*. Heredity 71: 193-198.

503. *Levin, D.A., Kerster, H.W. (1971)*: Neighborhood structure in plants under diverse reproductive methods. Amer. Naturalist 105: 345-354.

504. *Levin, D.A., Kerster, H.W. (1974)*: Gene flow in seed plants. In: Dobzhansky, T., Hecht, M.K., Steere, W.D., (eds.) Evol. Biol. 7: 139-220.

505. *Li, Z.J., Jarret, R.L., Cheng, M., Demski, J.W. (1995)*: Improved electroporation buffer enhances transient gene expression in *Arachis hypogaea* protoplasts. Genome 38: 858-863.

506. *Liang, X.-Y., Zhu, Y.-X., Mi, J.-J., Chen, Z.-L. (1994)*: Production of virus resistant and insect tolerant transgenic tobacco plants. Plant Cell Rep. 14: 141-144.

507. *Lin, W., Anuratha, C.S., Datta, K., Potrykus, I., Muthukrishnan, S., Datta, S.K. (1995)*: Genetic engineering of rice for resistance to sheath blight. Bio/Technology 13: 686-691.

508. *Lindbo, J.A., Dougherty, W.G. (1992)*: Pathogen-derived resistance to a potyvirus: Immune and resistant phenotypes in transgenic tobacco expressing altered forms of a potyvirus coat protein nucleotide sequence. Mol. Plant Microbe Interaction. 5: 144-153.

509. *Lindbo, J.A., Silva-Rosales, L., Proebsting, W.M., Dougherty, W.G. (1993)*: Induction of a highly specific antiviral state in transgenic plants: Implications for regulation of gene expression and virus resistance. Plant Cell 5: 1749-1759.

510. *Linde-Laursen, I. (1988)*: Giemsa C-banding of barley chromosomes. V. Localization of breakpoints in 70 reciprocal translocations. Hereditas 108: 65-76.

511. *Lipton, J., Galbraith, H., Burger, J., Wartenberg, D. (1993)*: A paradigm for ecological risk assessment. Environ. Manage. 17: 1-5.

512. *Liu, C.J., Witcombe, J.R., Pittaway, T.S., Nash, M., Hash, C.T., Busso, C.S., Gale, M.D. (1994)*: An RFLP-based genetic map of pearl millet (*Pennisetum glaucum*). Theor. Appl. Genet. 89: 481-487.

513. *Liu, Y., Silverstone, A.L., Wu, Y.M., Yang, S.F. (1995)*: Formation of N-malonyl-L-tryptophan in water-stressed tomato leaves. Phytochemistry 40: 691-697.

514. *Lloyd, D.G. (1992)*: Self- and cross-fertilization in plants. II. The selection of self-fertilization. Int. J. Plant Sci. 153: 370-380.

515. *Lössl, A., Frei, U., Wenzel, G. (1994)*: Interaction between cytoplasmic composition and yield parameters in somatic hybrids of *Solanum tuberosum* L. Theor. Appl. Genet. 89: 873-878.

516. *Logemann, J., Jach, G., Logemann, S., Leah, R., Wolf, G., Mundy, J., Oppenheim, A., Chet, I., Schell, J. (1993)*: Expression of a ribosome inhibiting protein (RIP) or a bacterial chitinase leads to fungal resistance in transgenic plants. In: Fritig, B., Legrand, M., (eds.) Developments in plant pathology. Mechanisms of plant defense responses. Kluwer Academic Publishers, Dordrecht, 2: 446-448.

517. *Lokar, L.C., Maurich, V., Poldini, L. (1986)*: Chemical aspect of floral biology in *Euphorbia fragifera*. Folia Geobot. Phytotaxon. 21: 277-285.

518. *Long, J.C. (1986)*: The allelic correlation structure of Gainj- and Kalam-speaking people. I. The estimation and interpretation of Wright's F-statistics. Genetics 112: 629-647.

519. *Loockermann, D.J., Jansen, R.K. (1996)*: The use of herbarium material for DNA studies. In: Stuessy, T.F., Sohmer, S.H., (eds.) Sampling the green world: Innovative concepts of collection, preservation, and storage of plant diversity. Columbia University Press, New York, 205-220.

520. *Lunau, K. (1991)*: Innate flower recognition in bumblebees (*Bombus terrestris, B. lucorum*; Apidae): Optical signals from stamens as landing reaction releasers. Ethology 88: 203-214.

521. *Lusardi, M.C., Neuhaus-Url, G., Potrykus, I., Neuhaus, G. (1994)*: An approach towards genetically engineered cell fate mapping in maize using the Lc gene as a visible marker: Transactivation capacity of Lc vectors in differentiated maize cells and microinjection of Lc vectors into somatic embryos and shoot apical meristems. Plant J. 5: 571-582.

522. *Lyons, E.E., Antonovics, J. (1991)*: Breeding system evolution in *Leavenworthia*: Breeding system variation and reproductive success in natural populations of *Leavenworthia crassa* (Cruciferae). Amer. J. Bot. 78: 270-287.

523. *Ma, J.K.-C., Hiatt, A., Hein, M., Vine, N.D., Wang, F., Stabila, P., Dolleweerd, C. van, Mostov, K., Lehner, T. (1995)*: Generation and assembly of secretory antibodies in plants. Science 268: 716-719.

524. *Ma, T.H., Cabrera, G.L., Cebulska-Wasilewska, A., Chen, R., Loarca, F., Vandenberg, A.L., Salamone, M.F. (1994)*: *Tradescantia* stamen hair mutation bioassay. Mutat. Res. - Fundam. Mol. Mech. Mut. 310: 211-220.

525. *Machado, M.L.D.C., Machado, A.D.C., Hanzer, V., Weiss, H., Regner, F., Steinkellner, H., Mattanovich, D., Plail, R., Knapp, E., Kalthoff, B., Katinger, H. (1992)*: Regeneration of transgenic plants of *Prunus armeniaca* containing the coat protein gene of Plum Pox Virus. Plant Cell Rep. 11: 25-29.

526. *MacIntosh, S.C., Kishore, G.M., Perlak, F.J., Marrone, P.G., Stone, T.B., Sims, S.R., Fuchs, R.L. (1990)*: Potention of *Bacillus thuringiensis* insecticidal activity by serine protease inhibitors. J. Agric. Food Chem. 38: 1145-1152.

527. *Maddox, G.D., Cook, R.E., Wimberger, P.H., Gardescu, S. (1989)*: Clone structure in four *Solidago altissima* (Asteraceae) populations: Rhizome connections within genotypes. Amer. J. Bot. 318-326.

528. *Madsen, K.H. (1994)*: Weed management and impact on ecology of growing glyphosate tolerant sugarbeets (*Beta vulgaris* L.). Ph.D. Thesis, The Royal Veterinary and Agricultural University, Copenhagen, 62 pp.

529. *Madsen, K.H., Blacklow, W.M., Jensen, J.E. (1996)*: Simulation of herbicide-use in a crop rotation with transgenic resistant sugarbeet. Second International Weed Control Congress 1387-1391.

530. *Madsen, K.H., Heitholt, J.J., Duke, S.O., Smeda, R.J., Streibig, J.C. (1995)*: Photosynthetic parameters in glyphosate-treated sugarbeet (*Beta vulgaris* L.). Weed Res. 35: 81-88.

531. *Maier, C.G.-A., Chapman, K.D., Smith, D.W. (1995)*: Differential estrogenic activities of male and female plant extracts from two dioecious species. Plant Sci. 109: 31-43.

532. *Maki, M., Masuda, M. (1994)*: Spatial genetic structure within two populations of a self-incompatible perennial, *Chionographis japonica* var. *japonica* (Liliaceae). J. Plant. Res. 107: 283-287.

533. *Malik, J., Barry, G., Kishore, G. (1989)*: The herbicide glyphosate. BioFactors 2: 17-25.

534. *Malléa, M., Soler, M. (1974)*: Les grains de pollen (Pollen grains). Laboratoires Sandoz S.A.R.L., Marseille, 229 pp.

535. *Manchenko, G.P. (1994)*: Handbook of detection of enzymes on electrophoretic gels. CRC Press, Boca Raton, 341 pp.

536. *Manly, B.F.J. (1985)*: The statistics of natural selection on animal populations. Chapman & Hall, London, 484 pp.

537. *Manly, B.F.J. (1986)*: Multivariate statistical methods, a primer. Chapman & Hall, London, 159 pp.

538. *Manteuffel, R., Panitz, R. (1993)*: *In situ* localization of faba bean and oat legumin-type proteins in transgenic tobacco seeds by a highly sensitive immunological tissue print technique. Plant Mol. Biol. 22: 1129-1134.

539. *Mariani, C., Beuckeleer, M. de, Truettner, J., Leemans, J., Goldberg, R.B. (1990)*: Induction of male sterility in plants by a chimaeric ribonuclease gene. Nature 347: 737-741.

540. *Mariani, C., Gossele, V., Beuckeleer, M. de, Block, M. de, Goldberg, R.B., Greef, W. de, Leemans, J. (1992)*: A chimaeric ribonuclease-inhibitor gene restores fertility to male sterile plants. Nature 357: 384-387.

541. *Markow, T.A. (1995)*: Evolutionary ecology and developmental instability. Annu. Rev. Ecol. Syst. 40: 105-120.

542. *Marsh, L.E., Pogue, G.P., Szybiak, U., Connell, J.P., Hall, T.C. (1991)*: Non-replicating deletion mutants of brome mosaic virus RNA-2 interfere with viral replication. J. Gen. Virol. 72: 2367-2374.

543. *Marshall, D.L., Ellstrand, N.C. (1988)*: Effective mate choice in wild radish: Evidence for selective seed abortion and its mechanism. Amer. Naturalist 131: 739-756.

544. *Mason, H.S., Lam, D.M.K., Arntzen, C.J. (1992)*: Expression of hepatitis B surface antigen in transgenic plants. Proc. Nat. Acad. Sci. USA 89: 11745-11749.

545. *Mason-Gamer, R.J., Holsinger, K.E., Jansen, R.K. (1995)*: Chloroplast DNA haplotype variation within and among populations of *Coreopsis grandiflora* (Asteraceae). Mol. Biol. Evol. 12: 371-381.

546. *Masoud, S.A., Johnson, L.B., White, F.F., Reeck, G.R. (1993)*: Expression of a cysteine protein-ase inhibitor (oryzacystatin-I) in transgenic tobacco plants. Plant Mol. Biol. 21: 655-663.

547. *Mathur, J., Koncz, C., Szabados, L. (1995)*: A simple method for isolation, liquid culture, transformation and regeneration of *Arabidopsis thaliana* protoplasts. Plant Cell Rep. 14: 221-226.

548. *Matousek, J., Schröder, A.R.W., Trnena, L., Reimers, M., Baumstark, T., Dedic, P., Vlasák, J., Becker, I., Kreuzaler, F., Fladung, M., Riesner, D. (1994)*: Inhibition of viroid infection by antisense RNA expression in transgenic plants. Biol. Chem. Hoppe-Seyler 375: 765-777.

549. *Matsumoto, T., Sakai, A., Yamada, K. (1994)*: Cryopreservation of *in vitro* grown apical meristems of wasabi (*Wasabia japonica*) by vitrification and subsequent high plant regeneration. Plant Cell Rep. 13: 442-446.

550. *Maughan, P.J., Saghai Maroof, M.A., Buss, G.R., Huestis, G.M. (1996)*: Amplified fragment length polymorphism (AFLP) in soybean: Species diversity, inheritance, and near-isogenic line analysis. Theor. Appl. Genet. 93: 392-401.

551. *Mayerhofer, R., Koncz-Kalman, Z., Nawrath, C., Bakkeren, G., Crameri, A., Angelis, K., Redei, G.P., Schell, J., Hohn, B., Koncz, C. (1991)*: T-DNA integration: A mode of illegitimate recombination in plants. EMBO J. 10: 697-704.

552. *Mazer, S.J., Schick, C.T. (1991)*: Constancy of population parameters for life history and floral traits in *Raphanus sativus* L. I. Norms of reaction and the nature of genotype by environment interactions. Heredity 67: 143-156.

553. *McCall, C., Mitchell-Olds, T., Waller, D.M. (1991)*: Distance between mates affects seedling characters in a population of *Impatiens capensis* (Balsaminaceae). Amer. J. Bot. 78: 964-970.

554. *McCall, C., Waller, D.M., Mitchell-Olds, T. (1994)*: Effects of serial inbreeding on fitness components in *Impatiens capensis*. Evolution 48: 818-827.

555. *McCarthy, B.C., Quinn, J.A. (1990)*: Reproductive ecology of *Carya* (Juglandaceae): Phenology, pollination, and breeding system of two sympatric tree species. Amer. J. Bot. 77: 261-273.

556. *McCauley, D.E. (1994)*: Contrasting the distribution of chloroplast DNA and allozyme polymorphism among local populations of *Silene alba*: Implications for studies of gene flow in plants. Proc. Nat. Acad. Sci. USA 91: 8127-8131.

557. *McCauley, D.E. (1995)*: The use of chloroplast DNA polymorphism in studies of gene flow in plants. Trend. Ecol. Evolut. 10: 198-202.

558. *McCrea, K.D., Levy, M. (1983)*: Photographic visualization of floral colors as perceived by honeybee pollinators. Amer. J. Bot. 70: 369-375.

559. *McGraw, J.B., Antonovics, J. (1983)*: Experimental ecology of *Dryas octopetala* ecotypes. I. Ecotypic differentiation and life-cycle stages of selection. J. Ecol. 71: 879-897.

560. *McKenna, M.A., Thomson, J.D. (1988)*: A technique for sampling and measuring small amounts of floral nectar. Ecology 69: 1306-1307.

561. *McKenzie, M.J., Jameson, P.E., Poulter, R.T.M. (1994)*: Cloning an *ipt* gene from *Agrobacterium tumefaciens*: Characterization of cytokinins in derivative transgenic plant tissue. Plant Growth Regul. 14: 217-228.

562. *McKersie, B.D. (1994)*: Environmental stress tolerance in genetically improved plants. Agri-food Research in Ontario 17: 2-6.

563. *McKone, M.J., Ostertag, R., Rauscher, J.T., Heiser, D.A., Russell, F.L. (1995)*: An exception to Darwin's syndrome: Floral position, protogyny, and insect visitation in *Besseya bullii* (Scrophulariaceae). Oecologia 101: 68-74.

564. *McNeilly, T., Antonovics, J. (1968)*: Evolution in closely adjacent plant populations. IV. Barriers to gene flow. Heredity 23: 205-218.

565. *McPherson, M.J., Quirke, P., Taylor, G.R. (1991)*: PCR: A practical approach. Oxford University Press, Oxford, 253 pp.

566. *Meagher, T.R. (1986)*: Analysis of paternity within a natural population of *Chamaelirium luteum*. 1. Identification of most-likely male parents. Amer. Naturalist 128: 199-215.

567. *Meagher, T.R. (1991)*: Analysis of paternity within a natural population of *Chamaelirium luteum*. II. Patterns of male reproductive success. Amer. Naturalist 137: 738-752.

568. *Medan, D. (1994)*: Reproductive biology of *Frangula alnus* (Rhamnaceae) in Southern Spain. Plant Syst. Evol. 193: 173-186.

569. *Mehta, H., Sarkar, K.R. (1992)*: Heterosis for leaf photosynthesis, grain yield and yield components in maize. Euphytica 61: 161-168.

570. *Meijden, R. van der (1994)*: European botanical files in relation to the safety of genetically modified plants: A case study in the Netherlands. In: OECD, Ottawa '92: The OECD workshop on methods for monitoring organisms in the environment. OECD Environment Monographs No. 90. OECD, Paris, 27-35.

571. *Meijer, E.G.M., Schilperoort, R.A., Rueb, S., Os-Ruygrok, P.E. van, Hensgens, L.A.M. (1991)*: Transgenic rice cell lines and plants: Expression of transferred chimeric genes. Plant Mol. Biol. 16: 807-820.

572. *Meins, F., Kunz, C. (1995)*: Gene silencing in transgenic plants: A heuristic autoregulation model. Curr. Top. Microbiol. Immunol. 197: 105-120.

573. *Meyer, P. (1995)*: DNA methylation and transgene silencing in *Petunia hybrida*. Curr. Top. Microbiol. Immunol. 197: 15-28.

574. *Meyer, P., Linn, F., Heidmann, I., Saedler, H. (1990)*: Engineering of a new flower colour variety of petunia. In: Lamb, C.J., Beachy, R.N., (eds.) Plant gene transfer: UCLA symposia on molecular and cellular biology. Wiley-Liss, New York, 129: 319-326.

575. *Miao, Z.H., Lam, E. (1995)*: Targeted disruption of the TGA3 locus in *Arabidopsis thaliana*. Plant J. 7: 359-365.

576. *Michaelson, M.J., Price, H.J., Ellison, J.R., Johnston, J.S. (1991)*: Comparison of plant DNA contents determined by Feulgen microspectrophotometry and laser flow cytometry. Amer. J. Bot. 78: 183-188.

577. *Michelmore, R. (1995)*: Molecular approaches to manipulation of disease resistance genes. Annu. Rev. Phytopathol. 33: 393-427.

578. *Mikkelsen, T.R., Andersen, B., Jørgensen, R.B. (1996)*: The risk of crop transgene spread. Nature 380: 31.

579. *Mikkelsen, T.R., Jensen, J., Jørgensen, R.B. (1996)*: Inheritence of oilseed rape (*Brassica napus*) RAPD markers in a backcross progeny with *Brassica campestris*. Theor. Appl. Genet. 92: 492-497.

580. *Milligan, B.G. (1991)*: Chloroplast DNA diversity within and among populations of *Trifolium pratense*. Curr. Genetics 19: 411-416.

581. *Misawa, N., Yamano, S., Linden, H., Felipe, M.R. de, Lucas, M., Ikenaga, H., Sandmann, G. (1993)*: Functional expression of the *Erwinia uredovora* carotenoid biosynthesis gene *crtI* in transgenic plants showing an increase of ß-carotene biosynthesis activity and resistance to the bleaching herbicide norflurazon. Plant J. 4: 833-840.

582. *Mitra, A., Que, Q.D. (1994)*: Ectopic expression of a viral adenine methyltransferase gene in tobacco. Biochim. Biophys. Acta 1219: 244-249.

583. *Møller, A.P. (1992)*: Female swallow preference for symmetrical male sexual ornaments. Nature 357: 228-240.

584. *Møller, A.P. (1993)*: Morphology and sexual selection in the barn swallow *Hirundo rustica* in Chernobyl, Ukraine. Proc. Roy. Soc. London Ser. B 252: 51-57.

585. *Møller, A.P. (1995)*: Bumblebee preference for symmetrical flowers. Proc. Nat. Acad. Sci. USA 92: 2288-2292.

586. *Møller, A.P. (1996)*: Developmental stability of flowers, embryo abortion, and developmental selection in plants. Proc. Roy. Soc. London Ser. B 263: 53-56.

587. *Møller, A.P., Eriksson, M. (1995)*: Pollinator preference for symmetrical flowers and sexual selection in plants. Oikos 73: 15-22.

588. *Möllers, C., Frei, U., Wenzel, G. (1994)*: Field evaluation of tetraploid somatic potato hybrids. Theor. Appl. Genet. 88: 147-152.

589. *Mogford, D.J. (1974)*: Flower colour polymorphism in *Cirsium palustre*. Heredity 33: 241-256.

590. *Mogie, M., Stamp, A.J. (1995)*: Pollen carryover and neighborhood in *Ranunculus bulbosus*. Experientia 51: 381-383.

591. *Mol, J.N.M., Holton, T.A., Koes, R.E. (1995)*: Floriculture: Genetic engineering of commercial traits. Trends Biotech. 13: 350-355.

592. *Molau, U., Prentice, H.C. (1992)*: Reproductive system and population structure in three arctic *Saxifraga* species. J. Ecol. 80: 149-161.

593. *Molina-Freaner, F., Jain, S.K. (1993)*: Inbreeding effects in a gynodioecious population of the colonizing species *Trifolium hirtum* All. Evolution 47: 1472-1479.

594. *Montalvo, A.M. (1994)*: Inbreeding depression and maternal effects in *Aquilegia caerulea*, a partially selfing plant. Ecology 75: 2395-2409.

595. *Mooring, J.S. (1994)*: A cytogenetic study of *Eriophyllum confertiflorum* (Compositae, Helenieae). Amer. J. Bot. 81: 919-926.

596. *Morgante, M., Olivieri, A.M. (1993)*: PCR-amplified microsatellites as markers in plant genetics. Plant J. 3: 175-182.

597. *Morse, D.H., Schmitt, J. (1991)*: Maternal and paternal effects on follicle production in the milkweed *Asclepias syriaca* (Asclepiadaceae). Amer. J. Bot. 78: 1304-1309.

598. *Motoyoshi, F. (1993)*: ToMV-resistant transgenic tomato as a material for the first field experiment of genetically engineered plants in Japan. In Vitro Cell. Dev. Biol. - Plant 29P: 13-16.

599. *Moya, S., Ackerman, J.D. (1993)*: Variation in the floral fragrance of *Epidendrum ciliare* (Orchidaceae). Nord. J. Bot. 13: 41-47.

600. *Mueller-Harvey, I., Dhanoa, M.S. (1991)*: Varietal differences among sorghum crop residues in relation to their phenolic HPLC fingerprints and responses to different environments. J. Sci. Food Agr. 57: 199-216.

601. *Mukai, Y., Nakahama, Y., Yamamoto, M. (1993)*: Simultaneous discrimination of the three genomes in hexaploid wheat by multicolor fluorescence *in situ* hybridisation using total genomic and highly repeated DNA probes. Genome 36: 489-494.

602. *Multani, D.S., Jena, K.K., Brar, D.S., Reyes, B.G. de los, Angeles, E.R., Khush, G.S. (1994)*: Development of monosomic alien addition lines and introgression of genes from *Oryza australiensis* Domin. to cultivated rice *Oryza sativa* L. Theor. Appl. Genet. 88: 102-109.

603. *Munson, A. (1995)*: Should a biosafety protocol be negotiated as part of the biodiversity convention? Global Environ. Change 5: 7-26.

604. *Muona, O., Moran, G.F., Bell, J.C. (1991)*: Hierarchical patterns of correlated mating in *Acacia melanoxylon*. Genetics 127: 619-626.

605. *Murata, N., Ishizaki-Nishizawa, O., Higashi, S., Hayashi, H., Tasaka, Y., Nishida, I. (1992)*: Genetically engineered alteration in the chilling sensitivity of plants. Nature 356: 710-713.

606. *Murray, B.G. (1990)*: Heterostyly and pollen-tube interactions in *Luculia gratissima* (Rubiaceae). Ann. Bot. 65: 691-698.

607. *Murray, B.G., Cameron, E.K., Standring, L.S. (1992)*: Chromosome numbers, karyotypes, and nuclear DNA variation in *Pratia* Gaudin (Lobeliaceae). N. Z. J. Bot. 30: 181-187.

608. *Musil, C.F. (1995)*: Differential effects of evaluated ultraviolet-B radiation on the photochemical and reproductive performances of dicotyledonous and moncotyledonous arid-environment ephemerals. Plant Cell Environ. 18: 844-854.

609. *Nabulsi, S.M., Page, N.W., Duval, A.L., Seabrook, Y.A., Scott, K.J. (1994)*: A gas-driven gene gun for microprojectile methods of genetic engineering. Meas. Sci. Technol. 5: 267-274.

610. *Nagy, F., Beffa, R., Meins Jr., F., Neuhaus, G., Metraux, J.-P. (1995)*: Cholera toxin induces defense reactions and pathogen resistance in transgenic plants. J. Cell. Biochem. 488-488.

611. *Nakajima, N., Ikada, Y. (1995)*: Effects of concentration, molecular weight, and exposure time of poly(ethylene glycol) on cell fusion. Polym. J. 27: 211-219.

612. *Nakamura, R.R., Stanton, M.L. (1987)*: Cryptic seed abortion and the estimation of ovule fertilization. Can. J. Bot. 65: 2463-2465.

613. *Namuth, D.M., Lapitan, N.L.V., Gill, K.S., Gill, B.S. (1994)*: Comparative RFLP mapping of *Hordeum vulgare* and *Triticum tauschii*. Theor. Appl. Genet. 89: 865-872.

614. *Nason, J.D., Ellstrand, N.C. (1995)*: Lifetime estimates of biparental inbreeding depression in the self-incompatible annual plant *Raphanus sativus*. Evolution 49: 307-316.

615. *Naumova, T., Nijs, A.P.M. den, Willemse, M.T.M. (1993)*: Quantitative analysis of aposporous parthenogenesis in *Poa pratensis* genotypes. Acta Bot. Neer. 42: 299-312.

616. *Neff, J.L., Simpson, B.B. (1990)*: The roles of phenology and reward structure in the pollination biology of wild sunflower (*Helianthus annuus* L., Asteraceae). Isr. J. Bot. 39: 197-216.

617. *Nei, M. (1972)*: Genetic distances between populations. Amer. Naturalist 106: 283-292.

618. *Nei, M. (1977)*: F-statistics and analysis of gene diversity in subdivided populations. Ann. Hum. Genet. 41: 225-233.

619. *Nei, M. (1978)*: Estimation of average heterozygosity and genetic distance from a small number of individuals. Genetics 89: 583-590.

620. *Nei, M. (1987)*: Molecular evolutionary genetics. Columbia University Press, New York, 512 pp.

621. *Nei, M., Chesser, R.K. (1983)*: Estimation of fixation indices and gene diversities. Ann. Hum. Genet. 47: 253-259.

622. *Neiland, M.R.M., Wilcock, C.C. (1995)*: Maximization of reproductive success by European Orchidaceae under conditions of infrequent pollination. Protoplasma 187: 39-48.

623. *Nejidat, A., Clark, W.G., Beachy, R.N. (1990)*: Engineered resistance against plant virus diseases. Physiol. Plant. 80: 662-668.

624. *Neuhaus, D., Kühl, H., Kohl, J.-G., Dörfel, P., Börner, T. (1993)*: Investigation on the genetic diversity of *Phragmites* stands using genomic fingerprinting. Aquat. Bot. 45: 357-364.

625. *Neuhaus, G., Kranz, E., Spangenberg, G., Schweiger, H.G. (1987)*: High efficiency transformation of *Nicotiana tabacum* and *Brassica napus* by intranuclear microinjection in combination with single cell culture technique. Eur. J. Cell Biol. 43: 39.

626. *Neuhaus, G., Spangenberg, G. (1990)*: Plant transformation by microinjection techniques. Physiol. Plant. 79: 213-217.

627. *Neuhaus, G., Spangenberg, G., Scheid, O.M., Schweiger, H.G. (1987)*: Transgenic rapeseed plants obtained by the microinjection of DNA into microspore-derived embryoids. Theor. Appl. Genet. 75: 30-36.

628. *Ney, B., Duthion, C., Turc, O. (1994)*: Phenological response of pea to water-stress during reproductive development. Crop Science 34: 141-146.

629. *Nida, D.L., Anjos, J.R., Lomonossoff, G.P., Ghabrial, S.A. (1992)*: Expression of cowpea mosaic virus coat protein precursor in transgenic tobacco plants. J. Gen. Virol. 73: 157-163.

630. *Nilsson, L.A. (1979)*: The pollination ecology of *Herminium monorchis* (Orchidaceae). Bot. Not. 132: 537-549.

631. *Nishizawa, Y. (1994)*: Transgenic tobacco plants with enhanced resistance to the fungal pathogens by chitinase gene transfer. National Institute of Agrobiological Resources - Annual Report 710: 4-5.

632. *Norelli, J.L., Aldwinckle, H.S., Destéfano-Beltrán, L., Jaynes, J.M. (1994)*: Transgenic "Malling 26" apple expressing the attacin E gene has increased resistance to *Erwinia amylovora*. Euphytica 77: 123-128.

633. *Norman, J.K., Sakai, A.K., Weller, S.G., Dawson, T.E. (1995)*: Inbreeding depression in morphological and physiological traits of *Schiedea lydgatei* (Caryophyllaceae) in two environments. Evolution 49: 297-306.

634. *NRC (National Research Concil of the National Academy of Sciences) (1989)*: Field testing genetically modified organisms: Framework for decisions. Committee on scientific evaluation of the genetically modified microorganisms and plants into the environment. Board on biology, Commission on life sciences, National Research Council. National Academy Press, Washington, 170 pp.

635. *Nunney, L. (1995)*: Measuring the ratio of effective population size to adult numbers using genetic and ecological data. Evolution 49: 389-392.

636. *Nybom, H. (1985)*: Active self-pollination in blackberries (*Rubus* subgen. *Rubus*, Rosaceae). Nord. J. Bot. 5: 521-525.

637. *Nybom, H. (1988)*: Apomixis versus sexuality in blackberries (*Rubus* subgen. *Rubus*, Rosaceae). Plant Syst. Evol. 160: 207-218.

638. *O'Brien, S.P. (1994)*: Andromonoecy and fruit set in *Leptospermum myrsinoides* and L. *continentale* (Myrtaceae). Aust. J. Bot. 42: 751-762.

639. *O'Brien, S.P., Calder, D.M. (1993)*: Reproductive biology and floral phenologies of the sympatric species *Leptospermum myrsinoides* and L. *continentale* (Myrtaceae). Aust. J. Bot. 41: 527-539.

640. *O'Connell, M.A. (1995)*: The role of drought-responsive genes in drought resistance. AgBiotech News Inf. 7: 143N-147N.

641. *O'Donoughue, L.S., Chong, J., Wight, C.P., Fedak, G., Molnar, S.J. (1996)*: Localization of stem rust resistance genes and associated molecular markers in cultivated oat. Phytopathology 86: 719-727.

642. *O'Neill, C., Horváth, G.V., Horváth, E., Dix, P.J., Medgyesy, P. (1993)*: Chloroplast transformation in plants: Polyethylene glycol (PEG) treatment of protoplasts is an alternative to biolistic delivery systems. Plant J. 3: 729-738.

643. *Oard, J.H. (1991)*: Physical methods for the transformation of plant cells. Biotechnol. Adv. 9: 1-11.

644. *OECD (1986)*: Recombinant DNA safety considerations: Safety considerations for industrial, agricultural and environmental applications of organisms derived by recombinant DNA techniques. OECD, Paris, 69 pp.

645. *OECD (1992)*: Safety considerations for biotechnology. OECD, Paris, 50 pp.

646. *OECD (1994)*: Ottawa '92: Report on the OECD workshop on the monitoring of organisms introduced into the environment. OECD Environment Monographs 90. OECD, Paris, 115 pp.

647. *OECD (1995)*: Commercialisation of agricultural products derived through modern biotechnology: Survey results. OECD Environment Monograph No. 99. OECD, Paris, 178 pp.

648. *Ohta, T. (1982)*: Linkage disequilibrium due to random genetic drift in finite subdivided populations. Proc. Nat. Acad. Sci. USA 79: 1940-1944.

649. *Okuda, T., Hatano, T., Agata, I., Nishibe, S. (1986)*: The components of tannic activites in *Labiatae* plants. I. Rosmarinic acid from *Labiatae* plants in Japan. Yakugaku Zasshi - J. Pharm. Soc. Jpn. 106: 1108-1111.

650. *Olesen, I., Warncke, E. (1992)*: Breeding system and seasonal variation in seed set in a population of *Potentilla palustris*. Nord. J. Bot. 12: 373-380.

651. *Olesen, J.M. (1987)*: Heterostyly, homostyly, and long-distance dispersal of *Menyanthes trifoliata* to Greenland. Can. J. Bot. 65: 1509-1513.

652. *Olesen, J.M. (1988)*: Floral biology of the Canarian *Echium wildpretii* - a bird-flower or a water resource to desert bees? Acta Bot. Neer. 37: 509-513.

653. *Olesen, J.M. (1996)*: From naiveté to experience: Bumblebee queens (*Bombus terrestris*) foraging on *Corydalis cava* (Fumariaceae). J. Kans. Entomol. Soc. 69: 274-286.

654. *Olesen, J.M., Knudsen, J.T. (1994)*: Scent profiles of flower colour morphs of *Corydalis cava* (*Fumariaceae*) in relation to foraging behaviour of bumblebee queens (*Bombus terrestris*). Biochem. Syst. Ecol. 22: 231-237.

655. *Olesen, J.M., Nielsen, N., Nielsen, L.M., Skov, F. (in preparation)*: Comparison of three Danish plant-pollinator communities.

656. *Olesen, J.M., Nielsen, N., Nielsen, L.M., Skov, F. (in preparation)*: Habitat determinants of reproductive success of plant invaders.

657. *Olesen, J.M., Warncke, E. (1989a)*: Flowering and seasonal changes in flower sex ratio and frequency of flower visitors in a population of *Saxifraga hirculus*. Holarctic Ecol. 12: 21-30.

658. *Olesen, J.M., Warncke, E. (1989b)*: Predation and transfer of pollen in a population of *Saxifraga hirculus* L. Holarctic Ecol. 12: 87-95.

659. *Olesen, J.M., Warncke, E. (1989c)*: Temporal changes in pollen flow and neighbourhood structure in a population of *Saxifraga hirculus* L. Oecologia 79: 205-211.

660. *Olin-Fatih M., Heneen, W.K. (1992)*: C-banded karyotypes of *Brassica campestris*, *B. oleracea*, and *B. napus*. Genome 35: 583-589.

661. *Oliveira, M.M., Barroso, J., Pais, M.S. S. (1991)*: Direct gene transfer into *Actinidia deliciosa* protoplasts: Analysis of transient expression of the *CAT* gene using TLC autoradiography and a GC-MS-based method. Plant Mol. Biol. 17: 235-242.

662. *Olivencia, A.O., Claver, J.P.C., Alcaraz, J.A.D. (1995)*: Floral and reproductive biology of *Drosophyllum lusitanicum* (L.) link (Droseraceae). Bot. J. Linn. Soc. 118: 331-351.

663. *Olivieri, I., Couvet, D., Slatkin, M. (1994)*: Allocation of reproductive effort in perennial plants under pollen limitation. Amer. Naturalist 144: 373-394.

664. *Olsen, F.L. (1991)*: Isolation and cultivation of embryogenic microspores from barley (*Hordeum vulgare* L.). Hereditas 115: 255-266.

665. *Olszewski, N.E., Martin, F.B., Ausubel, F.M. (1988)*: Specialized binary vector for plant transformation: Expression of the *Arabidopsis thaliana* AHAS gene in *Nicotiana tabacum*. Nucl. Acid. Res. 16: 765-782.

666. *Oparka, K.J., Murphy, R., Derrick, P.M., Prior, D.A.M., Smith, J.A.C. (1991)*: Modification of the pressure-probe technique permits controlled intracellular microinjection of fluorescent probes. J. Cell Sci. 98: 539-544.

667. *Opperman, C.H., Taylor, C.G., Conkling, M.A. (1994)*: Root-knot nematode-directed expression of a plant root-specfic gene. Science 263: 221-223.

668. *Orians, C.M. (1995)*: Preserving leaves for tannin and phenolic glycoside analyses: A comparison of methods using three willow taxa. J. Chem. Ecol. 21: 1235-1243.

669. *Orians, G.H. (1986)*: Site characteristics favoring invasions. In: Mooney, H.A., Drake, J.A., (eds.) Ecology of biological invasions of North America. Springer Verlag, Berlin, 133-148.

670. *Ouborg, N.J., Treuren, R. van (1994)*: The significance of genetic erosion in the process of extinction. IV. Inbreeding load and heterosis in relation to population size in the mint *Salvia pratensis*. Evolution 48: 996-1008.

671. *Ouborg, N.J., Treuren, R. van, Damme, J.M. M. van (1991)*: The significance of genetic erosion in the process of extinction. II. Morphological variation and fitness components in populations of varying size of *Salvia pratensis* L. and *Scabiosa columbaria* L. Oecologia 86: 359-367.

672. *Oud, J.S.N., Schneiders, H., Kool, A.J., Grinsven, M.Q.J.M. van (1995)*: Breeding of transgenic orange *Petunia hybrida* varieties. Euphytica 84: 175-181.

673. *Ouendeba, B., Ejeta, G., Nyquist, W.E., Hanna, W.W., Kumar, A. (1993)*: Heterosis and combining ability among African pearl millet landraces. Crop Sci. 33: 735-739.

674. *Ow, D.W., Medberry, S.L. (1995)*: Genome manipulation through site-specific recombination. Crit. Rev. Plant Sci. 14: 239-261.

675. *Owens, J.N., Colangeli, A.M., Morris, S.J. (1990)*: The effect of self-pollination, cross-pollination and no pollination on ovule, embryo, seed, and cone development in Western red cedar (*Thuja plicata*). Can. J. Forest Res. 20: 66-75.

676. *Oyama, K., Ito, M., Yahara, T., Ono, M. (1993)*: Low genetic differentiation among populations of *Arabis serrata* (Brassicaceae) along an altitudinal gradient. J. Plant. Res. 106: 143-148.

677. *Paige, K.N., Whitham, T.G. (1985)*: Individual and population shifts in flower color by scarlet *Gilia*: A mechanism for pollinator tracking. Science 227: 315-317.

678. *Palmer, A.R., Strobeck, C. (1986)*: Fluctuating asymmetry: Measurement, analysis, patterns. Annu. Rev. Ecol. Syst. 17: 391-421.

679. *Palmer, A.R., Strobeck, C. (1992)*: Fluctuating asymmetry as a measure of developmental stability: Implications of non-normal distributions and power of statistical tests. Acta Zool. Fenn. 191: 57-72.

680. *Palmer, J.D., Jansen, R.K., Michaels, H.E., Chase, M.W., Manhart, J.R. (1988)*: Chloroplast DNA variation and plant phylogeny. Ann. Mo. Bot. Gard. 75: 1180-1206.

681. *Palmer, T.M., Zimmerman, M. (1994)*: Pollen competition and sporophyte fitness in *Brassica campestris*: Does intense pollen competition result in individuals with better pollen? Oikos 69: 80-86.

682. *Pang, S.-Z., Slightom, J.L., Gonsalves, D. (1993)*: Different mechanisms protect transgenic tobacco against tomato spotted wilt and impatiens necrotic spot *Tospoviruses*. Bio/Technology 11: 819-824.

683. *Parker, I.M., Bartsch, D. (1996)*: Recent advances in ecological biosafety research on the risks of transgenic plants: A trans-continental perspective. In: Tomiuk, J., Wöhrmann, K., Sentker, A., (eds.) Transgenic organisms: Biological and social implications. Birkhäuser Verlag, Basel, 147-161.

684. *Parker, I.M., Nakamura, R.R., Schemske, D.W. (1995)*: Reproductive allocation and the fitness consequences of selfing in two sympatric species of *Epilobium* (Onagraceae) with contrasting mating systems. Amer. J. Bot. 82: 1007-1016.

685. *Parra, V., Vargas, C.F., Eguiarte, L.E. (1993)*: Reproductive biology, pollen and seed dispersal, and neighborhood size in the hummingbird-pollinated *Echeveria gibbiflora* (Crassulaceae). Amer. J. Bot. 80: 153-159.

686. *Pasteur, N., Pasteur, G., Bonhomme, F., Catalan, J., Britton-Davidian, J. (1987)*: Manuel technique de génétique par électrophorèse des protéines (Electrophoresis of proteins). Lavoisier, Technique et Documentation, Paris, 217 pp.

687. *Paszkowski, J., Baur, M., Bogucki, A., Potrykus, I. (1988)*: Gene targeting in plants. EMBO J. 7: 4021-4026.

688. *Peakall, R., Beattie, A.J. (1995)*: Does ant dispersal of seeds in *Sclerolaena diacantha* (Chenopodiaceae) generate local spatial genetic structure. Heredity 75: 351-361.

689. *Pedersen, C., Giese, H., Linde-Laursen, I. (1995)*: Towards integration of the physical and genetic chromosome map of barley by in situ hybridization. Hereditas 123: 77-88.

690. *Pedersen, C., Linde-Laursen, I. (1994)*: Chromosomal locations of four minor rDNA loci and a marker microsatellite sequence in barley. Chromosome Res. 2: 65-71.

691. *Pellmyr, O., Tang, W., Groth, I., Bergstrom, G., Thien, L.B. (1991)*: Cycad cone and angiosperm floral volatiles: Inferences for the evolution of insect pollination. Biochem. Syst. Ecol. 19: 623-627.

692. *Percival, M.S. (1961)*: Types of nectar in angiosperms. New Phytol. 60: 235-281.

693. *Percival, M.S. (1965)*: Floral biology. Pergamon Press, Oxford, 243 pp.

694. *Perl, A., Perl-Treves, R., Galili, S., Aviv, D., Shalgi, E., Malkin, S., Galun, E. (1993)*: Enhanced oxidative-stress defense in transgenic potato expressing tomato Cu,Zn superoxide dismutases. Theor. Appl. Genet. 85: 568-576.

695.   *Perry, D.J., Knowles, P. (1991)*: Spatial genetic structure within three sugar maple (*Acer saccharum* Marsh.) stands. Heredity 66: 137-142.

696.   *Petanidou, T., Ellis, W.N., Margaris, N.S., Vokou, D. (1995a)*: Constraints on flowering phenology in a phryganic (East Mediterranean shrub) community. Amer. J. Bot. 82: 607-620.

697.   *Petanidou, T., Nijs, J.C.M. den, Oostermeijer, J.G.B. (1995b)*: Pollination ecology and constraints on seed set of the rare perennial *Gentiana cruciata* L. in The Netherlands. Acta Bot. Neer. 44: 55-74.

698.   *Petanidou, T., Nijs, J.C.M. den, Oostermeijer, J.G.B., Ellis-Adam, A.C. (1995c)*: Pollination ecology and patch-dependent reproductive success of the rare perennial *Gentiana pneumonanthe* L. New Phytol. 129: 155-163.

699.   *Petanidou, T., Vokou, D. (1993)*: Pollination ecology of Labiatae in a Phryganic (East Mediterranean) ecosystem. Amer. J. Bot. 80: 892-899.

700.   *Peto, F.H. (1933)*: The cytology of certain intergeneric hybrids between *Festuca* and *Lolium*. J. Genet. 28: 113-156.

701.   *Pettersson, M.W. (1992)*: Advantages of being a specialist female in the gynodioecious *Silene vulgaris* S.L. (Caryophyllaceae). Amer. J. Bot. 79: 1389-1395.

702.   *Pettersson, M.W. (1994)*: Large plant size counteracts early seed predation during the extended flowering season of a *Silene uniflora* (Caryophyllaceae) population. Ecography 17: 264-271.

703.   *Philipp, M., Madsen, H.E.S., Siegismund, H.R. (1992)*: Gene flow and population structure in *Armeria maritima*. Heredity 69: 32-42.

704.   *Pigott, C.D., Warr, S.J. (1989)*: Pollination, fertilization and fruit development in sycamore (*Acer pseudoplatanus* L.). New Phytol. 111: 99-103.

705.   *Pilon-Smits, E.A.H., Ebskamp, M.J.M., Paul, M.J., Jeuken, M.J.W., Weisbeek, P.J., Smeekens, S.C.M. (1995)*: Improved performance of transgenic fructan-accumulating tobacco under drought stress. Plant Physiol. 107: 125-130.

706.   *Pleasants, J.M., Zimmerman, M. (1979)*: Patchiness in the dispersion of nectar resources: Evidence for hot and cold spots. Oecologia 41: 283-288.

707.   *Pleasants, J.M., Zimmerman, M. (1983)*: The distribution of standing crop of nectar: What does it really tell us? Oecologia 57: 412-414.

708.   *Plowright, R.C. (1987)*: Corolla depth and nectar concentration: An experimental study. Can. J. Bot. 65: 1011-1013.

709.   *Poirier, Y., Nawrath, C., Somerville, C. (1995)*: Production of polyhydroxyalkanoates, a family of biodegradable plastics and elastomers, in bacteria and plants. Bio/Technology 13: 142-150.

710.   *Potrykus, I. (1990)*: Gene transfer methods for plants and cell cultures. CIBA Found. Symp. 154: 198-212.

711.   *Potrykus, I. (1992)*: Micro-targeting of microprojectiles to target areas in the micrometer range. Nature 355: 568-569.

712.   *Potrykus, I., Spangenberg, G. (1995)*: Gene transfer to plants. Springer, Berlin, 361 pp.

713. *Potter, H. (1993)*: Application of electroporation in recombinant DNA technology. Meth. Enzymol. 217: 461-478.

714. *Poulet, P., Cahen, D., Malkin, S. (1983)*: Photoacoustic detection of photosynthetic oxygen evolution from leaves. Quantitative analysis by phase and amplitude measurements. Biochim. Biophys. Acta 724: 433-466.

715. *Prabhu, R.R., Gresshoff, P.M. (1994)*: Inheritance of polymorphic markers generated by DNA amplification fingerprinting and their use as genetic markers in soybean. Plant Mol. Biol. 26: 105-116.

716. *Pradhan, A.K., Sodhi, Y.S., Mukhopadhyay, A., Pental, D. (1993)*: Heterosis breeding in Indian mustard (*Brassica juncea* L. Czern and Coss): Analysis of component characters contributing to heterosis for yield. Euphytica 69: 219-229.

717. *Preiszner, J., Fehér, A., Veisz, O., Sutka, J., Dudits, D. (1991)*: Characterization of morphological variation and cold resistance in interspecific somatic hybrids between potato (*Solanum tuberosum* L.) and *Solanum brevidens* Phil. Euphytica 57: 37-49.

718. *Prentice, H.C., Lönn, M., Lefkovitch, L.P., Runyeon, H. (1995)*: Associations between allele frequencies in *Festuca ovina* and habitat variation in the alvar grasslands on the Baltic island of Öland. J. Ecol. 83: 391-402.

719. *Preston, R.E. (1991)*: The intrafloral phenology of *Streptanthus tortuosus* (Brassicaceae). Amer. J. Bot. 78: 1044-1053.

720. *Price, H.J. (1988)*: Nuclear DNA content variation within angiosperm species. Evol. Trend. Plant. 2: 53-60.

721. *Price, M.V., Waser, N.M. (1979)*: Pollen dispersal and optimal outcrossing in *Delphinium nelsonii*. Nature 277: 294-297.

722. *Pringle, G.J., Murray, B.G. (1991)*: Karyotype diversity and nuclear DNA variation in *Cyphomandra*. In: Hawkes, J.G., Lester, R.N., Nee, M., Estrada, N., (eds.) Solanaceae III: Taxonomy, chemistry, evolution. Royal Botanic Gardens, Kew, London, 247-252.

723. *Prins, T.W., Zadoks, J.C. (1994)*: Horizontal gene transfer in plants, a biohazard? Outcome of a literature review. Euphytica 76: 133-138.

724. *Proctor, M., Yeo, P., Lack, A. (1996)*: The natural history of pollination. Timber Press, Portland, 479 pp.

725. *Proença, C.E.B., Gibbs, P.E. (1994)*: Reproductive biology of eight sympatric Myrtaceae from Central Brazil. New Phytol. 126: 343-354.

726. *Pyke, G.H. (1978)*: Optimal foraging: Movement patterns of bumblebees between inflorescences. Theor. Pop. Biol. 13: 72-98.

727. *Pyke, G.H. (1984)*: Optimal foraging theory: A critical review. Ann. Rev. Ecol. Syst. 15: 523-575.

728. *Pyke, G.H. (1991)*: What does it cost a plant to produce floral nectar? Nature 350: 58-59.

729. *Pyke, G.H., Pulliam, H.R., Charnov, E.L. (1977)*: Optimal foraging: A selective review of theory and tests. Quart. Rev. Biol. 52: 137-154.

730. *Pyle, M.M., Adams, R.P. (1989)*: *In situ* preservation of DNA in plant specimens. Taxon 38: 576-581.

731. *Quarin, C.L. (1986)*: Seasonal changes in the incidence of apomixis of diploid, triploid, and tetraploid plants of *Paspalum cromyorrhizon*. Euphytica 35: 515-522.

732. *Queller, D.C., Goodnight, K.F. (1989)*: Estimating relatedness using genetic markers. Evolution 43: 258-275.

733. *Quesada, M., Bollman, K., Stephenson, A.G. (1995)*: Leaf damage decreases pollen production and hinders pollen performance in *Cucurbita texana*. Ecology 76: 437-443.

734. *Quiros, C.F., This, P., Laudie, M., Benet, A., Chevre, A.-M., Delseney, M. (1995)*: Analysis of a set of RAPD markers by hybridization and sequencing in Brassica: A note of caution. Plant Cell Rep. 14: 630-634.

735. *Rafalski, J.A., Tingey, S.V. (1993)*: Genetic diagnostics in plant breeding: RAPDs, microsatellites and machines. Trends Genet. 9: 275-280.

736. *Rahman, M.M., Nito, N. (1994)*: Use of glutamate oxaloacetate transaminase isozymes for detection of hybrids among genera of the true citrus fruit trees. Sci. Hort. - Amsterdam 58: 197-206.

737. *Ramakishana, W., Lagu, M.D., Gupta, V.S., Ranjekar, P.K. (1994)*: DNA fingerprinting in rice using oligonucleotide probes specific for simple repetitive DNA sequences. Theor. Appl. Genet. 88: 402-406.

738. *Ramsey, M. (1993)*: Floral morphology, biology and sex allocation in disjunct populations of christmas bells (*Blandfordia grandiflora*, Liliaceae) with different breeding systems. Aust. J. Bot. 41: 749-762.

739. *Ramsey, M. (1995)*: Causes and consequences of seasonal variation in pollen limitation of seed production in *Blandfordia grandiflora* (Liliaceae). Oikos 73: 49-58.

740. *Rantio-Lehtimäki, A. (1991)*: Sampling airborne pollen and pollen antigenes. In: D'Amato, G., Spieksma, F.T.M., Bonini, S., (eds.) Allergenic pollen and pollinosis in Europe. Blackwell Scientific Publications, Cambridge, 18-23.

741. *Rao, K.V., Rathore, K.S., Hodges, T.K. (1995)*: Physical, chemical and physiological parameters for electroporation-mediated gene delivery into rice protoplasts. Transgenic Res. 4: 361-368.

742. *Ravn, H., Pedersen, M.F., Borum, J., Andary, C., Anthoni, U., Christophersen, C., Nielsen, P.H. (1994)*: Seasonal variation and distribution of two phenolic compounds, rosmarinic acid and caffeic acid, in leaves and roots-rhizomes of eelgrass (*Zostera marina* L.). Ophelia 40: 51-61.

743. *Ray, S.D., Sen, S. (1992)*: Heterosis in sesame (*Sesamum indicum* L.). Trop. Agr. 69: 276-278.

744. *Raybould, A.F., Gray, A.J. (1994)*: Will hybrids of genetically modified crops invade natural communities? Trend. Ecol. Evol. 9: 85-89.

745. *Rayburn, A.L., Auger, J.A., Benzinger, E.A., Hepburn, A.G. (1989)*: Detection of intraspecific DNA content variation in *Zea mays* L. by flow cytometry. J. Exp. Bot. 40: 1179-1183.

746. *Raymond, M., Rousset, F. (1995)*: GENEPOP (Version 1.2): Population genetics software for exact tests and ecumeniscm. J. Hered. 86: 248-249.

747. *Reader, R.J. (1982)*: Variation in the flowering date of transplanted ericaceous shrubs in relation to their flowering phenology. J. Biogeogr. 9: 397-410.

748. *Real, L. (1983)*: Pollination biology. Academic Press, Orlando, 338 pp.

749. *Regal, P.J. (1993)*: The true meaning of "exotic species" as a model for genetically engineered organisms. Experientia 49: 225-234.

750. *Reich, T.J., Iyer, V.N., Miki, B.L. (1986)*: Efficient transformation of alfafa protoplasts by the intranuclear microinjection of Ti plasmids. Bio/Technology 4: 1001-1004.

751. *Reich, T.J., Iyer, V.N., Scobie, B., Miki, B.L. (1986)*: A detailed procedure for the intranuclear microinjection of plant protoplasts. Can. J. Bot. 64: 1255-1258.

752. *Reinhoud, P., Schrijnemakers, E., Pool, S., Iren, F. van, Kijne, J. (1993)*: Cryopreservation of tobacco suspension cells: A comparison of the vitrification method and the two-step freezing method. Cryobiology 30: 654.

753. *Reinisch, A.J., Dong, J., Brubaker, C.L., Stelly, D.M., Wendel, J.F., Paterson, A.H. (1994)*: A detailed RFLP map of cotton, Gossypium hirsutum X Gossypium barbadense: Chromosome organization and evolution in a disomic polyploid genome. Genetics 138: 829-847.

754. *Rejmánek, M. (1989)*: Invasibility of plant communities. In: Drake, J.A., Mooney, H.A., di Castri, F., Groves, R.H., Kruger, F.J., Rejmánek, M., Williamson, M., (eds.) Biological invasions: A global perspective. John Wiley & Sons, Chichester, 369-388.

755. *Rennenberg, H., Polle, A. (1994)*: Protection from oxidative stress in transgenic plants. Biochem. Soc. Trans. 22: 936-940.

756. *Reyment, R.A., Blackith, R.E., Campbell, N.A. (1984)*: Multivariate morphometrics. Academic Press, London, 233 pp.

757. *Rice, K.J., Mack, R.N. (1991)*: Ecological genetics of Bromus tectorum. III. The demography of reciprocally sown populations. Oecologia 88: 91-101.

758. *Richards, A.J. (1986)*: Plant breeding systems. Unwin Hyman, London, 529 pp.

759. *Rieseberg, L.H., Brunsfeld, S.J. (1992)*: Molecular evidence and plant introgression. In: Soltis, P.S., Soltis, D.E., Doyle, J.J., (eds.) Molecular systematics of plants. Chapman & Hall, New York, 151-176.

760. *Rieseberg, L.H., Choi, H., Chan, R., Spore, C. (1993)*: Genomic map of a diploid hybrid species. Heredity 70: 285-293.

761. *Rieseberg, L.H., Ellstrand, N.C. (1993)*: What can molecular and morphological markers tell us about plant hybridization. Crit. Rev. Plant Sci. 12: 213-241.

762. *Rieseberg, L.H., Fossen, C. van, Desrochers, A.M. (1995a)*: Hybrid speciation accompagnied by genomic reorganization in wild sunflowers. Nature 375: 313-316.

763. *Rieseberg, L.H., Linder, C.R., Seiler, G.J. (1995b)*: Chromosomal and genetic barriers to introgression in Helianthus. Genetics 141: 1163-1171.

764. *Rieseberg, L.H., Soltis, D.E. (1991)*: Phylogenetic consequences of cytoplasmic gene flow in plants. Evol. Trend. Plant. 5: 65-84.

765. *Rieseberg, L.H., Wendel, J.F. (1993)*: Introgression and its consequences in plants. In: Harrison, R.G., (ed.) Hybrid zones and the evolutionary process. Oxford University Press, Oxford, 70-109.

766. *Rigney, L.P. (1995)*: Postfertilization causes of differential success of pollen donors in Erythronium grandiflorum (Liliaceae): Nonrandom ovule abortion. Amer. J. Bot. 82: 578-584.

767. *Risseeuw, E., Offringa, R., Franke-van-Dijk, M.E.I., Hooykaas, P.J.J. (1995)*: Targeted recombination in plants using *Agrobacterium* coincides with additional rearrangements at the target locus. Plant J. 7: 109-119.

768. *Rissler, J., Mellon, M. (1993)*: Perils amidst the promise: Ecological risks of transgenic crops in a global market. Union of Concerned Scientists, Cambridge, 92 pp.

769. *Ritland, K. (1988)*: The genetic-mating structure of subdivided populations II. Correlated mating models. Theor. Pop. Biol. 34: 320-346.

770. *Ritland, K. (1989)*: Correlated matings in the partial selfer *Mimulus guttatus*. Evolution 43: 848-859.

771. *Ritland, K. (1990)*: A series of FORTRAN computer programs for estimating plant mating systems. J. Hered. 81: 235-237.

772. *Ritland, K., Jain, S. (1981)*: A model for the estimation of outcrossing rate and gene frequencies using *n* independent loci. Heredity 47: 35-52.

773. *Ritland, K., Soltis, D.E., Soltis, P.S. (1990)*: A two-locus model for the joint estimation of intergametophytic and intragametophytic selfing rates. Heredity 65: 289-296.

774. *Riveros, G.M., Barría, O.R., Humaña, P.A.M. (1995)*: Self-compatibility in distylous *Hedyotis salzmannii* (Rubiaceae). Plant Syst. Evol. 194: 1-8.

775. *Robert, T., Lespinasse, R., Pernès, J., Sarr, A. (1991)*: Gametophytic competition as influencing gene flow between wild and cultivated forms of pearl millet (*Pennisetum typhoides*). Genome 34: 195-200.

776. *Robertson, A.W., Diaz, A., MacNair, M.R. (1994)*: The quantitative genetics of floral characters in *Mimulus guttatus*. Heredity 72: 300-311.

777. *Robertson, A.W., MacNair, M.R. (1995)*: The effects of floral display size on pollinator service to individual flowers of *Myosotis* and *Mimulus*. Oikos 72: 106-114.

778. *Rodríguez-Robles, J.A., Meléndez, E.J., Ackerman, J.D. (1992)*: Effects of display size, flowering phenology, and nectar availability on effective visitation frequency in *Comparettia falcata* (Orchidaceae). Amer. J. Bot. 79: 1009-1017.

779. *Rogstad, S.H. (1994)*: The biosystematics and evolution of the *Polyalthia hypoleuca* species complex (Annonaceae) of Malesia. III. Floral ontogeny and breeding systems. Amer. J. Bot. 81: 145-154.

780. *Rood, S.B. (1995)*: Heterosis and the metabolism of gibberellin A_{20} in sorghum. Plant Growth Regul. 16: 271-278.

781. *Rosén, B., Halldén, C., Heneen, W.K. (1988)*: Diploid *Brassica napus* somatic hybrids: Characterization of nuclear and organellar DNA. Theor. Appl. Genet. 76: 197-203.

782. *Rosen, D., Barthlott, W. (1991)*: Ökologische Aspekte der Ultraviolett-Reflexion von Blumen in Mitteleuropa, besonders in der Eifel. Decheniana 144: 72-112.

783. *Rousset, F., Raymond, M. (1995)*: Testing heterozygote excess and deficiency. Genetics 140: 1413-1419.

784. *Rowland, L.J., Levi, A. (1994)*: RAPD-based genetic linkage map of blueberry derived from a cross between diploid species (*Vaccinium darrowi* and *V. elliottii*). Theor. Appl. Genet. 87: 863-868.

785. *Roy, B.A. (1995)*: The breeding systems of six species of *Arabis* (Brassicaceae). Amer. J. Bot. 82: 869-877.

786. *Russell, D.R., Wallace, K.M., Bathe, J.H., Martinell, B.J., McCabe, D.E. (1993)*: Stable transformation of *Phaseolus vulgaris* via electric-discharge mediated particle acceleration. Plant Cell Rep. 12: 165-169.

787. *Russell, J.A., Roy, M.K., Sanford, J.C. (1992)*: Physical trauma and tungsten toxicity reduce the efficiency of biolistic transformation. Plant Physiol. 98: 1050-1056.

788. *Ryan, C.A. (1990)*: Protease inhibitors in plants: Genes for improving defenses against insects and pathogens. Annu. Rev. Phytopathol. 28: 425-429.

789. *Saari, L.L., Cotterman, J.C., Thill, D.C. (1994)*: Resistance to acetolactate synthase inhibiting herbicides. In: Powles, S.B., Holtum, J.A., (eds.) Herbicide resistance in plants. Biology and biochemistry. Lewis Publishers, Boca Raton, 83-141.

790. *Saari, L.L., Mauvais, C.J. (1996)*: Sulfonylurea herbicide-resistant crops. In: Duke, S.O., (ed.) Herbicide resistant crops. Agricultural, environmental, economic, regulatory, and technical aspects. Lewis Publishers, Boca Raton, 127-139.

791. *Saez, F. (1995)*: Essential oil variability of *Thymus hyemalis* growing wild in Southeastern Spain. Biochem. Syst. Ecol. 23: 431-438.

792. *Sage, R.D., Selander, R.K. (1979)*: Hybridization between species of the *Rana pipiens* complex in central Texas. Evolution 33: 1069-1088.

793. *Sage, T.L., Webster, B.D. (1990)*: Seed abortion in *Phaseolus vulgaris* L. Bot. Gaz. 151: 167-175.

794. *Saito, K., Yamazaki, M., Murakoshi, I. (1992)*: Transgenic medicinal plants: *Agrobacterium*-mediated foreign gene transfer and production of secondary metabolites. J. Nat. Prod. - Lloydia 55: 149-162.

795. *Sakai, A., Kobayashi, S., Yamada, T., Uragami, A. (1991)*: Cryopreservation of cultured plant cells and meristems by a simple freezing method. Cryobiology 28: 549.

796. *Sakata, Y., Nishio, T., Narikawa, T., Monma, S. (1991)*: Cold and disease resistance of somatic hybrids between tomato (*Lycopersicon esculentum*) and *L. peruvianum*. J. Jpn. Soc. Hort. Sci. 60: 329-335.

797. *Sambrook, J., Fritsch, E.F., Maniatis, T. (1989)*: Molecular cloning: A laboratory manual. Cold Spring Harbor Laboratory Press, Cold Spring Harbor, 3 volumes.

798. *Sandhu, S.S., Serres, F.J. de, Gopalan, H.N.B., Grant, W.F., Svendsgaard, D., Veleminsky, J., Becking, G.C. (1994)*: Results and recommendations. Mutat. Res. - Fundam. Mol. Mech. Mut. 310: 257-263.

799. *Sanford, J.C., Johnston, S.A. (1985)*: The concept of parasite-derived resistance - deriving resistance genes from the parasite's own genome. J. Theor. Biol. 113: 395-405.

800. *Sangwan, R.S., Ducrocq, C., Sangwan-Norreel, B. (1993)*: *Agrobacterium*-mediated transformation of pollen embryos in *Datura innoxia* and *Nicotiana tabacum*: Production of transgenic haploid and fertile homozygous dihaploid plants. Plant Sci. 95: 99-115.

801. *Sarawat, P., Stoddard, F.L., Marshall, D.R., Ali, S.M. (1994)*: Heterosis for yield and related characters in pea. Euphytica 80: 39-48.

802.   *Sautter, C. (1993)*: Development of a microtargeting device for particle bombardment of plant meristems. Plant Cell Tissue Organ Cult. 33: 251-257.

803.   *Savolainen, O., Kärkkäinen, K., Harju, A., Nikkanen, T., Rusanen, M. (1993)*: Fertility variation in *Pinus sylvestris*: A test of sexual allocation theory. Amer. J. Bot. 80: 1016-1020.

804.   *Savolainen, V., Cuénoud, P., Spichiger, R., Martinez, M.D.P., Crèvecoeur, M., Manen, J.-F. (1995)*: The use of herbarium specimens in DNA phylogenetics: Evaluation and improvement. Plant Syst. Evol. 197: 87-98.

805.   *Savova, D., Rufener Al Mazyad, P., Felber, F. (1996)*: Cytogeography of *Medicago falcata* L. and *M. sativa* L. in Switzerland. Bot. Helv. 106: 197-207.

806.   *Sazima, M., Vogel, S., Cocucci, A., Hausner, G. (1993)*: The perfume flowers of *Cyphomandra* (Solanaceae): Pollination by euglossine bees, bellows mechanism, osmophores, and volatiles. Plant Syst. Evol. 187: 51-88.

807.   *Schaal, B.A., Learn, G.H. (1988)*: Ribosomal DNA variation within and among plant populations. Ann. Mo. Bot. Gard. 75: 1207-1216.

808.   *Schaal, B.A., O'Kane Jr., S.L., Rogstad, S.H. (1991)*: DNA variation in plant populations. Trend. Ecol. Evolut. 6: 329-333.

809.   *Schierwater, B., Streit, B., Wagner, G.P., DeSalle, R. (1994)*: Molecular ecology and evolution: Approaches and applications. Birkhäuser Verlag, Basel, 622 pp.

810.   *Schlüter, K., Fütterer, J., Potrykus, I. (1995)*: "Horizontal" gene transfer from a transgenic potato line to a bacterial pathogen (*Erwinia chrysanthemi*) occurs - if at all - at an extremely low frequency. Bio/Technology 13: 1094-1098.

811.   *Schmid, B., Dolt, C. (1994)*: Effects of maternal and paternal environment and genotype on offspring phenotype in *Solidago altissima* L. Evolution 48: 1525-1549.

812.   *Schmidt, J.M., Antlfinger, A.E. (1992)*: The level of agamospermy in a Nebraska population of *Spiranthes cernua* (Orchidaceae). Amer. J. Bot. 79: 501-507.

813.   *Schmidt-Rogge, T., Meixner, M., Srivastava, V., Guha-Mukherjee, S., Schieder, O. (1993)*: Transformation of haploid *Datura innoxia* protoplasts and analysis of the plasmid integration pattern in regenerated transgenic plants. Plant Cell Rep. 12: 390-394.

814.   *Schmitt, J., Gamble, S.E. (1990)*: The effect of distance from the parental site on offspring performance and inbreeding depression in *Impatiens capensis*: A test of the local adaptation hypothesis. Evolution 44: 2022-2030.

815.   *Schnabel, A., Asmussen, M.A. (1992)*: Comparative effects of pollen and seed migration on the cytonuclear structure of plant populations. II. Paternal cytoplasmic inheritance. Genetics 132: 253-267.

816.   *Schnepf, H.E., Whiteley, H.R. (1981)*: Cloning and expression of the *Bacillus thuringiensis* crystal protein gene in *Escherichia coli*. Proc. Natl. Acad. Sci. USA 78: 2893-2897.

817.   *Schoen, D.J. (1988)*: Mating system estimation via the one pollen parent model with the progeny array as the unit of observation. Heredity 60: 439-440.

818.   *Schoen, D.J., Brown, A.H.D. (1991)*: Intraspecific variation in population gene diversity and effective population size correlates with the mating system in plants. Proc. Nat. Acad. Sci. USA 88: 4494-4497.

819. *Schoen, D.J., Clegg, M.T. (1984)*: Estimation of mating system parameters when outcrossing events are correlated. Proc. Nat. Acad. Sci. USA 81: 5258-5262.

820. *Schou, O., Philipp, M. (1984)*: An unusual heteromorphic incompatibility system. 3. On the genetic control of distyly and self-incompatibility in *Anchusa officinalis* L. (Boraginaceae). New Phytol. 89: 693-703.

821. *Schroeder, H.E., Gollasch, S., Moore, A., Tabe, L.M., Craig, S., Hardie, D.C., Chrispeels, M.J., Spencer, D., Higgins, T.J.V. (1995)*: Bean α-amylase inhibitor confers resistance to the pea weevil (*Bruchus pisorum*) in transgenic peas (*Pisum sativum* L.). Plant Physiol. 107: 1233-1239.

822. *Schroeder, H.E., Khan, M.R.I., Knibb, W.R., Spencer, D., Higgins, T.J.V. (1991)*: Expression of a chicken ovalbumin gene in three lucerne cultivars. Aust. J. Plant Physiol. 18: 495-505.

823. *Schuster, W.S., Alles, D.L., Mitton, J.B. (1989)*: Gene flow in limber pine: Evidence from pollination phenology and genetic differentiation along an elevational transect. Amer. J. Bot. 76: 1395-1403.

824. *Schwarzacher, T., Anamthawat-Jónsson, K., Harrison, G.E., Islam, A.K.M.R., Jia, J.Z., King, I.P., Leitch, A.R., Miller, T.E., Reader, S.M., Rogers, W.J., Shi, M., Heslop-Harrison, J.S. (1992)*: Genomic in situ hybridization to identify alien chromosomes and chromosome segments in wheat. Theor. Appl. Genet. 84: 778-786.

825. *Scorza, R., Ravelonandro, M., Callahan, A.M., Cordts, J.M., Fuchs, M., Dunez, J., Gonsalves, D. (1994)*: Transgenic plums (*Prunus domestica* L.) express the plum pox virus coat protein gene. Plant Cell Rep. 14: 18-22.

826. *Scribailo, R.W., Barrett, S.C.H. (1994)*: Effects of prior self-pollination on outcrossed seed set in tristylous *Pontederia sagittata* (Pontederiaceae). Sex. Plant Reprod. 7: 273-281.

827. *Seavey, S.R., Carter, S.K. (1994)*: Self-sterility in *Epilobium obcordatum* (Onagraceae). Amer. J. Bot. 81: 331-338.

828. *Seberg, O., Linde-Laursen, I. (1996)*: *Eremium*, a new genus of the Triticeae (Poaceae) from Argentina. Syst. Bot. 21: 3-15.

829. *Sehnke, P.C., Pedrosa, L., Paul, A.L., Frankel, A.E., Ferl, R.J. (1994)*: Expression of active, processed ricin in transgenic tobacco. J. Biol. Chem. 269: 22473-22476.

830. *Shade, R.E., Schroeder, H.E., Pueyo, J.J., Tabe, L.M., Murdock, L.L., Higgins, T.J.V., Chrispeels, M.J. (1994)*: Transgenic pea seeds expressing the α-amylase inhibitor of the common bean are resistant to bruchid beetles. Bio/Technology 12: 793-796.

831. *Shah, D.M., Horsch, R.B., Klee, H.J., Kishore, G.M., Winter, J.A., Tumer, N.E., Hironaka, C.M., Sanders, P.R., Gasser, C.S., Aykent, S., Siegel, N.R., Rogers, S.G., Fraley, R.T. (1986)*: Engineering herbicide tolerance in transgenic plants. Science 233: 478-481.

832. *Shaner, D.L., Bascomb, N.F., Smith, W. (1996)*: Imidazolinone-resistant crops: Selection, characterization, and management. In: Duke, S.O., (ed.) Herbicide resistant crops. Agricultural, environmental, economic, regulatory, and technical aspects. Lewis Publishers, Boca Raton, 144-156.

833. *Shapcott, A. (1995)*: The spatial genetic structure in natural populations of the Australian temperate rainforest tree *Atherosperma moschatum* (Labill.) (Monimiaceae). Heredity 74: 28-38.

834. Sharon, D., Adato, A., Mhameed, S., Lavi, U., Hillel, J., Gomolka, M., Epplen, C., Epplen, J.T.
      (1995): DNA fingerprints in plants using simple-sequence repeat and minisatellite probes.
      Hortscience 30: 109-112.

835. Shaw, D.V., Brown, A.H.D. (1982): Optimum number of marker loci for estimating
      outcrossing in plant populations. Theor. Appl. Genet. 61: 321-325.

836. Shaw, D.V., Kahler, A.L., Allard, R.W. (1981): A multilocus estimator of mating system
      parameters in plant populations. Proc. Nat. Acad. Sci. USA 78: 1298-1302.

837. Sherwood, R.T., Berg, C.C., Young, B.A. (1994): Inheritance of apospory in buffelgrass. Crop
      Sci. 34: 1490-1494.

838. Shi, Y., Wang, M.B., Powell, K.S., Damme, E. van, Hilder, V.A., Gatehouse, A.M.R., Boulter,
      D., Gatehouse, J.A. (1994): Use of the rice sucrose synthase-1 promoter to direct phloem-
      specific expression of β-glucuronidase and snowdrop lectin genes in transgenic tobacco
      plants. J. Exp. Bot. 45: 623-631.

839. Shiba, H., Hinata, K., Suzuki, A., Isogai, A. (1995): Breakdown of self-incompatibility in
      Brassica by the antisense RNA of the SLG gene. Proc. Jpn. Acad. B 71: 81-83.

840. Shillito, R.D., Potrykus, I. (1987): Direct gene transfer to protoplasts of dicotyledonous and
      monocotyledonous plants by a number of methods, including electroporation. Meth.
      Enzymol. 153: 313-336.

841. Shiota, N., Nagasawa, A., Sakaki, T., Yabusaki, Y., Ohkawa, H. (1994): Herbicide-resistant
      tobacco plants expressing the fused enzyme between rat cytochrome P4501A1 (CYP1A1)
      and yeast NADPH-cytochrome P450 oxidoreductase. Plant Physiol. 106: 17-23.

842. Shmida, A., Kadmon, R. (1991): Within-plant patchiness in nectar standing crop in Anchusa
      strigosa. Vegetatio 94: 95-99.

843. Shonnard, G.C., Peloquin, S.J. (1991): Performance of true potato seed families. 1. Effect of
      level of inbreeding. Potato Res. 34: 397-407.

844. Shorrocks, B., Coates, D. (1993): The release of genetically-engineered organisms. Ecological
      Issues No. 4. Field Studies Council, Shrewsbury, UK, 45 pp.

845. Shriver, M.D., Jin, L., Boerwinkle, E., Deka, R., Ferrell, R.E., Chakraborty, R. (1995): A novel
      measure of genetic distance for highly polymorphic tandem repeat loci. Mol. Biol. Evol.
      12: 914-920.

846. Siegismund, H.R. (1995): G-stat, version 3.1: Genetical statistical programs for the analysis
      of population data. E-mail address of the author: hanss@bot.ku.dk

847. Siemens, D.H. (1994): Factors affecting regulation of maternal investment in an indeter-
      minate flowering plant (Cercidium microphyllum, Fabaceae). Amer. J. Bot. 81: 1403-1409.

848. Sigal, L.L. (1993): Sourcebook for the environmental assessment process. U.S. Environ-
      mental Protection Agency, Washington, 440 pp.

849. Silva, W.J. da, Prioli, L.M., Magalhaes, A.C.N., Pereira, A.C., Vargas, H., Mansanares, A.M.,
      Cella, N., Miranda, L.C.M., Alvarado-Gil, J. (1995): Photosynthetic O₂ evolution in maize
      inbreds and their hybrids can be differentiated by open photoacoustic cell technique.
      Plant Sci. 104: 177-181.

850. *Simmonds, J., Stewart, P., Simmonds, D. (1992)*: Regeneration of *Triticum aestivum* apical explants after microinjection of germ line progenitor cells with DNA. Physiol. Plant. 85: 197-206.

851. *Simonsen, V., Heneen, W.K. (1995)*: Genetic variation with in and among different cultivars and landraces of *Brassica campestris* L. and *B. oleracea* L. based on isozymes. Theor. Appl. Genet. 91: 346-352.

852. *Slatkin, M. (1985a)*: Gene flow in natural populations. Annu. Rev. Ecol. Syst. 16: 393-430.

853. *Slatkin, M. (1985b)*: Rare alleles as indicators of gene flow. Evolution 39: 53-65.

854. *Slatkin, M. (1987)*: Gene flow and the geographic structure of natural populations. Science 236: 787-792.

855. *Slatkin, M. (1993)*: Isolation by distance in equilibrium and non-equilibrium populations. Evolution 47: 264-279.

856. *Slatkin, M. (1995)*: A measure of population subdivision based on microsatellite allele frequencies. Genetics 139: 457-462.

857. *Slatkin, M., Arter, H.E. (1991)*: Spatial autocorrelation methods in population genetics. Amer. Naturalist 138: 499-517.

858. *Slatkin, M., Barton, N.H. (1989)*: A comparison of three indirect methods for estimating average levels of gene flow. Evolution 43: 1349-1368.

859. *Smith, D.N., Devey, M.E. (1994)*: Occurrence and inheritance of microsatellites in *Pinus radiata*. Genome 37: 977-983.

860. *Smith, H.A., Swaney, S.L., Parks, T.D., Wernsman, E.A., Dougherty, W.G. (1994)*: Transgenic plant virus resistance mediated by untranslatable sense RNAs: Expression, regulation, and fate of nonessential RNAs. Plant Cell 6: 1441-1453.

861. *Smith, O.S., Smith, J.S.C., Bowen, S.L., Tenborg, R.A., Wall, S.J. (1990)*: Similarities among a group of elite maize inbreds as measured by pedigree, F_1 grain yield, grain yield, heterosis and RFLPs. Theor. Appl. Genet. 80: 833-840.

862. *Smith, R.H., Hood, E.E. (1995)*: *Agrobacterium tumefaciens* transformation of monocotyledons. Crop Sci. 35: 301-309.

863. *Smith, S.E. (1989)*: Biparental inheritance of organelles and its implications in crop improvement. Plant Breed. Rev. 6: 361-393.

864. *Snow, A.A., Lewis, P.O. (1993)*: Reproductive traits and male fertility in plants: Empirical approaches. Annu. Rev. Ecol. Syst. 24: 331-351.

865. *Sokal, R.R., Oden, N.L. (1978a)*: Spatial autocorrelation in biology. 1. Methodology. Biol. J. Linn. Soc. 10: 199-228.

866. *Sokal, R.R., Oden, N.L. (1978b)*: Spatial autocorrelation in biology. 2. Some biological implications and four applications of evolutionary and ecological interest. Biol. J. Linn. Soc. 10: 229-254.

867. *Sokal, R.R., Rohlf, F.J. (1995)*: Biometry. The principles and practice of statistics in biological research.W.H. Freeman and Company, New York, 887 pp.

868.   *Soltis, D.E., Haufler C.H., Darrow, D.C., Gastony, G.J. (1983)*: Starch gel electrophoresis of ferns: A compilation of grinding buffers, gel and electrode buffers, and staining schedules. Amer. Fern J. 73: 9-27.

869.   *Soltis, D.E., Soltis, P.S. (1989)*: Isozymes in plant biology. Chapman & Hall, London, 268 pp.

870.   *Soltis, P.S., Soltis, D.E. (1991)*: Multiple origins of the allotetraploid *Tragopogon mirus* (Compositae): rDNA evidence. Syst. Bot. 16: 407-413.

871.   *Song, K.M., Osborn, T.C. (1994)*: A method for examining expression of homologous genes in plant polyploids. Plant Mol. Biol. 26: 1065-1071.

872.   *Songstad, D.D., Somers, D.A., Griesbach, R.J. (1995)*: Advances in alternative DNA delivery techniques. Plant Cell Tissue Organ Cult. 40: 1-15.

873.   *Southwick, E.E. (1982)*: Nectar biology and pollinator attraction in the north temperate climate. In: Breed, M.D., Michener, C.D., Evans, H.E., (eds.) The biology of social insects. Westview Press, Boulder, 19-23.

874.   *Southwick, E.E., Loper, G.M., Sadwick, S.E. (1981)*: Nectar production, composition, energetics and pollinator attractiveness in spring flowers of Western New York. Amer. J. Bot. 68: 994-1002.

875.   *Spalik, K., Woodell, S.R.J. (1994)*: Regulation of pollen production in *Anthriscus sylvestris*, an andromonoecious species. Int. J. Plant Sci. 155: 750-754.

876.   *Spangenberg, G., Gland, A., Neuhaus, G., Schweiger, H.G. (1987)*: Transgenic plants of *Brassica napus* obtained by microinjection of microspore derived embryoids. Eur. J. Cell Biol. 43: 56-56.

877.   *Spörlein, B., Streubel, M., Dahlfeld, G., Westhoff, P., Koop, H.U. (1991)*: PEG-mediated plastid transformation: A new system for transient gene expression assays in chloroplasts. Theor. Appl. Genet. 82: 717-722.

878.   *Stalker, D.M., Hiatt, W.R., Comai, L. (1985)*: A single amino acid substitution in the enzyme 5-enolpyruvylshikimate-3-phosphate synthase confers resistance to the herbicide glyphosate. J. Biol. Chem. 260: 4724-4728.

879.   *Stalker, D.M., Kiser, J.A., Baldwin, G., Coulombe, B., Houck, C.M. (1996)*: Cotton weed control using the BXN^TM system. In: Duke, S.O., (ed.) Herbicide-resistant crops. Agricultural, environmental, economic, regulatory, and technical aspects. Lewis Publishers, Boca Raton, 93-104.

880.   *Stalker, D.M., Malyj, L.D., McBride, K.E. (1988)*: Purification and properties of a nitrilase specific for the herbicide bromoxynil and corresponding nucleotide sequence analysis of the *bxn* gene. J. Biol. Chem. 263: 6310-6314.

881.   *Stalker, D.M., McBride, K.E., Malyj, L.D. (1988)*: Herbicide resistance in transgenic plants expressing a bacterial detoxification gene. Science 242: 419-423.

882.   *Stanton, M.L. (1987)*: Reproductive biology of petal color variants in wild populations of *Raphanus sativus*: II. Factors limiting seed production. Amer. J. Bot. 74: 188-196.

883.   *Steen, P., Pedersen, H.C. (1993)*: Gene transfer for herbcide resistance. J. Sugar Beet Res. 30: 267-274.

884. *Stenström, M., Molau, U. (1992)*: Reproductive ecology of *Saxifraga oppositifolia*: Phenology, mating system, and reproductive success. Arctic Alp. Res. 24: 337-343.

885. *Stephens, L.C., Hannapel, D.J., Krell, S.L., Shogren, D.R. (1996)*: *Agrobacterium* T-DNA mutation causes the loss of GUS expression in transgenic tobacco. Plant Cell Rep. 15: 414-417.

886. *Stephenson, A.G., Winsor, J.A. (1986)*: *Lotus corniculatus* regulates offspring quality through selective fruit abortion. Evolution 40: 453-458.

887. *Stevens, J.P., Bougourd, S.M. (1991)*: The frequency and meiotic behavior of structural chromosome variants in natural populations of *Allium schoenoprasum* L. (wild chives) in Europe. Heredity 66: 391-401.

888. *Stockey, A., Hunt, R. (1994)*: Predicting secondary succession in wetland mesocosms on the basis of autecological information on seeds and seedlings. J. Appl. Ecol. 31: 543-559.

889. *Stone, R. (1995)*: Taking a new look at life through a functional lens. Science 269: 316-317.

890. *Stratton, D.A. (1995)*: Spatial scale of variation in fitness of *Erigeron annuus*. Amer. Naturalist 146: 608-624.

891. *Streber, W.R., Kutschka, U., Thomas, F., Pohlenz, H.-D. (1994)*: Expression of a bacterial gene in transgenic plants confers resistance to the herbicide phenmedipham. Plant Mol. Biol. 25: 977-987.

892. *Streibig, J.C., Kudsk, P. (1993)*: Herbicide bioassays. CRC Press, Boca Raton, 270 pp.

893. *Sullivan, V.I., Neigel, J., Miao, B. (1991)*: Bias in inheritance of chloroplast DNA and mechanisms of hybridization between wind-pollinated and insect-pollinated *Eupatorium* (Asteraceae). Amer. J. Bot. 78: 695-705.

894. *Sumner, A.T. (1990)*: Chromosome banding. Unwin Hyman, London, 434 pp.

895. *Sumner, M.J., Caeseele, L. van (1988)*: Ovule development in *Brassica campestris*: A light microscope study. Can. J. Bot. 66: 2459-2469.

896. *Suter, G.W. (1993)*: Ecological risk assessment. Lewis Publishers, Boca Raton, 538 pp.

897. *Sutherland, S. (1986)*: Floral sex ratios, fruit set, and resource allocation in plants. Ecology 4: 991-1001.

898. *Swaddle, J.P., Witter, M.S., Cuthill, I.C. (1994)*: The analysis of fluctuating asymmetry. Anim. Behav. 48: 986-989.

899. *Tabashnik, B.E. (1994)*: Delaying insect adaptation to transgenic plants: Seed mixtures and refugia reconsidered. Proc. Roy. Soc. Lond.B. 255: 7-12.

900. *Tabashnik, B.E., Cushing, N.L., Finson, N., Johnson, M.W. (1990)*: Field development of resistance to *Bacillus thuringiensis* in diamondback moth (Lepidoptera: Plutellidae). J. Econ. Entomol. 83: 1671-1676.

901. *Tachibana, K., Watanabe, T., Sekizawa, Y., Takematsu, T. (1986)*: Accumulation of ammonia in plants treated with bialaphos. J. Pestic. Sci. 11: 33-37.

902. *Takahashi, H., Nishio, E., Hayashi, H. (1993)*: Pollination biology of the saprophytic species *Petrosavia sakuraii* (Makino) van Steenis in Central Japan. J. Plant Res. 106: 213-217.

903. *Takamizo, T., Spangenberg, G., Suginobu, K., Potrykus, I. (1991)*: Intergeneric somatic hybridization in *Gramineae*: Somatic hybrid plants between tall fescue (*Festuca arundinacea* Schreb.) and Italian ryegrass (*Lolium multiflorum* Lam.). Mol. Gen. Genet. 231: 1-6.

904. *Takeuchi, Y., Dotson, M., Keen, N.T. (1992)*: Plant transformation: A simple particle bombardment device based on flowing helium. Plant Mol. Biol. 18: 835-839.

905. *Talavera, S., Gibbs, P.E., Herrera, J. (1993)*: Reproductive biology of *Cistus ladanifer* (Cistaceae). Plant Syst. Evol. 186: 123-134.

906. *Tanhuanpää, P.K., Vilkki, J.P., Vilkki, H.J. (1996)*: Mapping of a QTL for oleic acid concentration in spring turnip rape (*Brassica rapa* ssp. *oleifera*). Theor. Appl. Genet. 92: 952-956.

907. *Tanksley, S.D., Young, N.D., Paterson, A.H., Bonierbale, M.W. (1989)*: RFLP mapping in plant breeding: New tools for an old science. Bio/Technology 7: 257-264.

908. *Tantikanjana, T., Nasrallah, M.E., Stein, J.C., Chen, C.-H., Nasrallah, J.B. (1993)*: An alternative transcript of the S locus glycoprotein gene in a class II pollen-recessive self-incompatibility haplotype of *Brassica oleracea* encodes a membrane-anchored protein. Plant Cell 5: 657-666.

909. *Tao, R., Uratsu, S.L., Dandekar, A.M. (1995)*: Sorbitol synthesis in transgenic tobacco with apple cDNA encoding NADP-dependent sorbitol-6-phosphate dehydrogenase. Plant Cell Physiol. 36: 525-532.

910. *Tao, Y.Z., Snape, J.W., Hu, H. (1991)*: The cytological and genetic characterisation of doubled haploid lines derived from triticale X wheat hybrids. Theor. Appl. Genet. 81: 369-375.

911. *Taylor, J.P. (1982)*: Carbon dioxide treatment as an effective aid to the production of selfed seeds in kale and brussel sprouts. Euphytica 31: 957-964.

912. *Taylor, J.W., Swann, E.C. (1994)*: DNA from herbarium specimens. In: Herrmann, B., Hummel, S., (eds.) Ancient DNA: Recovery an analysis of genetic material from paleontological, archaeological, museum, medical, and forensic specimens. Springer Verlag, Berlin, 166-181.

913. *Terauchi, R., Konuma, A. (1994)*: Microsatellite polymorphism in *Dioscorea tokoro*, a wild yam species. Genome 37: 794-801.

914. *Teso, B. (1993)*: OECD international principles for biotechnology safety. Agro Food Industry Hi-Tech 4: 27-31.

915. *Thanavala, Y., Yang, Y.-F., Lyons, P., Mason, H.S., Arntzen, C. (1995)*: Immunogenicity of transgenic plant-derived hepatitis B surface antigen. Proc. Nat. Acad. Sci. USA 92: 3358-3361.

916. *Thart, H., Sandbrink, J.M., Csikos, I., Ooyen, A. van, Brederode, J. van (1993)*: The allopolyploid origin of *Sedum rupestre* subsp. *rupestre* (Crassulaceae). Plant Syst. Evol. 184: 195-206.

917. *Thomas, J.C., Adams, D.G., Keppenne, V.D., Wasmann, C.C., Brown, J.K., Kanost, M.R., Bohnert, H.J. (1995)*: *Manduca sexta* encoded protease inhibitors expressed in *Nicotiana tabacum* provide protection against insects. Plant Physiol. Biochem. 33: 611-614.

918. *Thomas, J.C., Sepahi, M., Arendall, B., Bohnert, H.J. (1995)*: Enhancement of seed germination in high salinity by engineering mannitol expression in *Arabidopsis thaliana*. Plant Cell Environ. 18: 801-806.

919. *Thompson, C.J., Movva, N.R., Tizard, R., Crameri, R., Davies, J.E., Lauwereys, M., Botterman, J. (1987)*: Characterization of the herbicide-resistance gene *bar* from *Streptomyces hygroscopicus*. EMBO J. 6: 2519-2523.

920. *Thomson, D., Henry, R. (1993)*: Use of DNA from dry leaves for PCR and RAPD analysis. Plant Mol. Biol. Rep. 11: 202-206.

921. *Thomson, J.D., Plowright, R.C. (1980)*: Pollen carryover, nectar rewards, and pollinator behavior with special reference to *Diervilla lonicera*. Oecologia 46: 68-74.

922. *Thormann, C.E., Ferreira, M.E., Camargo, L.E.A., Tivang, J.G., Osborn, T.C. (1994)*: Comparison of RFLP and RAPD markers to estimating genetic relationships within and among cruciferous species. Theor. Appl. Genet. 88: 973-980.

923. *Thornhill, N.W. (1993)*: The natural history of inbreeding and outbreeding. Theoretical and emperical perspectives. The University of Chicago Press, Chicago, 575 pp.

924. *Thornthwaite, J.T., Sugarbaker, E.V., Temple, W.J. (1980)*: Preparation of tissues for DNA flow cytometric analysis. Cytometry 1: 229-237.

925. *Tiedje, J.M., Colwell, R.K., Grossman, Y.L., Hodson, R.E., Lenski, R.E., Mack, R.N., Regal, P.J. (1989)*: The planned introduction of genetically engineered organisms: Ecological considerations and recommendations. Ecology 70: 298-315.

926. *Tivang, J.G., Nienhuis, J., Smith, O.S. (1994)*: Estimation of sampling variance of molecular marker data using the bootstrap procedure. Theor. Appl. Genet. 89: 259-264.

927. *Toft, C.A. (1984)*: Resource shifts in bee flies (Bombyliidae): Interactions among species determine choice of resources. Oikos 43: 104-112.

928. *Tollsten, L., Knudsen, J.T. (1992)*: Floral scent in dioecious *Salix* (*Salicaceae*) - a cue determining the pollination system? Plant Syst. Evol. 182: 229-237.

929. *Tonsor, S.J., Kalisz, S., Fisher, J., Holtsford, T.P. (1993)*: A life-history based study of population genetic structure: Seed bank to adults in *Plantago lanceolata*. Evolution 47: 833-843.

930. *Torabinejad, J., Mueller, R.J. (1993)*: Genome constitution of the Australian hexaploid grass *Elymus scabrus* (Poaceae: Triticeae). Genome 36: 147-151.

931. *Touchell, D.H., Dixon, K.W., Tan, B. (1992)*: Cryopreservation of shoot-tips of *Grevillea scapigera* (Proteaceae): A rare and endangered plant from Western Australia. Aust. J. Bot. 40: 305-310.

932. *Traut, W. (1991)*: Chromosomen: Klassische und molekulare Cytogenetik. Springer Verlag, Berlin, 390 pp.

933. *Traveset, A. (1995)*: Reproductive ecology of *Cneorum tricoccon* L. (Cneoraceae) in the Balearic Islands. Bot. J. Linn. Soc. 117: 221-232.

934. *Treuren, R. van, Bijlsma, R., Ouborg, N.J., Delden, W. van (1993a)*: The effects of population size and plant density on outcrossing rates in locally endangered *Salvia pratensis*. Evolution 47: 1094-1104.

935. *Treuren, R. van, Bijlsma, R., Ouborg, N.J., Delden, W. van (1993b)*: The significance of genetic erosion in the process of extinction. IV. Inbreeding depression and heterosis effects caused by selfing and outcrossing in *Scabiosa columbaria*. Evolution 47: 1669-1680.

936. *Troyer, D.L., Goad, D.W., Xie, H., Rohrer, G.A., Alexander, L.J., Beattie, C.W. (1994)*: Use of direct in situ single-copy (DISC) PCR to physically map five porcine microsatellites. Cytogenet. Cell Genet. 67: 199-204.

937. *Troyer, M.E., Brody, M.S. (1994)*: Managing ecological risks at EPA: Issues and recommendations for progress. U.S. Environmental Protection Agency, Washington, 125 pp.

938. *Trudel, J., Potvin, C., Asselin, A. (1995)*: Secreted hen lysozyme in transgenic tobacco: Recovery of bound enzyme and in vitro growth inhibition of plant pathogens. Plant Sci. 106: 55-62.

939. *Truve, E., Aaspôllu, A., Honkanen, J., Puska, R., Mehto, M., Hassi, A., Terri, T.H., Kelve, M., Seppänen, P., Saarma, M. (1993)*: Transgenic potato plants expressing mammalian 2'-5' oligoadenylate synthetase are protected from potato virus X infection under field conditions. Bio/Technology 11: 1048-1052.

940. *Tsaftaris, S.A. (1995)*: Molecular aspects of heterosis in plants. Physiol. Plant. 94: 362-370.

941. *Tsuchiya, T., Toriyama, K., Yoshikawa, M., Ejiri, S., Hinata, K. (1995)*: Tapetum-specific expression of the gene for an endo-β-1,3-glucanase causes male sterility in transgenic tobacco. Plant Cell Physiol. 36: 487-494.

942. *Turner, B.L., Alston, R. (1959)*: Segregation and recombination of chemical constituents in a hybrid swarm of *Baptisia laevicaulis* x *B. viridis* and their taxonomic implications. Amer. J. Bot. 46: 678-686.

943. *Tuyl, J.M. van, Diën, M.P. van, Creij, M.G.M. van, Kleinwee, T.C.M. van, Franken, J., Bino, R.J. (1991)*: Application of in vitro pollination, ovary culture, ovule culture and embryo rescue for overcomming incongruity barriers in interspecific *Lilium* crosses. Plant Sci. 74: 115-126.

944. *Tzotzos, G.T. (1993)*: Biosafety considerations: An international perspective. Agro Food Industry Hi-Tech 4: 5-6.

945. *Tzotzos, G.T. (1995)*: Genetically modified organisms. A guide to biosafety. CAB International, Wallingford, 213 pp.

946. *Ulloa-G, M., Corgan, J.N., Dunford, M. (1994)*: Chromosome characteristics and behavior differences in *Allium fistulosum* L., *A. cepa* L., their F_1-hybrid, and selected backcross progeny. Theor. Appl. Genet. 89: 567-571.

947. *UNEP (1996)*: International technical guidelines for safety in biotechnology. United Nations Environmental Programme, Nairobi, 31 pp.

948. *Urwin, P.E., Atkinson, H.J., Waller, D.A., McPherson, M.J. (1995)*: Engineered oryzacystatin-I expressed in transgenic hairy roots confers resistance to *Globodera pallida*. Plant J. 8: 121-131.

949. *USDA (1991)*: User's guide for introducing genetically engineered plants and microorganisms. Technical bulletin No. 1783. U.S. Department of agriculture, animal and plant health inspection service, Hyattsville, various pagings.

950. *Utech, F.H., Kawano, S. (1975)*: Spectral polymorphisms in angiosperm flowers detemined by differential ultraviolet reflectance. Bot. Mag. 88: 9-30.

951. *Uyenoyama, M.K., Holsinger, K.E., Waller, D.M. (1993)*: Ecological and genetic factors directing the evolution of self-fertilization. In: Futuyma, D., Antonovics, J., (eds.) Oxford surveys in evolutionary biology. Oxford University Press, New York, 9: 327-381.

952. *Vaeck, M., Reynaerts, A., Höfte, H., Jansens, S., Beuckeleer, M. de, Dean, C., Zabeau, M., Montagu, M. van, Leemans, J. (1987)*: Transgenic plants protected from insect attack. Nature 328: 33-37.

953. *Valdes, A.M., Slatkin, M., Freimer, N.B. (1993)*: Allele frequencies at microsatellite loci: The stepwise mutation model revisited. Genetics 133: 737-749.

954. *Valk, P. van der, Vries, S.E. de, Everink, J.T., Verstappen, F., Vries, J.N. de (1991)*: Pre- and post-fertilization barriers to backcrossing the interspecific hybrid between *Allium fistulosum* L. and *A. cepa* L. with *A. cepa*. Euphytica 53: 201-209.

955. *Vasil, V., Castillo, A.M., Fromm, M.E., Vasil, I.K. (1992)*: Herbicide resistant fertile transgenic wheat plants obtained by microprojectile bombardment of regenerable embryogenic callus. Bio/Technology 10: 667-674.

956. *Vaughton, G., Ramsey, M. (1991)*: Floral biology and inefficient pollen removal in *Banksia spinulosa* var. *neoanglica* (Proteaceae). Aust. J. Bot. 39: 167-177.

957. *Veit, M., Bauer, K., Beckert, C., Kast, B., Geiger, H., Czygan, F.C. (1995)*: Phenolic characters of British hybrid taxa in *Equisetum* subgenus Equisetum. Biochem. Syst. Ecol. 23: 79-87.

958. *Verduin, B.J.M., Goldbach, R.W. (1990)*: Genetic engineering of viral resistance in plants. In: Jong, J. de, (ed.) Integration of *in vitro* techniques in ornamental plant breeding. Proceedings, symposium, 10-14 November 1990. Eucarpia, Wageningen, Netherlands, 162-172.

959. *Vermeire, T., Zandt, P. van der (1995)*: Procedures of hazard and risk assessment. In: Leeuwen, C.J. van, Hermens, J.M.L., (eds.) Risk assessment of chemicals: An introduction. Kluwer Academic Publishers, Dordrecht, 293-337.

960. *Vernon, D.M., Tarczynski, M.C., Jensen, R.G., Bohnert, H.J. (1993)*: Cyclitol production in transgenic tobacco. Plant J. 4: 199-205.

961. *Viegas-Péquignot, E. (1992)*: *In situ* hybridization to chromosomes with biotinylated probes. In: Wilkinson, D.G., (ed.) *In situ* hybridization: A practical approach. IRL Press, Oxford, 137-158.

962. *Vogel, S. (1990)*: The role of scent glands in pollination. Amerind Publishing Co., New Delhi, 202 pp.

963. *Vonhof, M.J., Harder, L.D. (1995)*: Size-number trade-offs and pollen production by papilionaceous legumes. Amer. J. Bot. 82: 230-238.

964. *Vries, F.T.F. de (1996)*: Cultivated plants and the wild flora. Effect analysis by dispersal codes. Rijksuniversiteit Leiden, Rijksherbarium, Hortus Botanicus, Leiden, 222 pp.

965. *Waddington, K.D. (1979a)*: Flight patterns of three species of sweat bees (Halictidae) foraging at *Convolvolus arvensis*. J. Kans. Entomol. Soc. 52: 751-758.

966. *Waddington, K.D. (1979b)*: Quantification of the movement patterns of bees: A novel method. Amer. Midland Naturalist 101: 278-285.

967. *Wakana, A., Uemoto, S. (1987)*: Adventive embryogenesis in citrus. I. The occurrence of adventive embryos without pollination or fertilization. Amer. J. Bot. 74: 517-530.

968. *Walden, R. (1988)*: Genetic transformation in plants. Oxford University Press, Manchester, 138 pp.

969. *Walden, R., Schell, J. (1990):* Techniques in plant molecular biology - progress and problems. Eur. J. Biochem. 192: 563-576.

970. *Walden, R., Wingender, R. (1995):* Gene-transfer and plant-regeneration techniques. Trends Biotech. 13: 324-331.

971. *Wang, G.J., Castiglione, S., Zhang, J., Fu, R.-Z., Ma, J.-S., Li, W.-B., Sun, Y.-R., Sala, F. (1994):* Hybrid rice (*Oryza sativa* L.): Identification and parentage determination by RAPD fingerprinting. Plant Cell Rep. 14: 112-115.

972. *Wang, G.-L., Paterson, A.H. (1994):* Assessment of DNA pooling strategies for mapping of QTLs. Theor. Appl. Genet. 88: 355-361.

973. *Wang, Z.Y., Legris, G., Nagel, J., Potrykus, I., Spangenberg, G. (1994):* Cryopreservation of embryogenic cell suspensions in *Festuca* and *Lolium* species. Plant Sci. 103: 93-106.

974. *Warwick, S.I., Bain, J.F., Wheatcroft, R., Thompson, B.K. (1989):* Hybridization and introgression in *Carduus nutans* and *C. acanthoides* reexamined. Syst. Bot. 14: 476-494.

975. *Warwick, S.I., Black, L.D. (1991):* Molecular systematics of *Brassica* and allied genera (subtribe Brassicinae, Brassiceae): Chloroplast genome and cytodeme congruence. Theor. Appl. Genet. 82: 81-92.

976. *Warwick, S.I., Thompson, B.K., Black, L.D. (1992):* Hybridization of *Carduus nutans* and *Carduus acanthoides* (Compositae): Morphological variation in F_1 hybrids and backcrosses. Can. J. Bot. 70: 2303-2309.

977. *Waser, N.M. (1982):* A comparison of distances flown by different visitors to flowers of the same species. Oecologia 55: 251-257.

978. *Waser, N.M. (1993):* Population structure, optimal outbreeding, and assortative mating in angiosperms. In: Thornhill, N.W., (ed.) The natural history of inbreeding and outbreeding. The University of Chicago Press, Chicago, 1: 173-199.

979. *Waser, N.M., Mitchell, R.J. (1990):* Nectar standing crops in *Delphinium nelsonii* flowers: Spatial autocorrelation among plants? Ecology 71: 116-123.

980. *Waser, N.M., Price, M.V. (1983):* Optimal and actual outcrossing in plants, and the nature of plant-pollinator interaction. In: Jones, C.E., Little, R.J., (eds.) Handbook of experimental pollination biology. Scientific and Academic Editions, New York, 341-359.

981. *Waser, N.M., Price, M.V. (1985):* The effect of nectar guides on pollinator preference: Experimental studies with a montane herb. Oecologia 67: 121-126.

982. *Waser, N.M., Price, M.V. (1989):* Optimal outcrossing in *Ipomopsis aggregata*: Seed set and offspring fitness. Evolution 43: 1097-1109.

983. *Waser, N.M., Price, M.V. (1994):* Crossing-distance effects in *Delphinium nelsonii*: Outbreeding and inbreeding depression in progeny fitness. Evolution 48: 842-852.

984. *Washitani, I., Osawa, R., Namai, H., Niwa, M. (1994):* Patterns of female fertility in heterostylous *Primula sieboldii* under severe pollinator limitation. J. Ecol. 82: 571-579.

985. *Watt, W.B. (1994):* Allozymes in evolutionary genetics: Self-imposed burden or extraordinary tool? Genetics 136: 11-16.

986. *Watt, W.B., Hoch, P.C., Mills, S.G. (1974):* Nectar resource use by *Colias* butterflies. Oecologia 14: 353-374.

987. *Webb, R.P., Wong-Vega, L., Allen, R.D. (1992)*: Superoxide dismutase gene expression in transgenic plants. Cell Biochem. Suppl. 16 Part F: 213.

988. *Weir, B.S. (1990)*: Genetic data analysis: Methods for discrete population genetic data. Sinauer Associates, Sunderland, 377 pp.

989. *Weir, B.S., Cockerham, C.C. (1984)*: Estimating F-statistics for the analysis of population structure. Evolution 38: 1358-1370.

990. *Weiss, J., Nerd, A., Mizrahi, Y. (1994)*: Flowering behavior and pollination requirements in climbing cacti with fruit crop potential. Hortscience 29: 1487-1492.

991. *Wells, H. (1980)*: A distance coefficient as a hybridization index: An example using *Mimulus longiflorus* and *M. flemingii* (Scrophulariaceae) from Santa Cruz Island, California. Taxon 29: 53-65.

992. *Welsh, J., Honeycutt, R.J., McClelland, M., Sobral, B.W.S. (1991)*: Parentage determination in maize hybrids using the arbitrarily primed polymerase chain reaction (Ap-PCR). Theor. Appl. Genet. 82: 473-476.

993. *Wendel, J.F., Weeden, N.F. (1989)*: Visualization and interpretation of plant isozymes. In: Soltis, D.E., Soltis, P.S., (eds.) Isozymes in plant biology. Chapman & Hall, London, 5-45.

994. *Wert, S.L. van, Saunders, J.A. (1992)*: Electrofusion and electroporation of plants. Plant Physiol. 99: 365-367.

995. *Wessels, J.G.H., Sietsma, J.H. (1981)*: Fungal cell walls: A survey. In: Tanner, W., Loewus, F.A., (eds.) Encyclopedia of plant physiology. Springer Verlag, Berlin, 13B: 352-394.

996. *Whittemore, A.T., Schaal, B.A. (1991)*: Interspecific gene flow in sympatric oaks. Proc. Nat. Acad. Sci. USA 88: 2540-2544.

997. *Widén, M. (1992)*: Sexual reproduction in a clonal, gynodioecious herb *Glechoma hederacea*. Oikos 63: 430-438.

998. *Widstrom, N.W., Bondari, K., McMillian, W.W. (1992)*: Hybrid performance among maize populations selected for resistance to insects. Crop Sci. 32: 85-89.

999. *Williams, C.F. (1994)*: Genetic consequences of seed dispersal in three sympatric forest herbs. II. Microspatial genetic structure within populations. Evolution 48: 1959-1972.

1000. *Williams, C.F., Guries, R.P. (1994)*: Genetic consequences of seed dispersal in three sympatric forest herbs. I. Hierarchical population-genetic structure. Evolution 48: 791-805.

1001. *Williams, J.G.K., Kubelik, A.R., Livak, K.J., Rafalski, J.A., Tingey, S.C. (1991)*: DNA polymorphisms amplified by arbitrary primers are useful as genetic markers. Nucl. Acid. Res. 18: 6531-6535.

1002. *Williamson, M. (1993)*: Risks from the release of GMOs: Ecological and evolutionary considerations. Environment Update 1: 5-9.

1003. *Williamson, M. (1994)*: Community response to transgenic plant release: Predictions from British experience of invasive plants and feral crop plants. Mol. Ecol. 3: 75-79.

1004. *Willis, J.H. (1993)*: Partial self-fertilization and inbreeding depression in two populations of *Mimulus guttatus*. Heredity 71: 145-154.

1005. *Willson, M.F. (1994)*: Sexual selection in plants: Perspective and overview. Amer. Naturalist 144: S13-S39.

1006. *Willson, M.F., Whelan, C.J. (1993)*: Variation of dispersal phenology in a bird-dispersed shrub, *Cornus drummondii*. Ecol. Monogr. 63: 151-172.

1007. *Wilson, P. (1992)*: On inferring hybridity from morphological intermediacy. Taxon 41: 11-23.

1008. *Wilson, T.M.A., Davies, J.W. (1994)*: New roads to crop protection against viruses. Outlook Agr. 23: 33-39.

1009. *Wolf, P.G., Campbell, D.R. (1995)*: Hierarchical analysis of allozymic and morphometric variation in a montane herb, *Ipomopsis aggregata* (Polemoniaceae). J. Hered. 86: 386-394.

1010. *Wolf, P.G., Soltis, P.S. (1992)*: Estimates of gene flow among populations, geographic races, and species in the *Ipomopsis aggregata* complex. Genetics 130: 639-647.

1011. *Wolf, S.J., Denford, K.E. (1984)*: *Arnica gracilis* (Compositae), a natural hybrid between *A. latifolia* and *A. cordifolia*. Syst. Bot. 9: 12-16.

1012. *Wolfe, A.D., Estes, J.R., Chissoe III, W.F. (1991)*: Tracking pollen flow of *Solanum rostratum* (Solanaceae) using backscatter scanning electron micriscopy and X-ray microanalysis. Amer. J. Bot. 78: 1503-1507.

1013. *Wolfe, L.M. (1993)*: Inbreeding depression in *Hydrophyllum appendiculatum*: Role of maternal effects, crowding, and parental mating history. Evolution 47: 374-386.

1014. *Woodward, J.R., Craik, D., Dell, A., Khoo, K.H., Munro, S.L.A., Clarke, A.E., Bacic, A. (1992)*: Structural analysis of the N-linked glycan chains from a stylar glycoprotein associated with expression of self-incompatibility in *Nicotiana alata*. Glycobiology 2: 241-250.

1015. *Wright, S. (1931)*: Evolution in Mendelian populations. Genetics 16: 97-158.

1016. *Wright, S. (1943)*: Isolation by distance. Genetics 28: 114-138.

1017. *Wright, S. (1946)*: Isolation by distance under diverse systems of mating. Genetics 31: 39-59.

1018. *Wright, S. (1969)*: Evolution and the genetics of populations. Vol. 2. The theory of gene frequencies. The University of Chicago Press, Chicago, 511 pp.

1019. *Wright, S. (1977)*: Evolution and the genetics of populations. Vol. 3. Experimental results and evolutionary deductions. The University of Chicago Press, Chicago, 613 pp.

1020. *Wu, G.S., Shortt, B.J., Lawrence, E.B., Levine, E.B., Fitzsimmons, K.C., Shah, D.M. (1995)*: Disease resistance conferred by expression of a gene encoding H_2O_2-generating glucose oxidase in transgenic potato plants. Plant Cell 7: 1357-1368.

1021. *Wyatt, R. (1990)*: The evolution of self pollination in granite outcrop species of *Arenaria* (Caryophyllaceae). V. Artificial crosses within and between populations. Syst. Bot. 15: 363-369.

1022. *Wyatt, R., Broyles, S.B. (1992)*: Hybridization in North American *Asclepias*. III. Isozyme evidence. Syst. Bot. 17: 640-648.

1023. *Xie, C.Y., Knowles, P. (1991)*: Spatial genetic substructure within natural populations of jack pine (*Pinus banksiana*). Can. J. Bot. 69: 547-551.

1024. *Xie, C.Y., Knowles, P. (1992)*: Associations between allozyme phenotypes and soil nutrients in a natural population of jack pine (*Pinus banksiana*). Biochem. Syst. Ecol. 20: 179-185.

1025. *Xu, G.-W., Magill, C.W., Schertz, K.F., Hart, G.E. (1994)*: A RFLP linkage map of *Sorghum bicolor* (L.) Moench. Theor. Appl. Genet. 89: 139-145.

1026. *Xue, B., Gonsalves, C., Provvidenti, R., Slightom, J.L., Fuchs, M., Gonsalves, D. (1994)*: Development of transgenic tomato expressing a high level of resistance to cucumber mosaic virus strains of subgroups I and II. Plant Dis. 78: 1038-1041.

1027. *Yamakawa, M., Shirata, A., Taniai, K., Miyamoto, K. (1990)*: Effects of the hemolymph from *Bombyx mori*, immunized with *Escherichia coli*, on the proliferation of plant pathogenic bacteria. Agr. Biol. Chem. 54: 2175-2176.

1028. *Yang, G.P., Maroof, M.A.S., Xu, C.G., Zhang, Q.F., Biyashev, R.M. (1994)*: Comparative analysis of microsatellite DNA polymorphism in landraces and cultivars of rice. Mol. Gen. Genet. 245: 187-194.

1029. *Yang, R.C., Yeh, F.C. (1993)*: Multilocus structure in *Pinus contorta* Dougl. Theor. Appl. Genet. 87: 568-576.

1030. *Yao, J.L., Cohen, D., Rowland, R.E. (1994)*: Inheritance and plastome-genome incompatibility in interspecific hybrids of *Zantedeschia* (Araceae). Theor. Appl. Genet. 88: 255-260.

1031. *Ye, G.N., Daniell, H., Sanford, J.C. (1990)*: Optimization of delivery of foreign DNA into higher plant chloroplasts. Plant Mol. Biol. 15: 809-819.

1032. *Young, A.M., Severson, D.W. (1994)*: Comparative analysis of steam distilled floral oils of cacao cultivars (*Theobroma cacao* L., Sterculiaceae) and attraction of flying insects: Implications for a Theobroma pollination syndrome. J. Chem. Ecol. 20: 2687-2703.

1033. *Young, H.J. (1988)*: Neighborhood size in a beetle pollinated tropical aroid: Effects of low density and asynchronous flowering. Oecologia 76: 461-466.

1034. *Young, J., Virmani, S.S. (1990)*: Heterosis in rice over environments. Euphytica 51: 87-93.

1035. *Yu, K.F., Pauls, K.P. (1993)*: Rapid estimation of genetic relatedness among heterogeneous populations of alfalfa by random amplification of bulked genomic DNA samples. Theor. Appl. Genet. 86: 788-794.

1036. *Yuan, C.-X., Bai, Y.-Y., Kuang, D.-R. (1993)*: Transfer of yeast ProB gene into tobacco plants. Acta Phytophysiol. Sin. 19: 306-312.

1037. *Yuan, S.-C., Zhang, Z.-G., He, H.-H., Zen, H.-L., Lu, K.-Y., Lian, J.-H., Wang, B.-X. (1993)*: Two photoperiodic-reactions in photoperiod-sensitive genic male-sterile rice. Crop Sci. 33: 651-660.

1038. *Yun, D.J., Hashimoto, T., Yamada, Y. (1992)*: Metabolic engineering of medicinal plants: Transgenic *Atropa belladonna* with an improved alkaloid composition. Proc. Nat. Acad. Sci. USA 89: 11799-11803.

1039. *Zhang, G., Angeles, E.R., Abenes, M.L.P., Khush, G.S., Huang, N. (1996)*: RAPD and RFLP mapping of the bacterial blight resistance gene *xa-13* in rice. Theor. Appl. Genet. 93: 65-70.

1040. *Zhu, Q., Maher, E.A., Masoud, S., Dixon, R.A., Lamb, C.J. (1994)*: Enhanced protection against fungal attack by constitutive co-expression of chitinase and glucanase genes in transgenic tobacco. Bio/Technology 12: 807-812.

1041. *Zimmerman, M. (1983)*: Calculating nectar production rates: Residual nectar and optimal foraging. Oecologia 58: 258-259.

12. Subject index

Bold page numbers refer to definitions, main discussions and method descriptions. Subject abbreviations (e.g., GMP, DNA, RFLP) and method numbers (e.g., M1, "Amplified restriction fragment polymorphism") listed in Section 7.2 are included in the index for convenience.

G. Kjellsson / V. Simonsen,
The National Environmental Research Institute, Silkeborg, Denmark (Eds)

Methods for Risk Assessment of Transgenic Plants

I. Competition, Establishment and Ecosystem Effects

1994. 224 pages. Hardcover.
ISBN 3-7643-5065-2

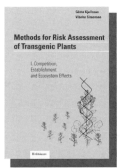

The present book is a compilation of current test methods useful in risk assessment of transgenic plants. It is intended to aid the environmental researcher in finding and comparing relevant methods quickly and easily. It may also be used as a general reference work for field-ecologists, laboratory biologists and others working in plant population biology and genetics.

The major processes affecting the fate of plants are covered with emphasis on invasion, competition and establishment, e.g., seed dispersal, density-dependent competition and plant growth. Ecosystem effects and genetic structure are also covered. For each process a number of relevant test methods have been selected; in total, 84 methods for field, greenhouse or laboratory research are included, employing 51 key processwords. Each method is described and evaluated briefly and succinctly, and there are comments on assumptions, restrictions, advantages and applications. An extensive bibliography provides entry into the scientific background, and cross references make it possible to find all relevant sources quickly.

Methods to study pollination and gene transfer will be considered in a future volume.

Birkhäuser Verlag • Basel • Boston • Berlin

J. Tomiuk / K. Wöhrmann, *University of Tübingen, Germany* /
A. Sentker, *Tübingen, Germany (Eds)*

Transgenic Organisms –
Biological and
Social Implications

1996. 280 pages. Hardcover
ISBN 3-7643-5262-0

This selected collection of contributions focuses on the modification of organisms through genetic manipulation. Scientists from various disciplines assess the quality of our knowledge on which risk assessment of genetechnology methods is currently based. Molecular biology and ecology, but also aspects of evolutionary and population genetics, human genetics and genetically modified food are among the topics covered.

The book analyzes the impetus behind, and progress in, research methods which have been introduced into genetechnology risk assessment procedures over the last three years, and, in so doing, reveals gaps in our understanding of evolutionary processes. The history of risk assessment and ethical implications with respect to the deliberate release of GMOs are considered. Finally, the transfer of knowledge from the laboratory to the public, and the role of the media in this process are discussed.

This monograph is of great interest to all those concerned with the risk assessment of genetechnology methods.

Birkhäuser Verlag • Basel • Boston • Berlin